# Lecture Notes in Electrical Engineering

## Volume 906

The book series *Lecture Notes in Electrical Engineering* (LNEE) publishes the latest developments in Electrical Engineering - quickly, informally and in high quality. While original research reported in proceedings and monographs has traditionally formed the core of LNEE, we also encourage authors to submit books devoted to supporting student education and professional training in the various fields and applications areas of electrical engineering. The series cover classical and emerging topics concerning:

- Communication Engineering, Information Theory and Networks
- Electronics Engineering and Microelectronics
- Signal, Image and Speech Processing
- Wireless and Mobile Communication
- Circuits and Systems
- Energy Systems, Power Electronics and Electrical Machines
- Electro-optical Engineering
- Instrumentation Engineering
- Avionics Engineering
- Control Systems
- Internet-of-Things and Cybersecurity
- Biomedical Devices, MEMS and NEMS

For general information about this book series, comments or suggestions, please contact leontina.dicecco@springer.com.

To submit a proposal or request further information, please contact the Publishing Editor in your country:

**China**

Jasmine Dou, Editor (jasmine.dou@springer.com)

**India, Japan, Rest of Asia**

Swati Meherishi, Editorial Director (Swati.Meherishi@springer.com)

**Southeast Asia, Australia, New Zealand**

Ramesh Nath Premnath, Editor (ramesh.premnath@springernature.com)

**USA, Canada:**

Michael Luby, Senior Editor (michael.luby@springer.com)

**All other Countries:**

Leontina Di Cecco, Senior Editor (leontina.dicecco@springer.com)

**** This series is indexed by EI Compendex and Scopus databases. ****

Sanjay Yadav · K. P. Chaudhary · Ajay Gahlot ·
Yogendra Arya · Aman Dahiya · Naveen Garg
Editors

# Recent Advances in Metrology

## Select Proceedings of AdMet 2021

Springer

*Editors*
Sanjay Yadav
CSIR-NPL
New Delhi, India

K. P. Chaudhary
Maharaja Surajmal Institute of Technology
New Delhi, India

Ajay Gahlot
Maharaja Surajmal Institute of Technology
New Delhi, India

Yogendra Arya
Maharaja Surajmal Institute of Technology
New Delhi, India

Aman Dahiya
Maharaja Surajmal Institute of Technology
New Delhi, India

Naveen Garg
CSIR-NPL
New Delhi, India

ISSN 1876-1100 ISSN 1876-1119 (electronic)
Lecture Notes in Electrical Engineering
ISBN 978-981-19-2467-5 ISBN 978-981-19-2468-2 (eBook)
https://doi.org/10.1007/978-981-19-2468-2



This Springer imprint is published by the registered company Springer Nature Singapore Pte Ltd.
The registered company address is: 152 Beach Road, #21-01/04 Gateway East, Singapore 189721, Singapore

# Organization

AdMet 2021 was jointly organized by CSIR-National Physical Laboratory, New Delhi and Metrology Society of India at Maharaja Surajmal Institute of Technology, Janakpuri, New Delhi, India, during March 5–6, 2021.

# Preface

This book presents proceedings of the 7th National Conference "Advances in Metrology" AdMet 2021 on Sensors & Advance Materials for Measurement & Quality Improvement organized by Maharaja Surajmal Institute of Technology, Janakpuri, New Delhi, India. The book highlights and discusses the technological developments in the areas of sensor technology, measurement, advance material for industrial application, automation and quality control. It provides significant advanced technological information and opportunities for metrologists, scientists, engineers, academicians and students from research institutes and industrial establishments to explore the future directions in the areas of sensors, advance materials, measurement and quality improvement. The book can be a valuable reference for beginners, science and engineering graduates and postgraduates, researchers, policy-makers and entrepreneurs for deeper understanding of the art of measurement and testing uncertainty in measurement. This book is aimed for all the personnel engaged in conformity assessment, quality system management, calibration and testing in all sectors of industry. The book is an asset for technical staff, researchers and scientists associated with the premier research institute like CSIR-NPL, NABL and MSI and all the NABL accredited calibration and testing laboratories in various disciplines.

The responses to the call for papers had been overwhelming from all over the India. AdMet 2021 ensured to be both a stimulating and an enlightening experience with numerous eminent keynote and invited speakers. The event consisted of invited talks, technical sessions, paper presentations, poster presentations and discussions with eminent speakers covering a wide range of topics in advances in metrology. This book contains the research papers presented at the conference.

We express our earnest gratitude to the eminent keynote speakers, authors and the participants. We are highly thankful to Surajmal Memorial Education Society (SMES) whose unconditional and unstinting support has made the dream of hosting AdMet 2021 conference a reality.

We express our thanks and appreciation to all the reviewers who helped us to maintain the high quality of manuscripts included in the proceedings. We are highly indebted to the entire Springer team for the excellent collaboration, assistance and patience during the production of this volume.

Organizing Committee
AdMet 2021
Metrology Society of India (MSI)
CSIR-National Physical Laboratory
(CSIR-NPL)
New Delhi, India

# Contents

# About the Editors

**Dr. Sanjay Yadav** born in 1962, obtained his master degree in science (M.Sc.) in 1985 and Ph.D. degree in Physics in 1990. Presently, he is Chief Scientist and Head, Physico Mechanical Metrology Division of NPL leading a group of 23 scientists working on various projects. In addition, he is also holding the post of Professor, Faculty of Physical Sciences, Academy of Scientific and Innovative Research (AcSIR), HRDG, Ghaziabad teaching 'Advanced Measurement Techniques & Metrology' course, taking practical classes and supervising graduate, master and Ph.D. students since 2011.

He is the recipient of research scholarships from Ministry of Home Affairs, India (1986); CSIR, India (1988); Col. G. N. Bajpayee Award of Institution of Engineers, India (1989); Commendation Certificates from Haryana Government (1991 and 1992); JICA Fellowship of Japan (1998), Commendation Certificates from SASO, Saudi Arabia (2003); 3 Appreciation Certificates from Director, NPL (2005); Managing Editor, MAPAN (2006–2014); nominated as Member of APMP Technical Committee of Mass Related Quantities (TCM), Australia (2013–2019); Nominated as Country Representative in APMP, China (2019); Vice President, Metrology Society of India (2020); Member, National Advisory Committee, NCERT, Delhi (2019); Members, Testing and Calibration Advisory Committee, BIS (2019, 2020 and 2021), and very recently received a prestigious International award i.e. APMP Award for Developing Economies, China (2020).

He is working as the Editor-in-Chief (EIC) of the *MAPAN: The Journal of Metrology Society of India*; Editorial Board of the *International Journal of Metrology and Quality Engineering* (IJMQE), France; Associate Editor-in-Chief of BGR Publications and reviewer of several SCI journals. He has extensively visited several countries i.e. France, Japan, SASO, Saudi Arabia, Indonesia, Bhutan and Malaysia on official assignments and deputations.

He has significantly contributed in the field of pressure metrology, biomedical instrumentation, ultrasonic transducers and instrumentation systems. His current research interests include research and developmental activities in physico mechanical measurements; establishment, realization, maintenance and up-gradation of national pressure and vacuum standards; dissemination of national practical pressure scale to users through apex level calibration, training and consultancy services; inter-laboratory comparisons, proficiency testing programme and key comparisons, implementation of Quality System in the laboratory as per ISO/IEC 17025 standard and Finite Element Analysis (FEA) and Monte Carlo Simulations for pressure balances.

He has more than 350 research publications to his credit, published in the national and international journals of repute and conferences besides contributing several books with Springer and MSI as well as in drafting several project, scientific and technical reports, documents and policy papers.

**Dr. K. P. Chaudhary** did his M.Tech in Instrumentation from NIT Kurukshetra and Ph.D. from IIT Delhi in Optical Instrumentation/Metrology. Dr. K. P. Chaudhary initially joined CSIO Chandigarh in 1980 and worked for about fifteen years in different capacities. He has designed and developed a no-contact microprocessor based Laser Scanning Micrometer. It's know how was transferred to CEL Sahibabad for commercialization and production. In addition to this, he was also responsible for the development of various opto-electronics instruments namely Digital Laser Power Meter, Laser Communication System, Optical Viscometer etc.

In 1996, he joined National Physical Laboratory (NPL), Delhi as Senior Scientist. He has published more than 90 research papers in journals and International conferences. He has also registered seven Patents to his credit and out of which six technologies have been transferred as consultancy projects. He has also been instrumental providing his technical expertise to the Indian industries as well as SAARC nations (Pakistan, Bangladesh, Bhutan, Nepal, Sri Lanka, Maldives, etc.) under SAARC-PTB cooperation project as Principle Investigator of International Collaboration of SAARC cooperation project. He has visited many countries on official deputations; to name few of them are Germany, France, USA, Indonesia, UK, Bhutan, Sri Lanka, Bangladesh etc. on various capacities/assignments. He is also a reviewer of CMCs of Asia Pacific Metrology Programme (APMP) as member of Technical Committee of Length (TCL).

He has also established a 3D Coordinate Measuring Machine (CMM), 4M Laser based Length Measuring Machine (LMM), Roundness Measuring Machine etc. He has chaired various prestigious committees of NPL. On a role of scientific societies, he is Fellow of IETE, MSI, USI, OSI, ISI and some International groups like TC213, WGDM, and APMP TCL, etc.

Since long, he is also associated with National Accreditation Board of Laboratory (NABL) and Bureau of Indian Standards (BIS) as Chairman of core/Supplementary committees and also as the Lead/Technical Assessor (NABL). He has delivered more than 100 invited talks and lectures in industries, laboratories and technical institutions. He has guided more than 15 students of M.Tech., B.Tech., and MDS for their project works He was working as Chief Scientist and Head at NPL before joining as Director at Maharaja Surajmal Institute of Technology, Janakpuri, New Delhi.

**Dr. Ajay Gahlot** is currently working as an Associate Professor in the Department of Applied Sciences, Maharaja Surajmal Institute of Technology (MSIT), Janakpuri, New Delhi. He received his M.Sc. in Physics from Delhi University (India), in 2001, M.Tech. (Solid State Technology) from Indian Institute of Technology (IIT), Chennai in 2004 and Doctorate in Plasma Physics from Guru Govind Singh Indraprastha University (GGSIPU), in 2011. His major areas of research interests include Plasma Physics, waves in Dusty & Strongly Coupled dusty plasma, Fusion plasma and Condensed Matter Physics. He has over 16 years of teaching experience and published 30 research papers in reputed National/International journals and conferences. He has published total 15 SCI research papers in highly reputed International journals.

**Dr. Yogendra Arya** is currently working as an Associate Professor with the Department of Electrical Engineering, J. C. Bose University of Science and Technology, YMCA, Faridabad, Haryana. He received his A.M.I.E. in Electrical Engineering from The Institution of Engineers (India), in 2008, M.Tech. (Instrumentation and Control) from Deenbandhu Chhotu Ram University of Science and Technology (DCRUST), Murthal, Haryana in 2010 and Ph.D. from Delhi Technological University (DTU), Delhi in 2018. His major areas of research interests include automatic generation control, load frequency control, automatic voltage control, control system, fuzzy control, optimization techniques, energy storage systems and renewable energy systems. He has published 50 research papers in reputed National/International journals and conferences. He has published 33 SCIE research papers in highly reputed International journals, out of which 13 are written as a sole author. He has received "MSIT Best Faculty Award" in 2018 and "MSIT Certificate of Excellence in Research" in 2018 and 2019. He has been placed among "Top 2% of Researchers in the World" for 2019 by Stanford University, USA. He is a regular reviewer of leading journals including IET and IEEE Transactions/journals. He is Associate Editor of *Journal of Electrical Engineering and Technology* (SCIE journal published by Springer), *International Journal of Numerical Modelling: Electronic Networks, Devices and Fields* (SCIE journal published by Wiley), and Academic Editor of *Plos One* (SCIE journal). He is Fellow of The Institution of Electronics and Telecommunication Engineers (I.E.T.E.), Associate Member of The Institution of Engineers (India), Life Member of Indian Society for Technical Education (I.S.T.E.) and Member of International Association of Engineers.

**Dr. Aman Dahiya** is currently working as an Assistant Professor in the Department of Electronics and Communication Engineering, Maharaja Surajmal Institute of Technology (MSIT), Janakpuri, New Delhi since 2009. She received Bachelor in Engineering (Electronics and Communication Engineering) with distinction, in 2005, from Maharashi Dayanand University, Rohtak, Haryana; MBA (Finance) in 2007 from Guru Jambheshwar University (GJU), Hissar; M.Tech. (ECE) in 2010 with distinction from Maharashi Dayanand University, Rohtak, Haryana and Ph.D. degree in Microwave Engineering in 2019. Her major areas of research include Design and

Analysis of Microwave and millimetre-wave Devices such as Rectangular Waveguides, Substrate Integrated Waveguides (SIW), Microstrip Patch Antennae, Reconfigurable antennae, IOT based applications etc. She has published 28 research papers in various national/ international journals and conferences of repute. She has authored three books in Microwave circuits and antenna designs. She has handled successfully IEEE MTT-s sponsored research project on Reconfigurable Antenna design for wireless communications. Her patent on "Antenna Designing" is published in 2020. She has been awarded "Certificate of Excellence" for best presentation, in 2006, "MSIT Certificate of Excellence in Research", in 2019. She is a reviewer of many leading journals. She is life member of Indian Society for Technical Education and associate member of IEEE.

**Dr. Naveen Garg** is Sr. Principal Scientist and Head, Acoustics and Vibration Standards in CSIR-National Physical Laboratory, New Delhi working in the field of Acoustics and Vibrations since past eighteen years. He is also guiding doctorate students in the field of noise control. He is a Mechanical Engineer specializing in Machine Design, Vibrations and Acoustics, Measurement science and working in development and up-gradation of primary standards of sound pressure and vibration amplitude and R&D in Applied Acoustics. He has done his Doctorate in Mechanical Engineering from Delhi Technological University, Delhi in January, 2016 in field of Noise Modelling and Control and M. Tech in Machine Design from IIT Delhi. He has been involved in many consultancy and sponsored projects pertaining to Environmental Impact Assessment studies in respect of noise and vibration for govt. bodies like Central Pollution Control Board (CPCB), Archaeological Survey of India (ASI), Delhi Metro Rail Corporation (DMRC), Bangalore Metro Rail Corporation Limited (BMRCL), India etc. He has published many papers in national and international journals in the field of acoustics and vibrations and has been actively involved in the Key Comparison exercises in Acoustics and Vibration metrology with other National Metrology Institutes (NMIs) of the world that has led to publishing of Calibration & Measurement Capabilities (CMCs) of CSIR-NPL, India in BIPM, Paris Key Comparison Database (KCDB). He has been Managing Editor of *MAPAN-Journal of Metrology Society of India* published by Springer and reviewer of many international journals and has enormously contributed towards evaluation, analysis and control of noise pollution. He was awarded APMP Iizuka Young Metrologist Prize for Developing Economies by Asia Pacific Metrology Programme (APMP) Secretariat, Japan in November, 2017 for his notable contributions to acoustics and vibration metrology.

# Energy Harvesting and Health Tracking Using Smart Shoe

**Mohammed Hammad Khan, Rakhi Kamra, Mohd Wasim, Kshitij Pathania, and Hritik**

**Abstract** Development in sensor technologies increasingly plays a key role in health care and fitness analysis, contributing to the development of new concepts of health maintenance like weight control and physical activity monitoring. These sensors allow us to study an individual's movement and mobility by collecting data such as step count and thus supporting a healthy living. In this paper, concept of a smart wearable system is presented called smart shoes. The designed system is based on Bluetooth technology to collect data using a smartphone in any environment. The system combines different sensor modules like a pedometer used for step counting, piezoelectric discs are used for electricity generation, Hall sensor is utilized for gesture detection, and load cell is used for measuring weight. The excess number of calories of energy utilized by an individual while walking provides enough motivation to implement a system with which energy can be generated that can further be used to recharge a battery. Also, designed smart shoes with sensors are able to send SMS alert in case of an emergency.

**Keywords** Sensor · Health care · Pedometer · Piezoelectric · Hall sensor · Load cell · Calories · Battery · SMS

## 1 Introduction

Wearable health devices are used in human body monitoring for fitness or medical purposes. Regardless of the purpose, the major challenge in designing these devices is to reduce power consumption to increase battery life, and to reduce short lifetime of wearable devices, the energy harvesting technology can be used. In this paper, concept of energy harvesting is used to design a system that observes the number of steps taken, calories of energy consumed and weight.

Thus, it can be utilized as a wearable health device [1] that can allow user to monitor the fitness level. The surplus number of calories consumed during walking

M. Hammad Khan (✉) · R. Kamra · M. Wasim · K. Pathania · Hritik
MSIT, Janakpuri, New Delhi, India
e-mail: hammad7khan@gmail.com

 
S. Yadav et al. (eds.), *Recent Advances in Metrology*, Lecture Notes in Electrical Engineering 906, https://doi.org/10.1007/978-981-19-2468-2_1

can be utilized for energy harvesting. Energy harvesting is to capture energy that is wasted or accessible free of cost from the environment. The excess of energy is often utilized to generate electricity by foot pressure. Due to exponential depletion of conventional sources of energy which are non-renewable resources, there is a need to shift focus on the non-conventional sources of energy, and energy harvesting is one of the options. The energy harvesters convert the energy available in surroundings in form of heat and vibrations into electrical energy [2]. Different types of energy harvesting methods are used, based on different phenomenon such as magnetism [3], electrostatic [4], thermoelectric [5], nano-triboelectric [6] and piezoelectric [7] strategies of energy harvesting. Among these methods, the piezoelectric structures are found to be more compact and easier to use than other energy harvesters.

This smart shoe can be used to monitor fitness levels, send SMS alert in case of emergency and uses piezoelectric effect for harvesting energy from the pressure and vibrations generated by walking. This energy is stored in a battery and is used to power the electronics in shoe. The background of the sensors used in the system is discussed in Sect. 2. An overview of the designed system is given in Sect. 3. The programming of microcontroller using flowchart is shown in Sect. 4. The result obtained after practical realization of the proposed system is discussed in Sect. 5. Section 6 concludes the paper and suggests the improvements in proposed system for future realizations.

## 2 Background

The various sensors used in the designed system are as follows:

### 2.1 *Accelerometer*

Accelerometer sensor detects the acceleration in any direction and accordingly gives a variable voltage output which can be used to determine the acceleration. The accelerometer generally works on the principle of capacitance [8] as due to movement, vibration occurs in a micro-spring-like structure which is used to vary the distance between two plates of capacitor, and thus, the capacitance varies, and this variation in capacitance gives a variable voltage (Fig. 1a).

### 2.2 *Load Cell*

Load cell is a type of transducer which converts force into electrical signals. In this system, strain gauge sensor is used as load cell; the strain gauge sensor works on the principle of resistance. The sensor has a microgrid of resistive wire (Fig. 1b), as the

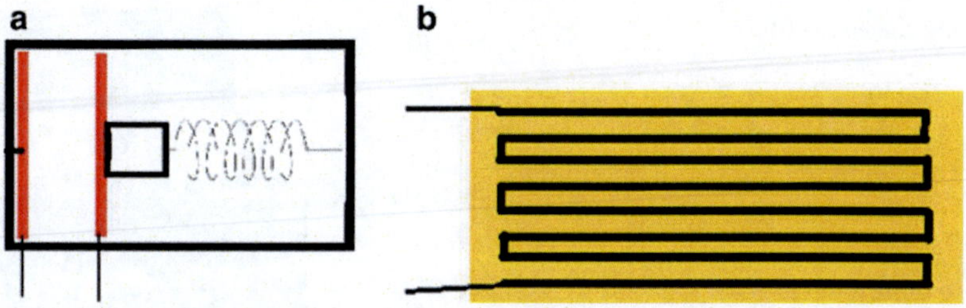


**Fig. 1** **a** Working of capacitive accelerometer. **b** Strain gauge

force is applied to the sensor, the area and length of the wire changes due to stress and strain, and this causes change in resistance. This change in resistance gives a change in voltage across it which can be measured and used to find the value of force. Load cells can be easily paired with a microcontroller to measure weight [9].

### 2.3 *Hall Sensor*

Hall sensors work on the phenomenon of Hall effect in which a potential difference is appeared across a current carrying conductor when a magnetic field is applied perpendicular to the direction of current, and the potential difference occurs is perpendicular to both the magnetic field and the direction of current. Due to this effect, the Hall sensors are generally used to detect magnetic fields.

### 2.4 *Piezoelectric Discs*

Piezoelectric discs are made up of piezoelectric materials, and these types of materials have special tendency to generate an electrical charge when they are subjected to mechanical stress. Thus, when these materials are subjected to a mechanical stress, a potential difference is generated across them (Fig. 2). Thus, these discs are generally used as sensors to detect vibrations, but they can also be utilized as energy harvesters as they can convert the energy generated due to vibrations into electrical energy.

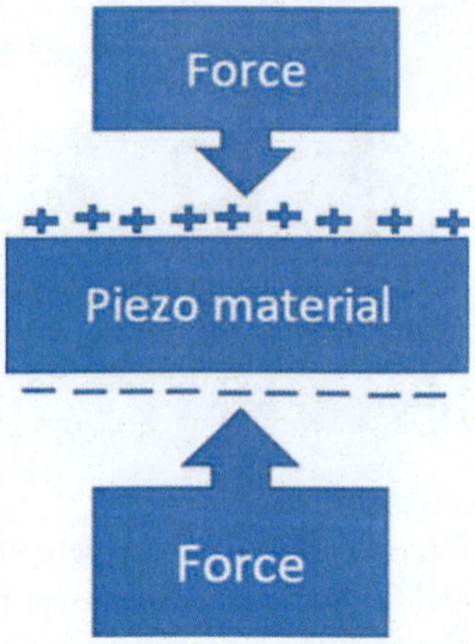


**Fig. 2** Piezoelectric effect

# 3 System Model

## *3.1 Health Tracking Unit*

Health tracking unit basically comprises of an Arduino microcontroller which takes data inputs from accelerometer (ADXL 335), load cell to determine step count, calories burnt and weight and sends it to smartphone companion app using Bluetooth module (HC-05). The circuit diagram of the health tracking unit is given (Fig. 3).

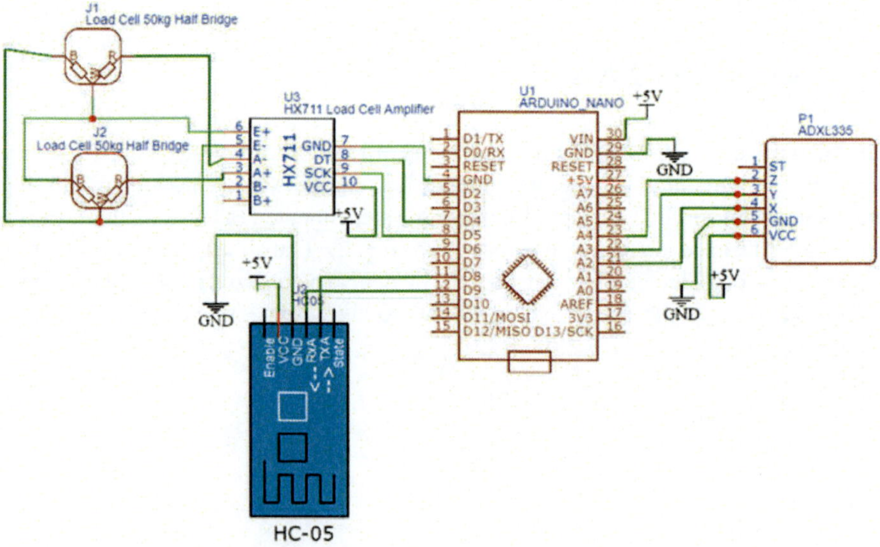


**Fig. 3** Circuit diagram of health tracking unit

### 3.2 *Emergency Alert Unit*

It detects a special gesture which user can make in an emergency situation by making an '*L*' shape by using two feet is shown in Fig. 4, systems detect the gesture [10] using a magnet and Hall sensor, and both are placed in opposite shoe. When gesture is detected, system sends a text message to predefined mobile number alerting that the user is in emergency. It utilizes SIM 800 l GSM module to send text messages. The circuit diagram of the emergency alert unit is given (Fig. 5).

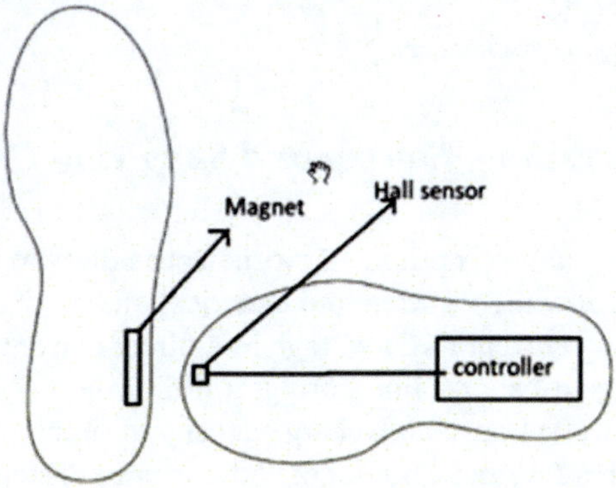


**Fig. 4** Gesture for emergency alert

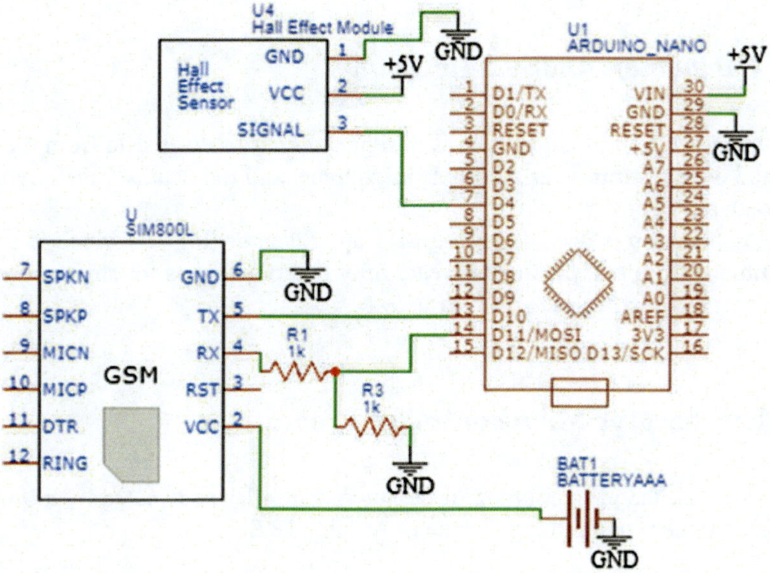


**Fig. 5** Circuit diagram of emergency alert unit

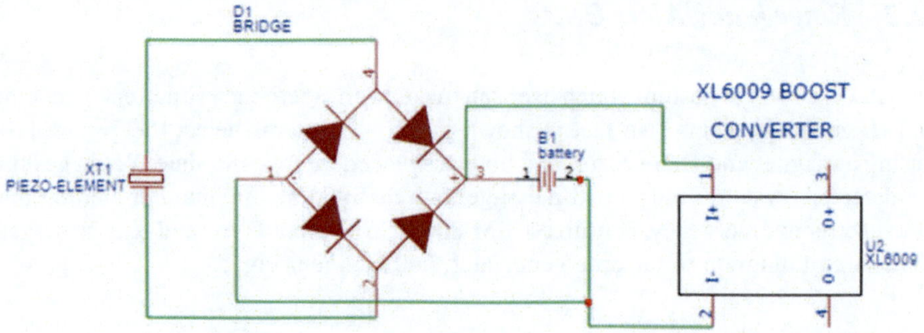


**Fig. 6** Circuit for energy harvesting unit

### 3.3 *Energy Harvesting, Storing and Supplying Unit*

The energy harvesting unit comprises 12 piezoelectric discs to convert the vibration and pressure produced while walking into electrical energy and a bridge rectifier to convert the alternating current (AC) voltage to a direct current (DC) voltage which can be used to charge a battery. For storing, a lithium ion (Li-ion) cell is used as they are compact as well as have high charge density to supply power to the modules most of which requires 5 V, and a boost converter is used to step up the 3.7 V battery voltage to 5 V.

The circuit diagram of the energy harvesting unit is given (Fig. 6).

### 3.4 *Companion Android Application*

A companion Android application is designed which receives data from the smart shoe and displays it in your Android smartphone, and the data is transferred using Bluetooth protocol.

The application is designed using MIT app inventor (Fig. 7), which is a visual programming tool that permits to create fully functional apps for smartphones and tablets.

## 4 Flowchart for Microcontroller Programme

In this section, the programming of the microcontroller is depicted by a flowchart (Fig. 8). All the programming is done in Arduino IDE.

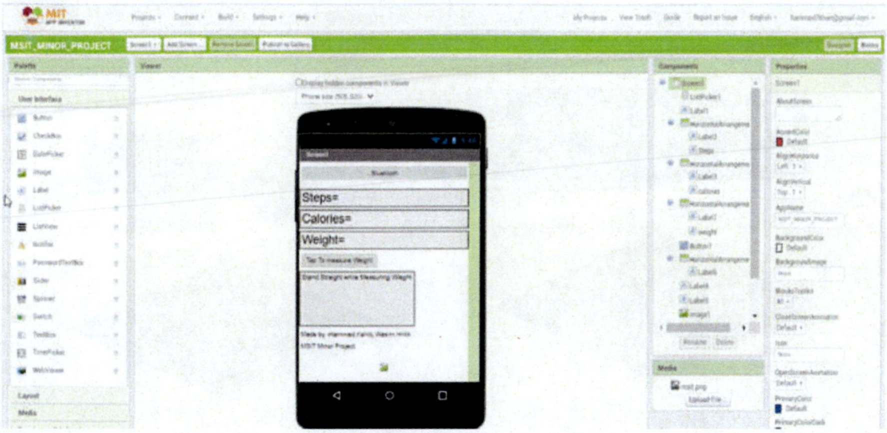


**Fig. 7** MIT APP inventor with application designed for this system

## 5 Results

All the components are assembled and tested successfully. The shoes are designed in such a way that the weight is calibrated only when the user presses the given button in the app so that the weight does not keep continuously changing as the user walks. The proposed system is found to be working, and steps count and weight values are precise, and with more polished algorithm, the values can be made more accurate.

The modules are attached to shoe in such a way that the whole assembly can be removed for adjustments and maintenance (Fig. 9).

The app is also installed in a smartphone and tested. The app is displaying all the data; also, the emergency alert system is able to detect the gesture and send the text message to the programmed number and the outcome (Fig. 10).

## 6 Conclusion

Smart shoe is an effective alternative of fitness bands as it can count the number of steps more accurately as compared to fitness band as they can sometime mistake hand movement for step, but since the shoe detects foot movements, they are more effective option; also, the weight of the body can be determined using a smart shoe which is difficult with any other smart wearable, and the energy harvesting also plays a significant role as it reduces the amount of time you need to recharge your device as most of the battery can be charged as you walk and run. The emergency alert system can be helpful for women security as well as kids who tend to go missing in public places and do not have a smartphone. The emergency alert system can be improved by installing a GPS module which can also provide the exact GPS coordinate to track the user.

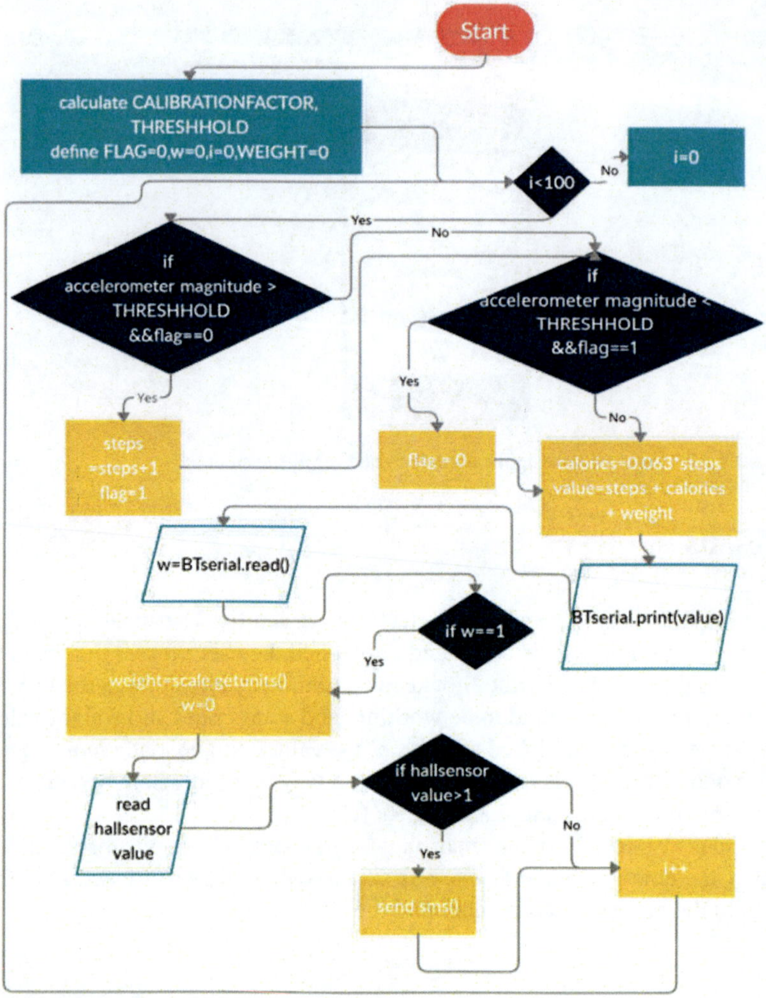


**Fig. 8** Flowchart for microcontroller programme

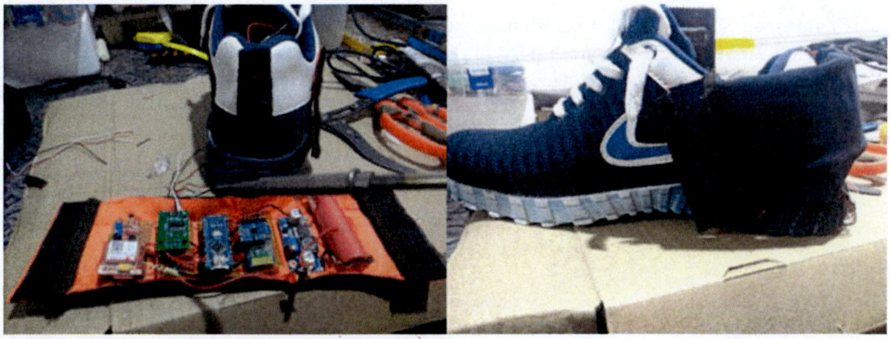

**Fig. 9** Prototype of the proposed system

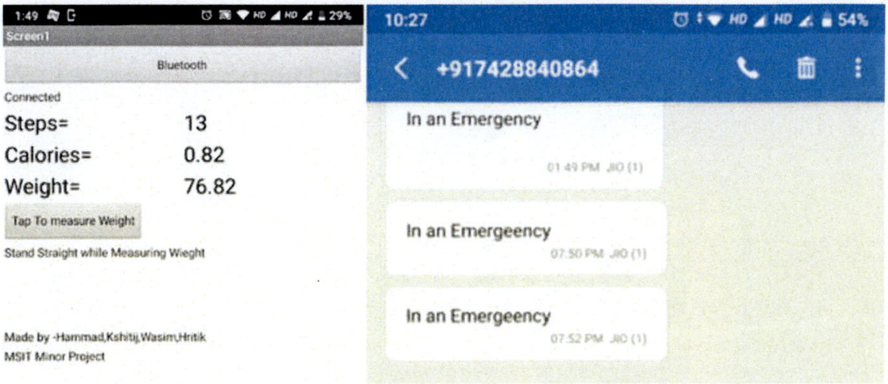


**Fig. 10** Companion application running on Android smartphone and alert message received when prototype detects gesture

## References

1. Dias D, Paulo silva cunha J (2018) Wearable health devices—vital sign monitoring, systems and technologies. Sensors 18(8):2414
2. Beeby SP, Tudor MJ, White NM (2006) Energy harvesting vibration sources for microsystems applications. Meas Sci Technol 17(12):R175
3. Yang B et al (2009) Electromagnetic energy harvesting from vibrations of multiple frequencies. J Micromech Microeng 19(3):035001
4. Basset P, Blokhina E, Galayko D (2016) Electrostatic kinetic energy harvesting. Wiley
5. Enescu D (2019) Thermoelectric energy harvesting: basic principles and applications. Green energy advances. IntechOpen
6. Wu C, Wang AC, Ding W, Guo H, Wang ZL (2019) Triboelectric nanogenerator: a foundation of the energy for the new era. Adv Energ Mat 9(1):1802906
7. Kim HS, Kim JH, Kim J (2011) A review of piezoelectric energy harvesting based on vibration. Int J Precis Eng Manuf 12(6):1129–1141
8. Hsu YW, Chen S, Chien HT (2012) Multi-axis capacitive accelerometer. U.S. Patent no. 8,205,498

9. Phulphagar V, Jaiswal R (2017) Arduino controlled weight monitoring with dashboard analysis. Int J Res Appl Sci Eng Technol 5:1164–1167
10. Walker DR (2017) Portable electronic device having a sensor arrangement for gesture recognition. U.S. Patent no. 9,569,002

# Uncertainty Evaluation Through Monte Carlo Simulation for Flatness Measurement of Optical Flats

**Girija Moona, Abhishek Singh, Anju, Guruvandra Singh, Sudha Yadav, Mukesh Jewariya, and Rina Sharma**

**Abstract** In this investigation, the flatness error of an optical flat has been measured using Fizeau laser interferometer, followed by measurement uncertainty evaluation. Various contributing error sources such as laser interferometer resolution, environmental conditions, mechanical distortions and reference flat uncertainty have been identified, and the uncertainty evaluation for flatness error measurement has been conducted through two approaches: law of propagation of uncertainty (LPU) and Monte Carlo simulation (MCS). Measurement uncertainty through LPU approach has been $\pm$ 20.8 nm ($k = 2$), whereas measurement uncertainty using MCS has been $\pm$ 20.3 nm ($k = 2$). Measurement uncertainties evaluated through both the techniques have been observed to be in good agreement.

**Keywords** Flatness · Optical flats · Interferometry · Uncertainty · Monte Carlo simulation

## 1 Introduction

Optical flats are used as reference standards in many advance engineering applications; hence, flatness measurement of optical flats should be carried out with significant accuracy and precision. Generally, flatness errors of various optical surfaces are determined through a popular non-contact measurement method, i.e. laser interferometry, by comparing the test surface with a superior reference surface of predetermined flatness deviations. The flatness error of optical flat under test is measured in the form of distance deviations between reference flat surface and test flat surface at various locations concurrently.

An interferogram is generated in accordance with the phase difference between the reflected and transmitted wavefronts, indicating the form deviation of the optical

G. Moona (✉) · A. Singh · Anju · G. Singh · S. Yadav · M. Jewariya · R. Sharma
Length, Dimension and Nanometrology, CSIR-National Physical Laboratory, Delhi, India
e-mail: moonag@nplindia.org

G. Moona · Anju · G. Singh · S. Yadav
Academy of Scientific & Innovative Research (AcSIR), Ghaziabad 201002, India

 
S. Yadav et al. (eds.), *Recent Advances in Metrology*, Lecture Notes in Electrical Engineering 906, https://doi.org/10.1007/978-981-19-2468-2_2

flat under test, in the form of p–v flatness (difference between the maximum perpendicular distance i.e. "peak" and minimum perpendicular distance, i.e. "valley" of the test surface from the best-fit plane) [1]. It has been conspicuously demonstrated that no measurement observations are perfect, and there shall always be inclusion of certain errors in every experimental outcome. Hence, to perform quality measurements, uncertainty evaluation is carried out and associated uncertainty is reported along with measurement observations [2].

## 2 Measurement Arrangement and Procedure

The laser interferometer (Zygo VerifireXP/D) used here for flatness error measurements functions on Fizeau configuration with phase shift capability. In this measurement setup as shown in Fig. 1, the collimated laser beam emitting from a low power Helium–Neon laser source by amplitude division exits the interferometer through aperture. First, the reference wavefront is created from the reflected laser beam by the reference/transmission flat, mounted on a receptacle attached to the main frame. Further, the transmitted laser light falls on the test flat, mounted on a separate adjustable receptacle and creates the measurement wave front after getting reflected from the test surface. The two receptacles are placed as close as possible in order to avoid air turbulence effects. The piezoelectric transducers in main frame move the reference flat forward and backward, maintaining constant phase deviations between reference and measurement wavefronts. The two wavefronts are combined and generate a fringe pattern, visible on live display window [1]. The interference pattern is transformed into electrical signals by a video camera allowing software acquisition, and analysis for flatness error is carried out being displayed on the primary monitor, as shown in Fig. 2.

In present experimental investigation, repeated flatness error measurement for optical flat of diameter 70 mm is conducted, followed by identification of various error sources contributing towards measurement uncertainties.

The measurement uncertainty evaluation is done using two techniques: law of propagation of uncertainties (LPU/GUM) and Monte Carlo simulation (MCS). In

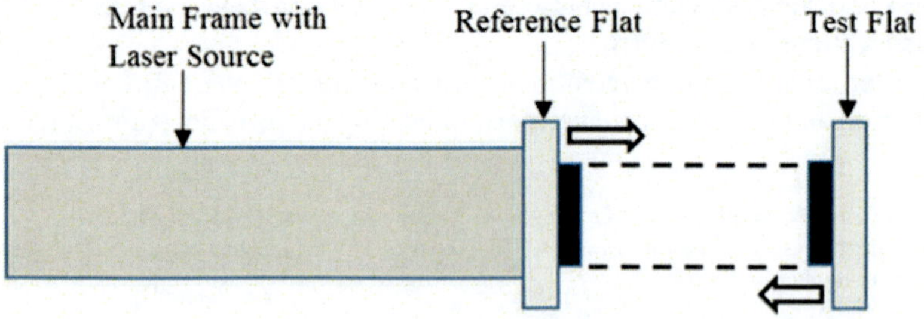


**Fig. 1** Flatness measurement setup

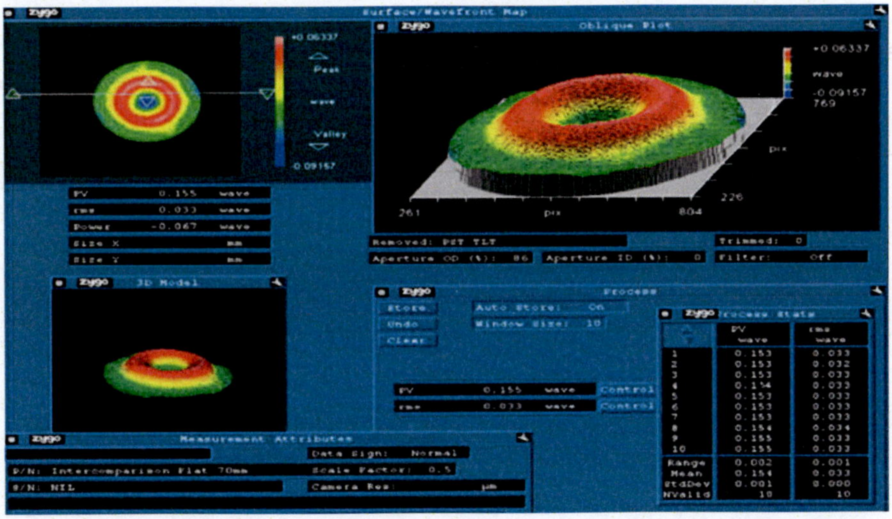


Fig. 2 Flatness measurement and analysis

LPU method, a model equation is formed for different input quantities and associated uncertainties, assuming that the measurement uncertainty evaluation is driven from input quantities. Estimated value of each input is considered for standard uncertainty computation [2]. Furthermore, sensitivity coefficient calculations are performed, and combined uncertainty evaluation is conducted. To surmount the challenges associated with LPU, such as dependence on central limit theorem and sensitivity coefficient calculation, Monte Carlo simulation being able to provide actual probability distribution of measurand and bypassing sensitivity coefficient and effective degree of freedom computation is projected as an alternative method of uncertainty evaluation [3]. Monte Carlo simulation approach for measurement uncertainty evaluation includes formulation of model function, probability density function estimation of input quantities, random number generation, conduction of Monte Carlo simulation and result compilation [4, 5].

## 3 Result and Discussion

Flatness measurement observations for test flat are demonstrated in Table 1. The model function for flatness measurement is displayed in Eq. 1. Various potential error sources existing during measurement and contributing in measurement uncertainty are shown in Fig. 3.

Model function for flatness measurement using Fizeau interferometer is established as:

**Table 1** Observation table

| Measurement set | PV wave | PV (nanometer) = PV wave × 632.8 nm |
|---|---|---|
| 1 | 0.153 | 96.82 |
| 2 | 0.153 | 96.82 |
| 3 | 0.153 | 96.82 |
| 4 | 0.154 | 97.45 |
| 5 | 0.153 | 96.82 |
| 6 | 0.153 | 96.82 |
| 7 | 0.153 | 96.82 |
| 8 | 0.154 | 97.45 |
| 9 | 0.155 | 98.08 |
| 10 | 0.155 | 98.08 |
| | Mean | 97.20 |
| | SD | 0.53 |
| | Type A uncertainty ua (nm) | 0.2 |

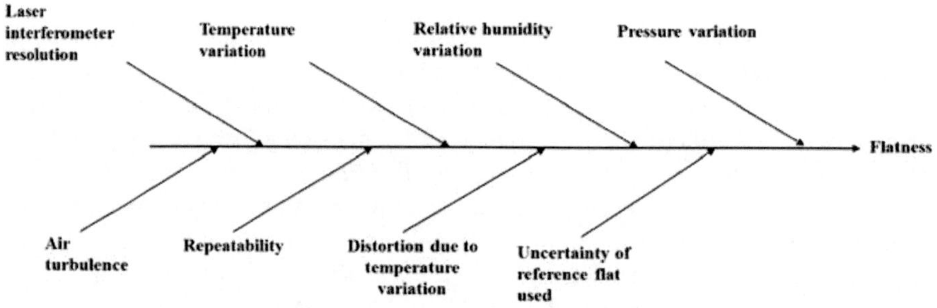


**Fig. 3** Fishbone diagram indicating dominant error sources [6, 7]

$$\text{Error in flatness measurement} = \delta_{LR} + \delta t + \delta_{RH} + \delta_{\mathrm{Pr}} + \delta_{Dt} + \delta_{AT} + \delta_{DM} + \delta_{TF} \quad (1)$$

Here:

$\delta_{LR}$ Laser interferometer resolution contribution.
$\delta t$ Temperature variation contribution.
$\delta_{RH}$ Relative humidity variation contribution.
$\delta_{Pr}$ Pressure variation contribution.
$\delta_{Dt}$ Distortion due to temperature variation contribution.
$\delta_{AT}$ Air turbulence contribution.
$\delta_{DM}$ Distortion due to mounting contribution.
$\delta_{TF}$ Transmission flat uncertainty contribution.

Uncertainty budget for flatness as per LPU is presented in Table 2

**Table 2** Uncertainty budget for flatness measurement as per LPU

| Sources of uncertainty | Limits | Units | Distribution | Standard uncertainty | Sensitivity coefficient | Uncertainty contribution (μm) |
|---|---|---|---|---|---|---|
| Laser resolution | 0.00001 | μm | Normal | 5.00E−06 | 1 | 5.00E−06 |
| Temperature variation | 1 | °C | Rectangular | 0.5774 | 1.00E−06 | 5.77E−07 |
| Relative humidity variation | 10 | % RH | Rectangular | 5.7735 | 1.00E−08 | 5.77E−08 |
| Pressure variation | 10 | mm of Hg | Rectangular | 5.7735 | 3.33E−07 | 1.92E−06 |
| Air turbulence effect | 0.002 | μm | Rectangular | 0.0012 | 1 | 1.15E−03 |
| Distortion due to temperature difference | 0.002 | μm | Rectangular | 0.0012 | 1 | 1.18E−03 |
| Distortion due to mounting | 0.0001 | μm | Rectangular | 0.000031 | 1 | 3.11E–-05 |
| Reference flat uncertainty | 0.020 | μm | Normal | 0.0102 | 1 | 1.02E−02 |
| Repeatability | 0.0002 | μm | Normal | 0.0002 | 1 | 1.78E−04 |
| Combined uncertainty | | | | | | ±10.4 nm |
| Expanded uncertainty | | | | | | ±20.8 nm |

The measurement uncertainty evaluation for flatness measurement of an optical flat has also been performed through Monte Carlo simulation. Random numbers are generated for each input quantity through 2, 00, 000 iterations, and subsequently, average measured value and standard deviation have been computed. The probability distribution histogram at $k = 2$ is obtained and is displayed in Fig. 4, and the terse summary of MCS result is given in Table 3.

Table 4 exhibits the comparison between LPU and MCS results.

## 4 Conclusion

Present paper describes the measurement uncertainty evaluation of an optical flat calibrated against a reference flat through Fizeau interferometry. Major error sources influencing the measurement procedure are identified, and uncertainty evaluation is performed through two different techniques: law of propagation of uncertainty and Monte Carlo simulation as per the JCGM recommendations.

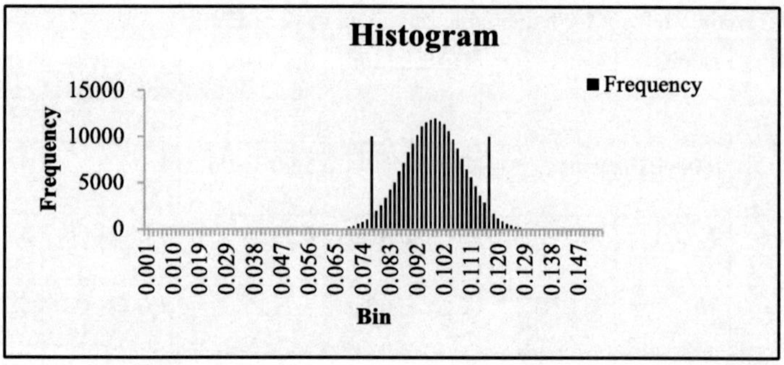


**Fig. 4** MCS histogram for flatness measurement

**Table 3** Monte Carlo simulation summary

| Parameters | Monte Carlo simulation outcomes for flatness measurement |
|---|---|
| Mean measured flatness value (nm) | 97.3 |
| Expanded uncertainty (nm) at k = 2 | ±20.3 |
| Low-end point (nm) | 77.4 |
| High-end point (nm) | 117.2 |

**Table 4** LPU versus MCS

| Technique | Mean flatness value (nm) | Expanded uncertainty (nm) at $k = 2$ |
|---|---|---|
| LPU | 97.2 | ±20.8 |
| MCS | 97.3 | ±20.3 |

Measurement uncertainties evaluated by both the approaches are noticed to be agreeable. According to LPU, the mean flatness error value is 97.2 nm with expanded uncertainty ±20.8 nm, whereas as per MCS, the mean flatness error value is 97.3 nm mm with expanded uncertainty ±20.3 nm

**Acknowledgements** Authors are thankful to the Director, National Physical Laboratory, for his continuous support and encouragement

## References

1. Moona G, Sharma R, Kiran U, Chaudhary KP (2014) Evaluation of measurement uncertainty for absolute flatness measurement, by using fizeau interferometer with phase-shifting capability. Mapan 29(4):261–267
2. JCGM 101 (2008) Evaluation of measurement data—Supplement 1 to the Guide to the expression of uncertainty in measurement—Propagation of distributions using a Monte Carlo method, First (2008)
3. Moona G, Sharma R, Kumar H (2018) Evaluation of uncertainty of measurement of shadow mask dot pitch using different approaches. Trans Inst Meas Control 40(7):2428–2435
4. Kumar H, Moona G, Arora PK, Haleem A, Singh J, Kumar R, Kumar A (2017) Monte Carlo method for evaluation of uncertainty of measurement in brinell hardness scale. Indian J Pure Appl Phys 55:445–453
5. Moona G, Kumar V, Jewariya M, Sharma R, Kumar H (2019) Measurement uncertainty evaluation using Monte Carlo simulation for newly established line scale calibration facility at CSIR-NPLI. Mapan 34(3):325–331
6. Azpurua M, Tremola C, Paez E (2011) Comparison of the gum and Monte Carlo simulations for the uncertainty estimation in electromagnetic compatibility testing. Prog Electromagnet Res B 34:125–144
7. Chen A, Chen C (2016) Comparison of GUM and Monte Carlo methods for evaluating uncertainty of measurement of perspiration measurement systems. Meas 87:27–37

# Internet of Things and Cognitive Radio Networks: Applications, Challenges and Future

**V. Dankan Gowda, B. S. Puneeth Kumar, P. Ramesh Naidu, and S. B. Sridhara**

**Abstract** Intelligent technology to cope with bandwidth shortage problems has arisen as cognitive radio (CR). The CR is supposed to utilise the unoccupied spectrum band if the approved person may not use it. Since this technology was introduced in 1999, comprehensive research has taken place, addressing diverse challenges including spectrum sensing, coordination among CR users and the applicability of CR networks. In this article, we introduce new applications of Internet of Things (IoT) CR technology and provide adequate solutions for the real CR technology problems that render IoT more accessible and more useful and also provide an overview of CR-based IoT systems. We illustrate feasible implementations of IoT systems focused on CR.

**Keywords** Cooperative communication · Pervasive computing · Internet of Things · Cognitive radio

## 1 Introduction

The radio spectrum is the most powerful resource for wireless communications [1]. Latest experiments have shown that radio spectrum is still underused [2]. Cognitive (CR) technology [3–5] has appeared as a potential approach for addressing spectrum

V. Dankan Gowda (✉)
Department of Electronics and Communication Engineering, BMS Institute of Technology and Management, Bangalore, Karnataka, India
e-mail: dankanies@gmail.com

B. S. Puneeth Kumar
Department of Computer Science, St. Joseph's College, Bangalore, Karnataka, India

P. Ramesh Naidu
Department of Computer Science and Engineering, Nitte Meenakshi Institute of Technology, Bangalore, Karnataka, India

S. B. Sridhara
Department of Electronics and Communication Engineering, Vijaya Vittala Institute of Technology, Bangalore, Karnataka, India

 
S. Yadav et al. (eds.), *Recent Advances in Metrology*, Lecture Notes in Electrical Engineering 906, https://doi.org/10.1007/978-981-19-2468-2_3

depletion and its inefficiency of wireless networking. CR nodes are intelligent wireless machines that interpret the world, experience changing the network and make wise choices to use the information obtained from the previous encounter with the network. This spectrum scanning method (*S*), control-sharing knowledge (*E*), white space agreement (*A*) and network data transfer (*T*) are continuously replicated in a SEAT cycle [6]. Figure 1 displays the contact model for CR.

CR technology dramatically enhances bandwidth performance by providing unlicenced consumers opportunistic access to frequency bands (also called CR users or secondary users, SUs). Licenced consumers (also known as main users, PUs) build spectrum gaps or white spaces in the spectrum when not being broadcast. SUs utilise certain PU-owned white spaces to enter these spectrum gaps opportunistically without intervening with PUs.

Over time, the technology advances to make it simpler for users, contributing to an explosion in gadgets connecting to the Internet. Owing to the vast number of devices impacts data transmission and receiving speed and productivity and creates interference in the network architecture. So, CR may effectively use the unused bandwidth, which already belongs to the primary consumer to allow IoT objects (PU). CR is an insightful technique of wireless communication that, in all situations, realises its environment. The first to suggest CR is Joseph Motila, using cognitive radio to optimise the bandwidth usable and boost efficiency. It is called an intelligent radio which knows the state of the spectrum. It is made up of an extra consumer (SU) and a PU. The bandwidth is used by SUs so long as it does not interact with PUs. The SU must quit the channel and jump on to another channel if PU wants its channel [1–4]. Notice that the key roles of CR are spectrum sensing, spectrum exchange, spectrum mobility and spectrum control 2 [5] and [6] (Fig. 2).

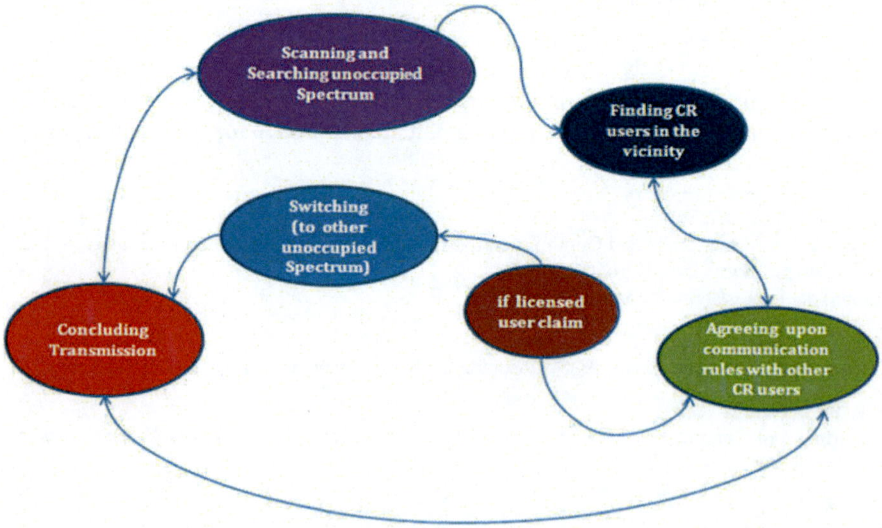


**Fig. 1** Communication model for cognitive radio

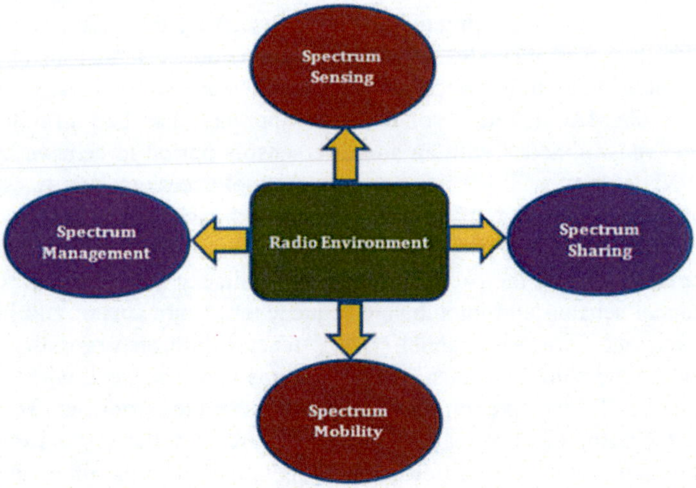


**Fig. 2** Functions of cognitive radio

The latest definition of IoT implies that different objects can be linked to the Internet through Internet IPv4 or IPv6. It helps people to engage with, connect with and serve without being in a single area. The official IoT declaration is by Kevin Ashton in 1999. There are various patterns of IoT relations such as: Things-to-There, Human-to-Thing and Human-to-Human (H2H). The protocols that may be used for IoT connectivity are: telephone, Wi-Fi, Bluetooth, ZigBee, detection of radio frequencies and close field communication [7–11].

## 2 Literature Survey

In order to increase spectrum access and usage of CR-IoT, this segment carries out a detailed survey relevant to the different current methods offered. In addition, the study void for achieving an appropriate spectrum resource distribution model for CR-IoT describes state-of-the-art work.

Opportunistic spectrum access model for CR-IoT: in [6], reciprocal regulation of sample rates and channel connectivity of sensor nodes under energy usage, channel bandwidth and interference constraints was considered as an optimiser network utility. With fluctuating rates of energy harvesting and channel transition costs, we formulate network efficiency maximisation as a nonlinear programming problem with a mixer and cope with it effectively and coupled with double decline. A joint channel access and sampling rate control framework called JASC would then be presented, taking account of the results of real-time channel sensing and energy recovery rates. If nodes have data to relay, spectrum sensing on demand is initiated and a subset of spectrum sensing nodes is often used to gather information on spectrum availability across all nodes. In addition, CRSN nodes establish an adaptive duty period that enables them to sleep consistently and stay awake if transmission data is necessary. CAMAC is thus a sophisticated approach that uses a limited number of spectrum sensing nodes with an adaptive sensing period to achieve low energy consumption. Aslam et al. [7] the complex channel access model was introduced in light of the energy usage of channel sensing and exchange, and the state of the sensor nodes and the transfer to a licenced channel in order to increase energy efficiency are calculated by the packet loss rate on the licence-free channel. In order to define channel sensing and interchange sequences for intracluster and intercluster data transfer, two complex channel control schemes are also proposed. In [8], a branch-and-bound solution focused on algorithms was proposed, with a reduction in search space. Their approach constructs the best routes from each source to the sink node, including the best set of hops in each direction, the best variety of relays and the maximum allotment of power for cooperative transmission connections. In [9], a QoS-conscious packet scheduling approach is implemented to boost secondary consumer transmission efficiency. A QoS-based priority model is specifically structured to resolve data classification within this framework. And then, channel efficiency and channel transmission impact are implemented in the packet scheduling process on the basis of priority.

Opportunistic spectrum access employing evolutionary computing model for CR-IoT: increased channel sensing learning was carried out in the cluster-based wireless sensor network [10]. The efficiency technique is built to reduce sensing by adding the problem of the Markov decision. The result indicates a rise in the identification of primary customers and the cost of energy over conventional greedy searches. The spectrum sensing time period model has been developed [10] to provide estimated data for the secondary consumer channel sensing sources. They also introduced a complex sensing paradigm for the Markov chain, which facilitates listening before talking. The simulation involves various time scales. The findings indicate that the model for the Markov chain enhances cognitive radio WSN energy efficiency, but the network density is known to differ. The dilemma of sequential multi-user channel sensing links to complex neural networks, in which the active user collection is discussed [11]. In addition, each user has just his or her particular details without information sharing between users.

## 3 Motivations for Using CR in IoT

With the continuing growth of CRNs and IoT, it is anticipated that potential requirements will be fulfilled through CR-based IoT frameworks. In order to read, think and make choices from an awareness of both social and physical environments, the IoT objects are fitted with cognition [3, 6, 8]. Intelligent decision-making, a perception action loop, large data processing, provision of on-demand resources, semanticised derivation and information exploration are also needed. A CR-based IoT is thus a predictable requirement in future; this may be for the following purposes.

The primary incentive is the distribution of bandwidth for IoT artefacts. The IoT number objects are anticipated to evolve in significant numbers, and these objects would be challenging to grant spectrum bands. The number of Pus would also rise, posing issues for unlicenced consumers. The strategy of fixed spectrum assignment involves spectrum purchasing costs; thus, spectrum assignment will generate needless expenses for a large range of items. In all these cases, CRNs may facilitate. Orthodox networking strategies do not accept multiple users' spectrum sharing. With its spectrum-sharing benefit, CRN can in future improve as artefacts grow and spectrum access is pursued [8, 9]. Therefore, CRNs will interpret the spectrum landscape and provide consumers with on-demand resources through smart decision-making.

A new paradigm of cloud servers is gaining prominence to help big data produced by a variety of IoT artefacts and semantic derivation problems in future. CR-based IoT artefacts are required to autonomously scan for accessible storage places on cloud servers and submit data from them through spectrum sensing [2, 4].

## 4 CR-based IoT Applications

Given the fact that cognitive radio technology is at an early age and there is no clear illustration of completely functioning, usable, cognitive radio, because of its amazing and accessible capabilities, there is a great deal of promise and applicability for cognitive radio. Cognitive radio is also an exciting application for developing current software-defined radio (SDR) technology. We will analyse cognitive IoT implementations in this portion.

### *4.1 Health Care*

There are IoT health technologies that are feasible. Shrewd sensors are distributed silently to track essential details, such as temperature, blood pressure, glucose and other information. The restaurant team constantly tracks criteria [12, 13] with remote control. We believe that few low-energy IoT devices (sensors) are attached to the body of the patient, as seen in the Fig. 3. IoT devices like phones for patients, portable devices and RFIDs are responsible for transmitting data packets to a strong node named the cluster head (CH). The CH conducts several activities such as data analysis, spectrum sensing, transmitting control messages to PU operation IoT computers. We presume that the CH has a double antenna, one antenna for spectrum detection and another for sharing of data with the member nodes. Once each active acknowledgement or beacon is obtained, the IoT devices must relay data.

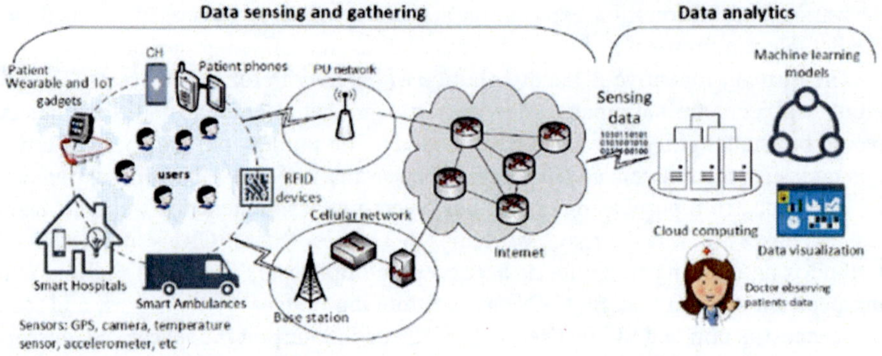


**Fig. 3** Application scenario of the CR-IoT

## *4.2 In-home Applications*

IoT-based arrangements are now inside the operating room within a house/building, and IoT would be a must in future as creativity progresses. The administration of domestic fertility is now reflected in the frame of such photographs such as a keen fridge and keen lamps [14].

## *4.3 Smart Grid*

Buyers like to recognise everything and anything relevant to their energy use at all moment. This demonstrates the need for IoT in future in a strong network. One big drawback here is the sharing of vast quantities of data at a small pace without obstructing large distances from many meters/devices [15].

## *4.4 Smart Cities*

It is an urban planning model that involves convergence of ICT and IoT structures and creativity. The motivation behind shrewd cities is to offer e-services to consumers in an eco-friendly fashion for better living. In order to enable this, an infinite network would be necessary, which would render a data and connectivity system the backbone. In addition, knowledge processing and consumer engagement would be important. The continuous network problem can be sponsored by CRNs of cognitive radio networks [16].

## 5 Issues Associated with CR-based IoT

Proper CRN hardware architecture results in optimal usage of the whole network. Another key concern is that the antennas used with CRNs are not the same in scale as the antennas are used with one frequency range (say cellular) (ISM 2.4 GHz). In addition, the transmitting power levels often differ nature with the spot. Some gateways are necessary to link IoT artefacts to networks. In construction procedures, versatility, scalability, protection and energy efficiency of the entrances are truly essential. Effective spectrum usage should be adequately managed in multi-user situations. CR users typically scan individually for links to a continuum. Properly built gateways will sensor their faithful spectrum. Geolocation-based spectrum analysis may be a great assistance with previous quest figures. There may also be two other issues, versatility and interoperability [17, 18]. The IoT is a mixture of heterogeneous items. Normally, the CR technology involves the node-based design with proper control strategies. Networks. However, this impact on certain protection concerns is not relevant on all heterogeneous networks as consistent security requirements.

## 6 Spectrum-related Functions

The sensing of spectrum is a preliminary and simple phase. In a complex setting, IoT artefacts like CRs have to search for spectrum in the presence of many Pus, and the issue is amplified when Pus are in the same band. At the same time, it is also necessary to distinguish between licenced and unlicenced signals. The sensing of spectrum requires time and needs electricity. Quick and energy-efficient algorithms should therefore be created. CRs do not conduct a role while spectrum sensing. We think different radios can.

Usage when a radio detects the spectrum, while another transfers the details. However, it is efficient when paired with sophisticated geolocation and historical algorithms. It can be effective. In addition, we cannot compromise on the likelihood of misdetection, but we can compromise on the probability of incorrect detection. Misdetection is a case when a PU is not detected and a PU is detected, where there is no PU [5].

If the range is ideally sensed, the search bands may settle on the transition of data.

A problem. A issue. Multiple bands cannot satisfy application-specific QoS specifications in some bands. Real-time responses are needed for sensitive applications, whereas non-critical applications may sacrifice response time. In addition, spectrum assignment among many CRs is also a challenge and involves shared resource and application-specific algorithms.

## 7 Standardisation Activities and Challenges

Standardisation of IoT is still in its early stages, and CR-based IoT racticesation is a meagre issue to explore. The first and most critical issue is the creation of semantic norms, since it will speed up the pathway towards interoperable solutions. CR-based IoT frames will enable a wide variety of various devices for effective scenarios. The actual appearance of a wide number of instruments is related to a set of application specifications. This needs continuing sémantic exchange of knowledge [4]. Other main fields of standardisation include communication protocols, collection of artefacts and transmitting capacity on layer. When standardisation practices are conducted simultaneously across the world, the cumulative gains will be accomplished. In addition, these efforts could be condensed and saved easily, which may be a more concern in future [19].

## 8 Conclusion

In this article, we have mentioned the need for cognitive IoT radio networks. Both systems are in the early stages, but IoT has achieved relatively little testing. The realisation of CR-based IoT remains thus a modern paradigm and must be analysed in depth. IoT is meant to be a drain on current network without cognition. We believe that CR and IoT have a great deal of potential and applicability, especially if both technologies constitute a coalition. Any of the applications for CR was presented in this paper along with the difficulties and problems currently addressed by CR technology.

## References

1. Kakkavas G, Tsitseklis K, Karyotis V Papavassiliou S (2019) A software defined radio cross-layer resource allocation approach for cognitive radio networks: from theory to practice. IEEE Trans Cogn Commun Network. https://doi.org/10.1109/TCCN.2019.2963869
2. Ren J, Zhang Y, Deng R, Zhang N; Zhang D, Shen X (2016 April 21) Joint channel access and sampling rate control in energy harvesting cognitive radio sensor networks. IEEE Trans Emerg Top Comput PP(99) pp1–1
3. Shah G, Akan O (2015) Cognitive adaptive medium access control in cognitive radio sensor networks. IEEE Trans Veh Tech 64(2):757–767
4. Li N, Xiao M, Rasmussen LK (2017)Spectrum sharing with network coding for multiple cognitive users. IEEE Internet Things J. https://doi.org/10.1109/JIOT.2017.2728626
5. Aijaz A, Aghvami AH (2015) Cognitive machine-to-machine communications for internet-of-things: a protocol stack perspective. IEEE Internet Things J 2(2):103–112
6. Chiwewe TM, Hancke GP (2017) Fast convergence cooperative dynamic spectrum access for cognitive radio networks. IEEE Trans Indust Inf https://doi.org/10.1109/TII.2017.2783973
7. Aslam S, Ejaz W, Ibnkahla M (2018) Energy and spectral efficient cognitive radio sensor networks for internet of things. IEEE Internet Things J https://doi.org/10.1109/JIOT.2018.2837354

8. Ejaz W, Shah GA, Kim HS et al (2015) Energy and throughput efficient cooperative spectrum sensing in cognitive radio sensor networks. Trans Emerg Telecommun Technol 26(7):1019–1030
9. Maghsudi S, Stanczak S (2016) Hybrid centralized-distributed resource allocation for device-to-device communication underlaying cellular networks. IEEE Trans Veh Technol
10. Xu Y, Wang J, Wu Q, Anpalagan A, Yao YD (2012) Opportunistic spectrum access in unknown dynamic environment: a game-theoretic stochastic learning solution. IEEE Trans Wirel Commun 1380–1391
11. Anandkumar A, Michael N, Tang A, Swami A (2012) Distributed algorithms for learning and cognitive medium access with logarithmic regret. IEEE J Sel Areas Commun Adv Cogn Radio Network Commun 29:731–745
12. Vakili S, Liu K, Zhao Q (2013) Deterministic sequencing of exploration and exploitation for multi-armed bandit problems. IEEE J Sel Top Sig Process 59(3):1902–1916
13. Gai Y, Krishnamachari B (2011) Decentralized online learning algorithms for opportunistic spectrum access. Proc IEEE Global Commun Conf (GLOBECOM)pp 1–6
14. Xu Y, Wu Q, Wang J, Shen L, Anpalagan A (2015) Robust multiuser sequential channel sensing and access in dynamic cognitive radio networks: potential games and stochastic learning. IEEE Trans Veh Technol 64(8):3594–3607
15. Zhou F, Beaulieu NC, Cheng J, Chu Z, Wang Y (2017) Robust Max–Min fairness resource allocation in sensing-based wideband cognitive radio with SWIPT: imperfect channel sensing. IEEE Syst J https://doi.org/10.1109/JSYST.2017.2698502
16. Jang H, Yun SY, Shin J, Yi Y (2018) IEEE Trans Wirel Commun 17(1):194–209. https://doi.org/10.1109/TWC.2017.2764081
17. Lu Y, Duel-Hallen A (2018) A sensing contribution-based two-layer game for channel selection and spectrum access in cognitive radio ad-hoc networks. IEEE Trans Wirel Commun 17(6):3631–3640
18. Ren J, Zhang Y, Zhang N, Zhang D, Shen X (2016) Dynamic channel access to improve energy efficiency in cognitive radio sensor networks. IEEE Trans Wireless Commun 15(5):3143–3156
19. Mansourkiaie F, Ahmed MH (2016) Optimal and near-optimal cooperative routing and power allocation for collision minimization in wireless sensor networks. IEEE Sens J 16(5):1398–1411

# Carbon Nanotubes-Based Biosensors

**Shashi Chawla, Prateek Rai, and Tanya Garain**

**Abstract** The use of carbon nanotubes (CNTs) in the field of biosensing technology is considered a revolution, as it aims to produce faster and more sensitive biosensor devices. Both the physical and mechanical properties of CNTs make them highly suitable to be utilized as biosensors. Surface functionalizations and modifications of CNTs make them even more versatile, giving rise to various types of CNT-based biosensors. Though CNT-based biosensors find wide operations in the field of biomedical sciences, some major issues regarding their toxicity and biocompatibility need to be acknowledged before commercializing them completely. This review article focuses on the properties of CNTs which make them one of the prominent candidates to be used for biosensing purposes, chemical modifications of CNTs, and types and working of CNT-based biosensors.

**Keyword** Carbon nanotubes · Chemical functionalization · Biosensors · Toxicity · Biocompatibility

## 1 General Characteristics of Carbon Nanotubes (CNTs)

Carbon is one of the core elements for the survival of living organisms and exists in three allotropic forms, namely: graphite, diamond, and amorphous carbon [1]. Discovered by Iijima in 1991 [2, 3] carbon nanotubes, also called Bucky tubes, have been the topic of interdisciplinary research, owing to its distinctive electrical, optical, thermal, and mechanical properties like high tensile strength, high thermal and electrical conductivity, and the scope of conjugating different functional entities on their surface [3–10]. They can be structurally defined as sheets of graphene which are rolled upon themselves that make them look like a tube consisting of $sp^2$ hybridized carbon atoms. They have a hollow structure, and their diameter is in nanometers scale [1, 11–17]. Carbon nanotubes are classified on the basis of the number of walls present in them into mainly two types: single-walled carbon nanotubes (SWCNTs)

S. Chawla (✉) · P. Rai · T. Garain
Department of Chemistry, Amity Institute of Applied Sciences, Amity University, Noida, Uttar Pradesh, India
e-mail: schawla5@amity.edu

 
S. Yadav et al. (eds.), *Recent Advances in Metrology*, Lecture Notes in Electrical Engineering 906, https://doi.org/10.1007/978-981-19-2468-2_4

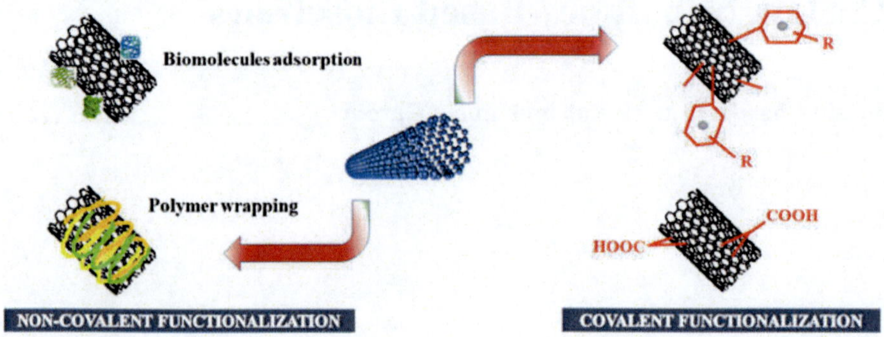


**Fig. 1** Chemical functionalization of CNTs

and multi-walled carbon nanotubes (MWCNTs). MWCNTs are formed by several homocentric layers of rolled graphene sheets, whereas SWCNTs are made up of single graphene sheets rolled upon themselves [18–23].

Chemical functionalization is a crucial step to make the carbon nanomaterials more versatile and adjustable in biological conditions. Besides these, functionalization helps in minimizing the toxicity and increasing its solubility in an aqueous condition. Functionalization is based on different chemical operations like oxidation, esterification, amidation, and halogenation [24–30]. Although CNTs show exceptional behavior for their use as theranostic aid, the problem of agglomeration due to the van der Waals attraction makes it tough to control them during their interaction with biochemical molecules. Hence, functionalization is usually required to control the functions and properties of carbon nanotubes to make them biocompatible and to improve their solubility in water [31–35]. There are two ways in which this functionalization can take place: (See Fig. 1).

(i) *Covalent functionalization*:

In this, covalent bonds are formed between the functional entities like −OH, −COOH, $-CONH_2$ and the carbon skeleton of the CNTs.

(ii) *Non-Covalent functionalization*:

Another type of functionalization, i.e., non-covalent functionalization employs proteins, carbohydrates, DNA, etc., which are connected with the CNTs that make their interaction with biomolecules more active in biological conditions. Also, coating the CNTs with surfactants and polymers like ethyleneglycol comes under the non-covalent functionalization.

## 2 Some Common Methods to Synthesize Carbon Nanotubes

Several methods are mentioned in the literature that can be employed for the manufacturing of carbon nanotubes. The *carbon arc method* [36] is one such method that is used prominently. Two carbon rods, positioned end to end and filled with a low-pressure inert gas, are taken, and their arc vaporization is done [37]. Another method to develop a carbon nanotube is the *laser method* in which carbon rods with a catalyst mixture of nickel and cobalt (50:50) at about 1200 °C are taken [38]. This is followed by heat treatment to remove impurities. *Catalyzed chemical vapor deposition* [39] is also a common method of synthesis in which acetylene is deposited over Fe and Co catalysts, producing huge amounts of CNTs [40]. Another simpler method of CNT production is *Ball Milling* [41] and subsequent annealing, which involves placing graphite powder in a container of stainless steel [42], containing four hardened balls of steel.

Some other methods can be used for CNT synthesis like *diffusion flame synthesis* [37] and *heat treatment electrolysis* [43].

## 3 CNTs-Based Biosensors

Biosensors find their extensive utilization in the diagnose of biochemical molecules, pathogenic microorganisms, or any other agents causing diseases [18]. In 1906, a biosensor was discovered for the first time. However, a full-fledged biosensor was manufactured in 1956 by Clark et al. [44]. In 1969, Guilbault and Montalvo used glass electrodes and discovered a potential biometric biosensor [44].

Biosensing devices can be made of carbon nanotubes as CNTs have an exceptionally low detection limit. Carbon nanotubes have proved to be ideal for highly sensitive biosensor devices, owing to their extraordinary responsiveness toward their surface. Biosensors composed of carbon nanotubes were developed in the early 2000s [45–48].

In 1962, Clark constructed an enzyme electrode. The idea of a biosensor, made with CNTs, was based on this [49]. The surface area of CNTs is large and this enables incapacitation of many functionalized units such as receptors for biosensing [50]. CNTs have distinctive optical properties like photoluminescence, shown in the near-infrared region [51], as well as strong resonance scattering. This makes them perfect for detecting biological moieties [1, 52–54]. A CNT-based biosensor generally consists of two components—a transducer and a biologically sensitive element (See Fig. 2) [55].

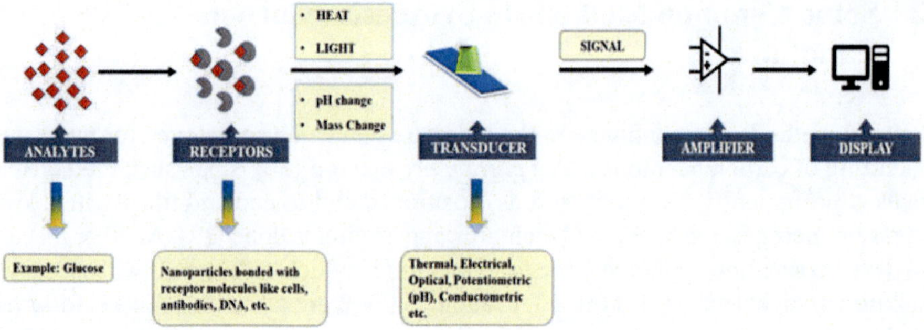


**Fig. 2** Basic components and working of CNTs-based biosensors

Many biological molecules, for example, enzymes, cell receptors, proteins, antibodies, oligo or polynucleotides, microorganisms, or tissues can be used to functionalize CNTs, and this makes them effective to be used as a biosensor [44, 56–58]. A transducer, in a biosensor, converts the concentration of analyte to different signals which can be detected, for example, current, absorbance, mass, acoustic variables, etc. CNT-based biosensors are categorized into two categories: chemical and physical. This classification is based on the interactivity between the analyte and the biologically sensitive materials [18, 55, 59–62].

## 4 Classification of Biosensors

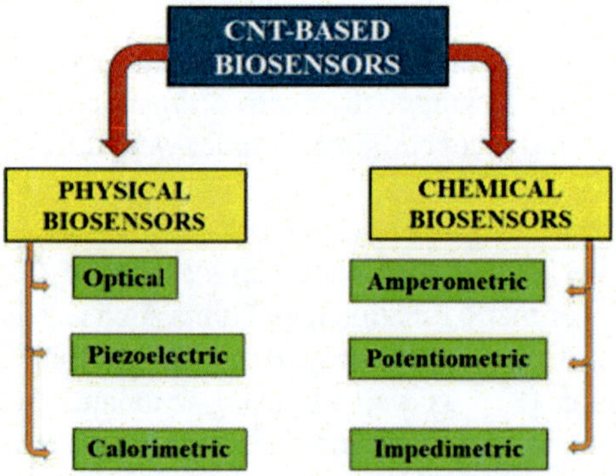

## 4.1 Chemical Biosensors

There are three types of electrochemical biosensors [47, 63]:

- *Amperometric based*
- *Potentiometric based*
- *Impedimetric based*

Most commonly used biosensors are those which operate through amperometric mode [64–68].

The working principle of chemical biosensors involves the conversion of chemical signals to electrical signals [69]. Analytes come into the enzymatic layer through the effect of diffusion, after immersing an enzyme electrode in a test solution. This analyte then undergoes an immediate enzymatic reaction [1].

CNTs have a hollow structure, due to which it is considered ideal for absorption of an enzyme. Hence, an enzyme-CNT electrode is generated generally by functionalizing the surface of the amperometric CNT by an enzyme [70, 71]. Oxidation or reduction reactions are catalyzed by an enzyme-fixed CNT electrode. This forms the basis of the sensing mechanism of an amperometric CNT-based biosensor [72–76]. Various types of enzymes are selected for surface modifications depending upon the different types of analytes used. They can be NADH [70], glucose oxidase [77], aflatoxin oxidase [66], cholesterol oxidase [77], urease, lactic acid oxidase, and horseradish peroxidase [78] and so on [79].

## 4.2 Physical Biosensors

CNT-based physical biosensors are of the following types [49]:

- *Optical biosensors*

  Detection of molecules is done at 0.9–1.3 eV, using near-infrared light [80]. This light can penetrate the tissues deeply, and hence, it can be used for several biomedical applications [18, 51, 81–83]. Some optical biosensors were manufactured, which were based on SWCNTs [84] and had their emission modulated, responding to adsorption of diverse biomolecules [49, 56, 85, 86]. Optical biosensors have many advantages like:

  (a) remarkable luminescence
  (b) good wavelength conversion function, and
  (c) most importantly, NIR (Near-infrared region) emission takes place, which is tunable and responsive towards local dielectric function [84].

- *Piezoelectric biosensors*

  When an external force is applied, then the resistance change observed is known as piezoresistive [79, 87–93]. CNTs are being considered as an ideal candidate for piezoelectric biosensors because of the following advantages [93–99]:

(a) Tough and high tensile strength.
(b) Young's modulus is high (> 1 TPa), which is greater than steel by 100 times.
(c) Density is 1/6th of that of steel.

- *Calorimetric biosensors*

  The working principle of the calorimetric biosensor is based on the fact that biological entities react with a particular biomolecule, which is being measured and hence producing a heating effect [96, 98, 100]. The temperature changes alter the size of the carbon nanotube, creating a strain on the network of CNT, which can be detected [72].

## 5 Limitations of CNT-Based Biosensors

Due to their extraordinary structure and characteristic properties, CNTs are being part of a huge amount of research, particularly in the biomedical field. But, there are certain issues related to it that are to be taken care of for improving its biocompatibility. One of them is toxicity which is associated with all the types of carbon nanomaterials, and studies are being done to recognize their cytotoxic behavior and measures to reduce them [23, 25, 101–107]. Another important factor for improving the efficacy of CNTs is their biodegradability. So, it can be said that there are three major complications for the use of CNTs in biological conditions which are as follows: solubility, toxicity, and biocompatibility.

## 6 Conclusion

Biosensors, when fabricated using nanomaterials, show enhanced properties and performance. There are numerous advantages related to the use of carbon nanotubes as biosensors, and CNTs also provide an ample number of opportunities to develop a new generation of nanotools that can be applied for diagnosis as well as disease probing devices. However, before developing and using biosensors as an immunological aid, it is important to consider the fact that despite the small size of a nanoparticle, they may remain accumulated inside the body and do not get metabolized. There can be a significant number of impurities in a CNT like certain metals or compounds that can have adverse effects. So, the purification of CNTs before their use is also a major concern. Therefore, proper evaluation should be done so that various parameters like strength, length, stability, biocompatibility, durability, and toxicity are analyzed. Nevertheless, CNTs show versatility and adaptability in biological conditions which are vital for their use in biosensing devices.

## References


1. Tîlmaciu CM, Morris MC (2015) Carbon nanotube biosensors. Front Chem 3:1–21. https://doi.org/10.3389/fchem.2015.00059
2. Iijima S, Ichihashi T (1993) Single-shell carbon nanotubes of 1-nm diameter. Nature 363:603–605. https://doi.org/10.1038/363603a0
3. Aqel A, El-Nour KMMA, Ammar RAA, Al-Warthan A (2012) Carbon nanotubes, science and technology part (I) structure, synthesis and characterisation. Arab J Chem 5:1–23. https://doi.org/10.1016/j.arabjc.2010.08.022
4. Ouyang M, Huang JL, Lieber CM (2002) One-dimensional energy dispersion of single-walled carbon nanotubes by resonant electron scattering. Phys Rev Lett 88:4. https://doi.org/10.1103/PhysRevLett.88.066804
5. Ibrahim KS (2013) Carbon nanotubes-properties and applications: a review. Carbon Lett 14:131–144. https://doi.org/10.5714/CL.2013.14.3.131
6. Popov V (2004) Carbon nanotubes: properties and application. Mater Sci Eng R Reports 43:61–102. https://doi.org/10.1016/j.mser.2003.10.001
7. Eatemadi A, Daraee H, Karimkhanloo H et al (2014) Carbon nanotubes: properties, synthesis, purification, and medical applications. Nanoscale Res Lett 9:393. https://doi.org/10.1186/1556-276X-9-393
8. Hirlekar R, Yamagar M, Garse H et al (2009) Carbon nanotubes and its applications: a review. Asian J Pharm Clin Res 2:17–27
9. He H, Pham-Huy LA, Dramou P et al (2013) Carbon nanotubes: applications in pharmacy and medicine. Biomed Res Int 2013:1–12. https://doi.org/10.1155/2013/578290
10. Saifuddin N, Raziah AZ, Junizah AR (2013) Carbon nanotubes: a review on structure and their interaction with proteins. J Chem 2013:1–18. https://doi.org/10.1155/2013/676815
11. Sajid MI, Jamshaid U, Jamshaid T et al (2016) Carbon nanotubes from synthesis to in vivo biomedical applications. Int J Pharm 501:278–299. https://doi.org/10.1016/j.ijpharm.2016.01.064
12. Bianco A, Kostarelos K, Prato M (2005) Applications of carbon nanotubes in drug delivery. Curr Opin Chem Biol 9:674–679. https://doi.org/10.1016/j.cbpa.2005.10.005
13. Stout D (2015) Recent advancements in carbon nanofiber and carbon nanotube applications in drug delivery and tissue engineering. Curr Pharm Des 21:2037–2044. https://doi.org/10.2174/1381612821666150302153406
14. Coleman JN, Khan U, Blau WJ, Gun'ko YK, (2006) Small but strong: A review of the mechanical properties of carbon nanotube-polymer composites. Carbon N Y 44:1624–1652. https://doi.org/10.1016/j.carbon.2006.02.038
15. Sun Y-P, Fu K, Lin Y, Huang W (2002) Functionalized carbon nanotubes: properties and applications. Acc Chem Res 35:1096–1104. https://doi.org/10.1021/ar010160v
16. Xie S, Li W, Pan Z et al (2000) Mechanical and physical properties on carbon nanotube. J Phys Chem Solids 61:1153–1158. https://doi.org/10.1016/S0022-3697(99)00376-5
17. Endo M, Strano MS, Ajayan PM (2007) Potential applications of carbon nanotubes, pp 13–62
18. Kour R, Arya S, Young S-J et al (2020) Review—recent advances in carbon nanomaterials as electrochemical biosensors. J Electrochem Soc 167:037555. https://doi.org/10.1149/1945-7111/ab6bc4
19. Richard C, Doan BT, Beloeil JC et al (2008) Noncovalent functionalization of carbon nanotubes with amphiphilic Gd 3+ chelates: toward powerful T1 and T2 MRI contrast agents. Nano Lett 8:232–236. https://doi.org/10.1021/nl072509z
20. Qiu H, Yang J (2017) Structure and properties of carbon nanotubes. Elsevier Inc.
21. Jain KK (2012) Advances in use of functionalized carbon nanotubes for drug design and discovery. Expert Opin Drug Discov 7:1029–1037. https://doi.org/10.1517/17460441.2012.722078
22. Rao CNR, Muller A AKC (2004) The chemistry of nanomaterials: synthesis, properties and applications. Wiley

23. Beg S, Rizwan M, Sheikh AM et al (2011) Advancement in carbon nanotubes: basics, biomedical applications and toxicity. J Pharm Pharmacol 63:141–163. https://doi.org/10.1111/j.2042-7158.2010.01167.x
24. Raval JP, Joshi P, Chejara DR (2018) Carbon nanotube for targeted drug delivery. Elsevier Inc.
25. Firme CP, Bandaru PR (2010) Toxicity issues in the application of carbon nanotubes to biological systems. Nanomed Nanotechnol, Biol Med 6:245–256. https://doi.org/10.1016/j.nano.2009.07.003
26. Sanginario A, Miccoli B, Demarchi D (2017) Carbon nanotubes as an effective opportunity for cancer diagnosis and treatment. Biosensors 7:1–23. https://doi.org/10.3390/bios7010009
27. Liu, Speranza (2019) Functionalization of carbon nanomaterials for biomedical applications. C—J Carbon Res 5:72. https://doi.org/10.3390/c5040072
28. Zhou Y, Fang Y, Ramasamy RP (2019) Non-covalent functionalization of carbon nanotubes for electrochemical biosensor development. Sensors (Switzerland) 19. https://doi.org/10.3390/s19020392
29. Mata D, Amaral M, Fernandes AJS et al (2015) Diels-Alder functionalized carbon nanotubes for bone tissue engineering: In vitro/in vivo biocompatibility and biodegradability. Nanoscale 7:9238–9251. https://doi.org/10.1039/c5nr01829c
30. Merum S, Veluru JB, Seeram R (2017) Functionalized carbon nanotubes in bio-world: applications, limitations and future directions. Mater Sci Eng B Solid-State Mater Adv Technol 223:43–63. https://doi.org/10.1016/j.mseb.2017.06.002
31. Risoluti R, Gullifa G, Carcassi E et al (2020) TGA/chemometrics addressing innovative preparation strategies for functionalized carbon nanotubes. J Pharm Anal. https://doi.org/10.1016/j.jpha.2020.02.009
32. Holzinger M, Vostrowsky O, Hirsch A et al (2001) Sidewall functionalization of carbon nanotubes. Angew Chemie Int Ed 40:4002–4005. https://doi.org/10.1002/1521-3773(20011105)40:21%3c4002::AID-ANIE4002%3e3.0.CO;2-8
33. Hirsch A (2002) Functionalization of single-walled carbon nanotubes. Angew Chemie Int Ed 41:1853–1859. https://doi.org/10.1002/1521-3773(20020603)41:11%3c1853::AID-ANIE1853%3e3.0.CO;2-N
34. Liu Z, Tabakman S, Welsher K, Dai H (2009) Carbon nanotubes in biology and medicine: in vitro and in vivo detection, imaging and drug delivery. Nano Res 2:85–120. https://doi.org/10.1007/s12274-009-9009-8
35. Kang HS (2004) Organic functionalization of sidewall of carbon nanotubes. J Chem Phys 121:6967–6971. https://doi.org/10.1063/1.1775783
36. Sharma R, Sharma AK, Sharma V (2015) Synthesis of carbon nanotubes by arc-discharge and chemical vapor deposition method with analysis of its morphology, dispersion and functionalization characteristics. Cogent Eng 2:1094017. https://doi.org/10.1080/23311916.2015.1094017
37. Medjo E (2011) Carbon nanotube synthesis. Carbon Nanotub Appl Electron Devices 1–11. https://doi.org/10.5772/23030
38. Chrzanowska J, Hoffman J, Małolepszy A et al (2015) Synthesis of carbon nanotubes by the laser ablation method: effect of laser wavelength. Phys status solidi 252:1860–1867. https://doi.org/10.1002/pssb.201451614
39. Wojtera K, Walczak M, Pietrzak L et al (2020) Synthesis of functionalized carbon nanotubes for fluorescent biosensors. Nanotechnol Rev 9:1237–1244. https://doi.org/10.1515/ntrev-2020-0096
40. Koziol K, Boskovic BO, Yahya N (2010) Synthesis of carbon nanostructures by CVD method. Adv Struct Mater pp 23–49
41. Soares OSGP, Rocha RP, Gonçalves AG et al (2015) Easy method to prepare N-doped carbon nanotubes by ball milling. Carbon N Y 91:114–121. https://doi.org/10.1016/j.carbon.2015.04.050
42. Munir KS, Li Y, Qian M, Wen C (2016) Identifying and understanding the effect of milling energy on the synthesis of carbon nanotubes reinforced titanium metal matrix composites. Carbon N Y 99:384–397. https://doi.org/10.1016/j.carbon.2015.12.041

43. Awasthi K, Srivastava A, Srivastava ON (2005) Synthesis of Carbon Nanotubes. J Nanosci Nanotechnol 5:1616–1636. https://doi.org/10.1166/jnn.2005.407
44. Sireesha M, Jagadeesh Babu V, Kranthi Kiran AS, Ramakrishna S (2018) A review on carbon nanotubes in biosensor devices and their applications in medicine. Nanocomposites 4:36–57. https://doi.org/10.1080/20550324.2018.1478765
45. Holzinger M, Le GA, Cosnier S (2014) Nanomaterials for biosensing applications: a review. Front Chem 2:1–10. https://doi.org/10.3389/fchem.2014.00063
46. Malhotra BD, Ali MA (2018) Nanomaterials in biosensors
47. Ijeomah G, Obite F, Rahman O (2016) Development of carbon nanotube-based biosensors. Int J Nano Biomater 6:83–109. https://doi.org/10.1504/IJNBM.2016.079682
48. Erol O, Uyan I, Hatip M et al (2018) Recent advances in bioactive 1D and 2D carbon nanomaterials for biomedical applications. Nanomedicine Nanotechnology Biol Med 14:2433–2454. https://doi.org/10.1016/j.nano.2017.03.021
49. Yang N, Chen X, Ren T et al (2015) Carbon nanotube based biosensors. Sens Actuators B Chem 207:690–715. https://doi.org/10.1016/j.snb.2014.10.040
50. Lewis K, Choi A, Chou J, BB-N (2007) Bioceramics—The changing role of ceramics and nanoceramics in medical applications. pp 1–9
51. Boghossian AA, Zhang J, Barone PW et al (2011) Near-infrared fluorescent sensors based on single-walled carbon nanotubes for life sciences applications. Chemsuschem 4:848–863. https://doi.org/10.1002/cssc.201100070
52. Malik P, Katyal V, Malik V et al (2013) Nanobiosensors: concepts and variations. ISRN Nanomater 2013:1–9. https://doi.org/10.1155/2013/327435
53. Zhu Z (2017) An overview of carbon nanotubes and graphene for biosensing applications. Nano-Micro Lett 9:1–24. https://doi.org/10.1007/s40820-017-0128-6
54. Mohajeri M, Behnam B, Sahebkar A (2018) Biomedical applications of carbon nanomaterials: drug and gene delivery potentials. J Cell Physiol 234:298–319. https://doi.org/10.1002/jcp.26899
55. Gupta S, Murthy CN, Prabha CR (2018) Recent advances in carbon nanotube based electrochemical biosensors. Int J Biol Macromol 108:687–703. https://doi.org/10.1016/j.ijbiomac.2017.12.038
56. Yemini M, Reches M, Gazit E, Rishpon J (2005) Peptide nanotube-modified electrodes for enzyme-biosensor applications. Anal Chem 77:5155–5159. https://doi.org/10.1021/ac050414g
57. Luo XL, Xu JJ, Wang JL, Chen HY (2005) Electrochemically deposited nanocomposite of chitosan and carbon nanotubes for biosensor application. Chem Commun 2169. https://doi.org/10.1039/b419197h
58. Gergeroglu H, Yildirim S, Ebeoglugil MF (2020) Nano-carbons in biosensor applications: an overview of carbon nanotubes (CNTs) and fullerenes (C60). SN Appl Sci 2:603. https://doi.org/10.1007/s42452-020-2404-1
59. Umasankar Y, Chen SM (2008) Recent trends in the application of carbon nanotubes-polymer composite modified electrodes for biosensors: a review. Anal Lett 41:210–243. https://doi.org/10.1080/00032710701792638
60. Trojanowicz M (2006) Analytical applications of carbon nanotubes: a review. TrAC Trends Anal Chem 25:480–489. https://doi.org/10.1016/j.trac.2005.11.008
61. Pasinszki T, Krebsz M, Tung TT, Losic D (2017) Carbon nanomaterial based biosensors for non-invasive detection of cancer and disease biomarkers for clinical diagnosis. Sensors (Switzerland) 17:1–32. https://doi.org/10.3390/s17081919
62. Gao M, Dai L, Wallace GG (2003) Biosensors based on aligned carbon nanotubes coated with inherently conducting polymers. Electroanalysis 15:1089–1094. https://doi.org/10.1002/elan.200390131
63. Lawal AT (2016) Synthesis and utilization of carbon nanotubes for fabrication of electrochemical biosensors. Mater Res Bull 73:308–350. https://doi.org/10.1016/j.materresbull.2015.08.037

64. Musameh M, Wang J, Merkoci A, Lin Y (2002) Low-potential stable NADH detection at carbon-nanotube-modified glassy carbon electrodes. Electrochem commun 4:743–746. https://doi.org/10.1016/S1388-2481(02)00451-4
65. Liu X, Shi L, Niu W et al (2008) Amperometric glucose biosensor based on single-walled carbon nanohorns. Biosens Bioelectron 23:1887–1890. https://doi.org/10.1016/j.bios.2008.02.016
66. chuan Li S, hua Chen J, Cao H, et al (2011) Amperometric biosensor for aflatoxin B1 based on aflatoxin-oxidase immobilized on multiwalled carbon nanotubes. Food Control 22:43–49. https://doi.org/10.1016/j.foodcont.2010.05.005
67. Sirkar K, Pishko MV (1998) Amperometric biosensors based on oxidoreductases immobilized in photopolymerized poly(ethylene glycol) redox polymer hydrogels. Anal Chem 70:2888–2894. https://doi.org/10.1021/ac971236x
68. Hou S-F, Fang H-Q, Chen H-Y (1997) An amperometric enzyme electrode for glucose using immobilized glucose oxidase in a ferrocene attached poly(4-vinylpyridine) multilayer film. Anal Lett 30:1631–1641. https://doi.org/10.1080/00032719708001682
69. Kim M, Jang J, Cha C (2017) Carbon nanomaterials as versatile platforms for theranostic applications. Drug Discov Today 22:1430–1437. https://doi.org/10.1016/j.drudis.2017.05.004
70. Thévenot DR, Toth K, Durst RA, Wilson GS (2001) Electrochemical biosensors: recommended definitions and classification. Biosens Bioelectron 16:121–131. https://doi.org/10.1016/S0956-5663(01)00115-4
71. Bartlett PN, Birkin PR (1993) Enzyme switch responsive to glucose. Anal Chem 65:1118–1119. https://doi.org/10.1021/ac00056a029
72. Allen BL, Kichambare PD, Star A (2007) Carbon nanotube field-effect-transistor-based biosensors. Adv Mater 19:1439–1451. https://doi.org/10.1002/adma.200602043
73. Kwon OS, Song HS, Park TH, Jang J (2019) Conducting nanomaterial sensor using natural receptors. Chem Rev 119:36–93. https://doi.org/10.1021/acs.chemrev.8b00159
74. Barone PW, Strano MS (2009) Single walled carbon nanotubes as reporters for the optical detection of glucose. J Diabetes Sci Technol 3:242–252. https://doi.org/10.1177/193229680900300204
75. Wang Y, Liu L, Li M et al (2011) Multifunctional carbon nanotubes for direct electrochemistry of glucose oxidase and glucose bioassay. Biosens Bioelectron 30:107–111. https://doi.org/10.1016/j.bios.2011.08.038
76. Harper A, Anderson MR (2010) Electrochemical glucose sensors—developments using electrostatic assembly and carbon nanotubes for biosensor construction. Sensors 10:8248–8274. https://doi.org/10.3390/s100908248
77. Zhu L, Xu L, Tan L et al (2013) Direct electrochemistry of cholesterol oxidase immobilized on gold nanoparticles-decorated multiwalled carbon nanotubes and cholesterol sensing. Talanta 106:192–199. https://doi.org/10.1016/j.talanta.2012.12.036
78. Moyo M, Okonkwo JO, Agyei NM (2014) An amperometric biosensor based on horseradish peroxidase immobilized onto maize tassel-multi-walled carbon nanotubes modified glassy carbon electrode for determination of heavy metal ions in aqueous solution. Enzyme Microb Technol 56:28–34. https://doi.org/10.1016/j.enzmictec.2013.12.014
79. Yun Y, Dong Z, Shanov V et al (2007) Nanotube electrodes and biosensors This article reviews the state of the art in carbon nanotube electrode. Nano Today 2:30–37
80. Barone PW, Baik S, Heller DA, Strano MS (2005) Near-infrared optical sensors based on single-walled carbon nanotubes. Nat Mater 4:86–92. https://doi.org/10.1038/nmat1276
81. Jain A, Homayoun A, Bannister CW, Yum K (2015) Single-walled carbon nanotubes as near-infrared optical biosensors for life sciences and biomedicine. Biotechnol J 10:447–459. https://doi.org/10.1002/biot.201400168
82. Ziegler KJ (2005) Developing implantable optical biosensors. Trends Biotechnol 23:440–444. https://doi.org/10.1016/j.tibtech.2005.07.006
83. García-Aljaro C, Cella LN, Shirale DJ et al (2010) Carbon nanotubes-based chemiresistive biosensors for detection of microorganisms. Biosens Bioelectron 26:1437–1441. https://doi.org/10.1016/j.bios.2010.07.077

84. O'Connell MJ, Bachilo SH, Huffman CB et al (2002) Band gap fluorescence from individual single-walled carbon nanotubes. Science 297 (80- ):593–596. https://doi.org/10.1126/science.1072631
85. Kelley K, Pehrsson PE, Ericson LM, Zhao W (2005) Optical pH response of dna wrapped hipco carbon nanotubes. J Nanosci Nanotechnol 5:1041–1044. https://doi.org/10.1166/jnn.2005.155
86. Yoo SM, Lee SY (2016) Optical biosensors for the detection of pathogenic microorganisms. Trends Biotechnol 34:7–25. https://doi.org/10.1016/j.tibtech.2015.09.012
87. Cullinan MA, Culpepper ML (2010) Carbon nanotubes as piezoresistive microelectromechanical sensors: theory and experiment. Phys Rev B 82:115428. https://doi.org/10.1103/PhysRevB.82.115428
88. Ramaratnam A, Jalili N (2006) Reinforcement of piezoelectric polymers with carbon nanotubes: pathway to next-generation sensors. J Intell Mater Syst Struct 17:199–208. https://doi.org/10.1177/1045389X06055282
89. Alamusi HuN, Fukunaga H et al (2011) Piezoresistive strain sensors made from carbon nanotubes based polymer nanocomposites. Sensors 11:10691–10723. https://doi.org/10.3390/s111110691
90. Obitayo W, Liu T (2012) A review: carbon nanotube-based piezoresistive strain sensors. J Sensors 2012:1–15. https://doi.org/10.1155/2012/652438
91. Lekawa-Raus A, Koziol KKK, Windle AH (2014) Piezoresistive effect in carbon nanotube fibers. ACS Nano 8:11214–11224. https://doi.org/10.1021/nn503596f
92. Wan-Lu W, Ke-Jun L, Yong L, Yong-Tian W (2003) Piezoresistive effect of doped carbon nanotube/cellulose films. Chin Phys Lett 20:1544–1547. https://doi.org/10.1088/0256-307X/20/9/339
93. Li X, Levy C, Elaadil L (2008) Multiwalled carbon nanotube film for strain sensing. Nanotechnol 19:045501. https://doi.org/10.1088/0957-4484/19/04/045501
94. Vemuru SM, Wahi R, Nagarajaiah S, Ajayan PM (2009) Strain sensing using a multiwalled carbon nanotube film. J Strain Anal Eng Des 44:555–562. https://doi.org/10.1243/03093247JSA535
95. Bu L, Steitz J, Kanoun O (2010) Influence of processing parameters on properties of strain sensors based on carbon nanotube films. In: 2010 7th International multi- conference on systems, signals and devices. IEEE, pp 1–6
96. Kang I, Schulz MJ, Kim JH et al (2006) A carbon nanotube strain sensor for structural health monitoring. Smart Mater Struct 15:737–748. https://doi.org/10.1088/0964-1726/15/3/009
97. Gloor TA, Mila F (2004) Strain induced correlation gaps in carbon nanotubes. Eur Phys J B Condens Matter 38:9–12. https://doi.org/10.1140/epjb/e2004-00092-2
98. Dharap P, Li Z, Nagarajaiah S, Barrera EV (2004) Nanotube film based on single-wall carbon nanotubes for strain sensing. Nanotechnol 15:379–382. https://doi.org/10.1088/0957-4484/15/3/026
99. Grow RJ, Wang Q, Cao J et al (2005) Piezoresistance of carbon nanotubes on deformable thin-film membranes. Appl Phys Lett 86:093104. https://doi.org/10.1063/1.1872221
100. Campbell JD, Ferguson WG (1970) The temperature and strain-rate dependence of the shear strength of mild steel. Philos Mag 21:63–82. https://doi.org/10.1080/14786437008238397
101. Cifuentes-Rius A, Boase NRB, Font I et al (2017) In vivo fate of carbon nanotubes with different physicochemical properties for gene delivery applications. ACS Appl Mater Interfaces 9:11461–11471. https://doi.org/10.1021/acsami.7b00677
102. Bianco A, Kostarelos K, Prato M (2011) Making carbon nanotubes biocompatible and biodegradable. Chem Commun 47:10182–10188. https://doi.org/10.1039/c1cc13011k
103. Hubbs AF, Mercer RR, Benkovic SA et al (2011) Nanotoxicology—a pathologist's perspective. Toxicol Pathol 39:301–324. https://doi.org/10.1177/0192623310390705
104. Dumortier H, Lacotte S, Pastorin G et al (2006) Functionalized carbon nanotubes are non-cytotoxic and preserve the functionality of primary immune cells. Nano Lett 6:1522–1528. https://doi.org/10.1021/nl061160x

105. Muller J, Huaux F, Moreau N et al (2005) Respiratory toxicity of multi-wall carbon nanotubes. Toxicol Appl Pharmacol 207:221–231. https://doi.org/10.1016/j.taap.2005.01.008
106. Schipper ML, Nakayama-Ratchford N, Davis CR et al (2008) A pilot toxicology study of single-walled carbon nanotubes in a small sample of mice. Nat Nanotechnol 3:216–221. https://doi.org/10.1038/nnano.2008.68
107. Smart SK, Cassady AI, Lu GQ, Martin DJ (2006) The biocompatibility of carbon nanotubes. Carbon N Y 44:1034–1047. https://doi.org/10.1016/j.carbon.2005.10.011

# Design and Analysis of Lever Type Chainless Drive Mechanism of a Bicycle

P. Srinivasa Rao, M. Muzibul, V. Naveen, R. Ravikumar, S. Subbarao, and P. Yaswanth

**Abstract** The conventional bicycle employs the chain drive to transmit power from pedal to rear wheel and it requires accurate mounting and alignment for proper working. This paper focuses on the design of a chainless drive mechanism by using lever principle for the bicycle. This mechanism also seeks the usage of gravitational effect while pedaling because of the oscillatory pedaling instead of rotary. Using the mechanical advantage of lever principle we tried to decrease the effort which should require propelling a bicycle. This drive mechanism consists of majorly 3 components: bell crank lever, link, and drive wheel. The theoretical values of forces and all are calculated initially and after that, the three components are modeled in design software SOLID WORKS and analyzed further and the results were compared. The results obtained are optimized with this new mechanism.

**Keywords** Chain drive · Crank lever · Link · Lever · Chainless drive mechanism

## 1 Introduction

### *1.1 Definition of Bicycle*

A bicycle, is a human-powered or motor-powered, pedal-driven, single-track vehicle, having two wheels attached to a frame, one behind the other. A bicycle rider is called a cyclist, or bicyclist. The dandy horse, also called Draisienne or laufmaschine, was the first human means of transport to use only two wheels in tandem and was invented by the German Baron Karl von Drais. It is regarded as the modern bicycle's forerunner; Drais introduced it to the public in Mannheim in summer 1817 and in Paris in 1818. The first mechanically-propelled, two-wheeled vehicle may have been built by Kirkpatrick MacMillan, a Scottish blacksmith, in 1839, although the claim is often disputed. In the early 1860s, Frenchmen Pierre Michaux and Pierre Lallement

P. S. Rao (✉) · M. Muzibul · V. Naveen · R. Ravikumar · S. Subbarao · P. Yaswanth
Department of Mechanical Engineering, Lendi Institute of Engineering and Technology, Vizianagram, Andhra Pradesh, India
e-mail: srinivas.p@lendi.org



S. Yadav et al. (eds.), *Recent Advances in Metrology*, Lecture Notes in Electrical Engineering 906, https://doi.org/10.1007/978-981-19-2468-2_5

took bicycle design in a new direction by adding a mechanical crank drive with pedals on an enlarged front wheel (the velocipede). In that same year, bicycle wheels with wire spokes were patented by Eugene Meyer of Paris. Further innovations increased comfort and ushered in a second bicycle craze, the 1890s Golden Age of Bicycles.

## *1.2 Design of Chainless Drive Mechanism of a Bicycle*

This paper [1] illustrates the characteristics of the two alternate drive mechanisms, chain drive and shaft drive. After carefully examining the two alternatives, the conventional shaft drive was selected for the project since its cost and flexibility were determined to be better suited for the project. In this journal [2] Chase discloses about A bicycle of the type comprising a frame and a front and rear wheel rotatable mounted to the frame is provided with improved means for rotatable driving the rear wheel. The driving means comprises a right and left foot lever pivotally connected at one end to the frame and adapted to be alternately depressed by the feet of the bicyclist. This paper [3] work discloses about a driving device without chain for bicycle comprises a driving wheel mounted to a crank shaft, a rear gear mounted to a wheel hub for mounting a rear wheel of the bicycle, an input gear coupled with the driving wheel by means of an interconnecting link and a gear set having two identical gears, respectively, engaging the rear gear and the input gear.

In the work [4] discloses about the reciprocating rectilinear motion using a system of levers, gears, and chains. The mechanism relies upon crank levers whose fulcrum lies at the base of the bicycle frame. The motion of the crank levers pushes connecting rods that drive a number of gears located on an extension to the bicycle frame above the rear wheel.

This paper [5] says that invention relates to an improved lever driven bicycle including a shifting mechanism for controlling a drive ratio between the oscillating lever and the drive wheel. The drive lever arrangement may be located about the front or rear wheels of the bicycle and, when about the rear wheel, may include a Sole lever or dual levers. This paper describes [6] invention relates to a bicycle having a driving force transmitting mechanism with oscillating fulcrum point. A sliding part is provided with a dust cover or a brush to prevent dust and the like from being entered therein and to remove the dust, if entered, there from. When a rider steps on the pedals alternatively, force of the pedal levers generated by the oscillation of the pedal levers is transmitted to the sprocket through the crank lever. This paper [1] invention relates to a bicycle and like, having a rear wheel or flywheel, and provided with a pedal group comprising a central shaft and a gear wheel integrally coupled each other, a left crank and a right crank, coupled on the two ends of the central Shaft, characterized in that a pedal crank is rotatable coupled with each one of said left and right cranks by mechanical means creating a freewheel like mechanism between the crank and the corresponding pedal crank, a pedal being coupled with one end of each one of the two pedal cranks. This [3] gives the information about the lever driven bicycle consists of the following parts: mounting plate, torsion spring, and oscillating lever. The lever is pivoted at a point on the mounting plate which is fixed to the bicycle frame and a torsion spring is present in between the mounting plate and the lever. The end of the lever contains a gear sector which is in mesh with the free wheel. This paper [5] illustrates the characteristics of the two alternate drive mechanisms, chain drive and shaft drive. After carefully examining the two alternatives, the conventional shaft drive was selected for the project since its cost and flexibility were determined to be better suited for the project.

## 2 Methodology

### 2.1 Design Process

#### 2.1.1 Lever

A lever is simple machine consisting of a beam or rigid rod pivoted at a fixed end. A lever is a rigid body capable of rotating on a point on itself. On the basis of the locations of fulcrum, load, and effort, the lever is divided into three types. A lever amplifies an input force to provide a greater output force, which is said to provide leverage. The ratio of the output force to the input force is the mechanical advantage of the lever. As such, the lever is a mechanical advantage device, trading off force against movement.

#### 2.1.2 Law of Lever

The lever is a movable bar that pivots on a fulcrum attached to a fixed point. The lever operates by applying forces at different distances from the fulcrum, or a pivot. Assuming the lever does not dissipate or store energy, the power into the lever must equal the power out of the lever. As the lever rotates around the fulcrum, points farther from this pivot move faster than points closer to the pivot. Therefore, a force applied to a point farther from the pivot must be less than the force located at a point closer in, because power is the product of force and velocity.

If $a$ and $b$ are distances from the fulcrum to points $A$ and $B$ and the force $F_A$ applied to $A$ is the input and the force $F_B$ applied at $B$ is the output, the ratio of the velocities of points $A$ and $B$ is given by $a/b$, so we have the ratio of the output force to the input force, or mechanical advantage, is given by [2]

$$\mathrm{MA} = F_B/F_A = a/b.$$

This is the law of the lever, which was proven by Archimedes using geometric reasoning. It shows that if the distance $a$ from the fulcrum to where the input force is applied (point A) is greater than the distance $b$ from fulcrum to where the output force is applied (point B), then the lever amplifies the input force. On the other hand, if the distance a from the fulcrum to the input force is less than the distance b from the fulcrum to the output force, then the lever reduces the input force. The use of velocity in the static analysis of a lever is an application of the principle of virtual work.

### 2.1.3 Solid Work Models

The design of lever crank, lever arm and drive arm has been done by using SOLID WORKS 2020 SP 2.0 software. The design is done in such a way that there should not be any damage to lever crank, lever arm, and drive arm. SOLID WORKS started as an in-house development in 1977 by French aircraft manufacture Avions Marcel Dassault, at that time adopted in the automotive, shipbuilding, aerospace, and other industries. SOLID WORKS enables the creation of 3-D sketches, sheet metal, forged or tooling parts up to the definition of mechanical assemblies. The software provides advanced technologies for mechanical surfacing and BIW. It has wide range of applications for tooling design, for both generic tooling and mold and die. SOLID WORKS offer solution to shape design, styling, surfacing workflow and visualization to create, modify and validate complex innovative shapes from industrial design to Class-A surfacing with the ICEM surfacing technologies. SOLID WORKS supports multiple stages of product design whether started from scratch or from 2-D sketches. SOLID WORKS is able to read and produce STEP format files for reverse engineering and surface reuse (Fig. 1).

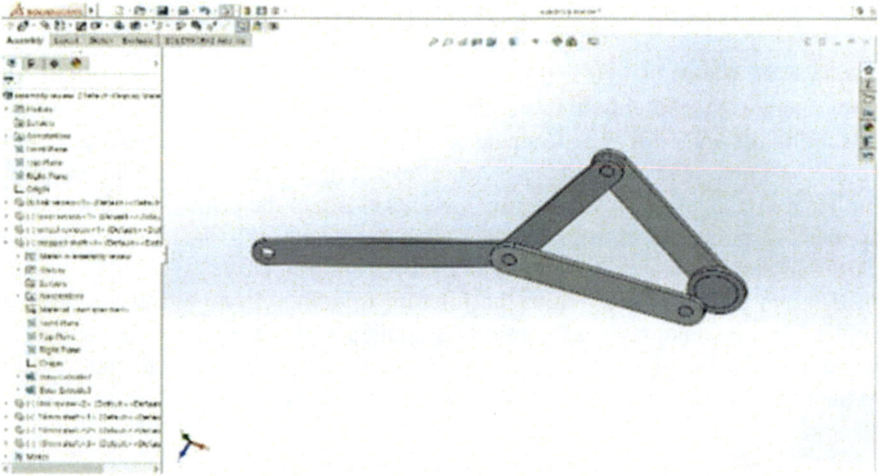

**Fig. 1** Solid works model

## 2.2 *Design Criteria*

### 2.2.1 Force Required to Move Bicycle

Let us consider the design parameters from book [1–3]

Weight of the bicycle = 20 kg = 200 N.

Weight of the human bicycle = 70 kg = 700 N.

**Torque required to propel the bicycle**

$$\text{Total tractive force(TTE)} = \text{RR}(N).$$

where

RR = Force required to overcome rolling resistance.

Rolling Resistance (RR) is the force necessary to propel a vehicle over a particular surface worst possible surface type, if we consider

$$\text{RR} = \text{GVW} * \text{Crr} \tag{1}$$

where

GVW = Gross vehicle weight and

Crr = Surface friction between road and tyer, typical value is 0.055

$$\begin{aligned} \text{GVW} &= \text{Man weight} + \text{cycle weight} \\ &= 70\,\text{kg} + 20\,\text{kg} = 90\,\text{kg}. \end{aligned} \tag{2}$$

From Eq. (1) RR = 900 * 0.055 = 49.5 N.

**Torque Required**

$$\begin{aligned} T_w &= F * r \\ &= 49.5 * 0.3 = 14.85\,N - m \end{aligned} \tag{3}$$

$$\begin{aligned} F &= T_w / r \\ &= 14.85/0.05 = 297\,N \end{aligned} \tag{4}$$

### 2.2.2 Bending Design Criteria for Lever

The max bending moment on the lever applied is,

$$M_b = \text{load} * \text{perpendicular distance from pivot to load point} \tag{5}$$

Therefore, $M_b = 297 * 125 = 37{,}125$ N-mm.
From bending stress formula,

$$M/I = f/y \tag{6}$$

where

$M =$ bending moment.
$I =$ moment of inertia about neutral axis.
$f =$ bending stress.
$y =$ distance from outermost layer to neutral axis.

For aluminum value of yield strength, $S_{yt} = 207$Mpa.
Considering F.S = 4

$$\begin{aligned} \text{Bending stress, } f &= \left(S_{yt}/F.S\right) \\ &= (207/4) = 51.74\,\text{Mpa} \\ y &= 25\,\text{mm} \end{aligned} \tag{7}$$

Therefore from Eq. (6), $I = ((37{,}125*25)/51.75) = 17{,}938.24\text{mm}^4$.
We considered the b and d values of the lever as, $b = 10$ mm and $d = 50$ mm.
Therefore $I = (\text{bd}^3/12) = 104{,}166.67\text{mm}^4$.

## 2.3 *Simulation*

### 2.3.1 Cad Design

Considering the design calculations we have done in the design criteria, we have designed the virtual CAD model of the individual parts with the required dimensions in the CAD software SOLIDWORKS 2016 and assembled the individual design parts (Figs. 2 and 3).

### 2.3.2 Motion Study

After all, motion study was performed on the complete assembly of the design and hence confirmed that the dimensions are giving the required motion as per our requirement (Fig. 4).

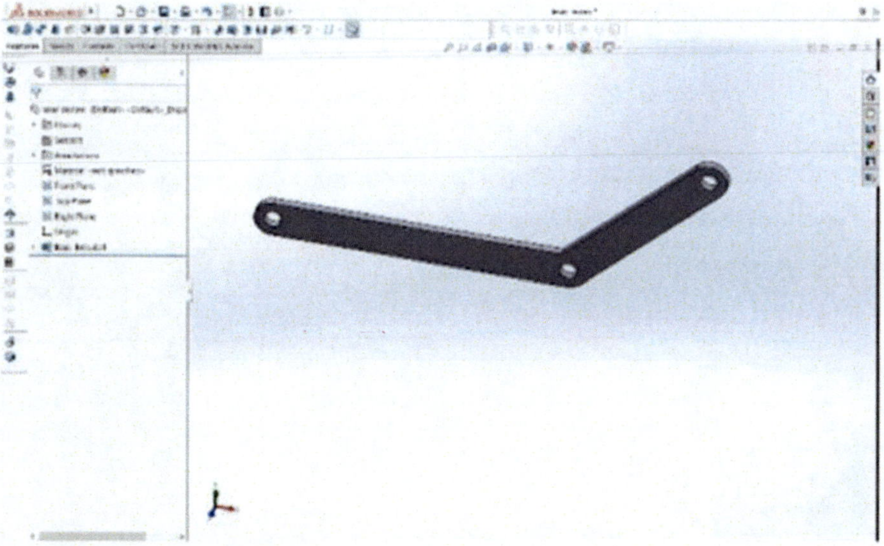

**Fig. 2** Lever Arm design model

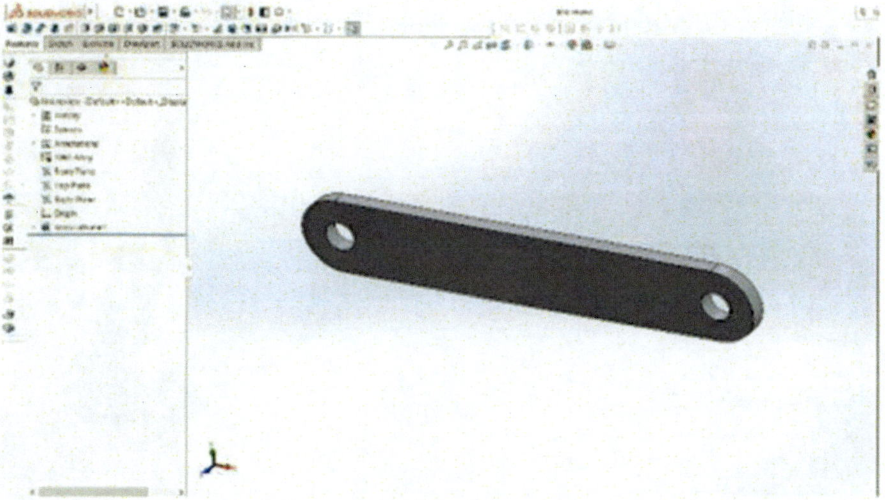

**Fig. 3** Drive Arm design model

### 2.3.3 Analysis of Components

Also the simulation of the analyzing the forces was done in the same software and contours and profiles of the stresses, strains and displacements were developed (Figs 5, 6 and 7).

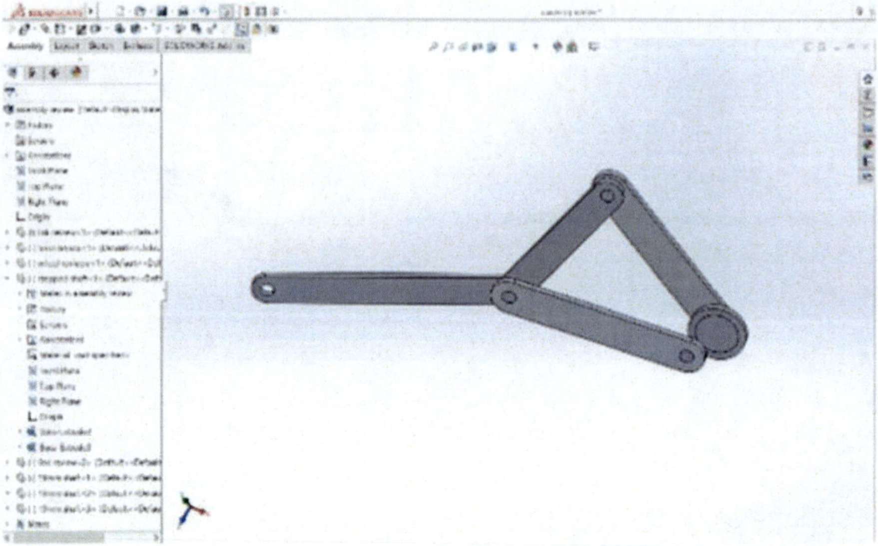

**Fig. 4** Four bar mechanism

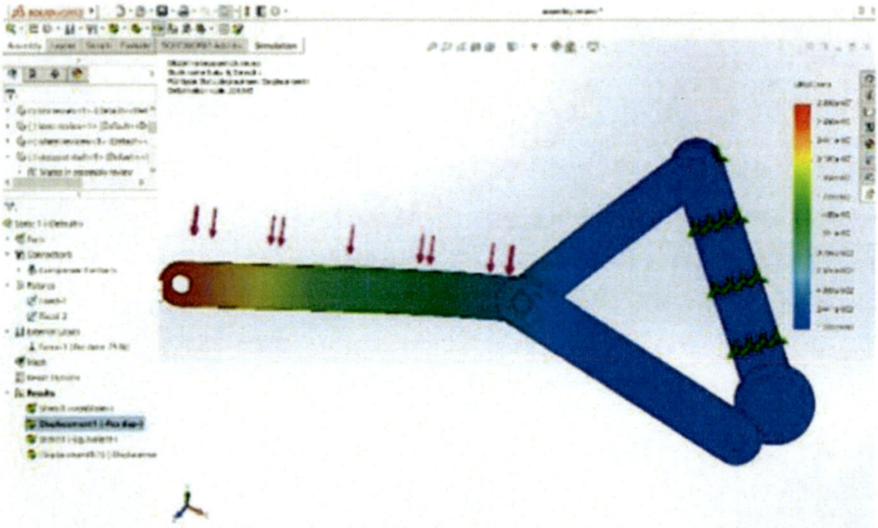

**Fig. 5** Principal of stresses analysis

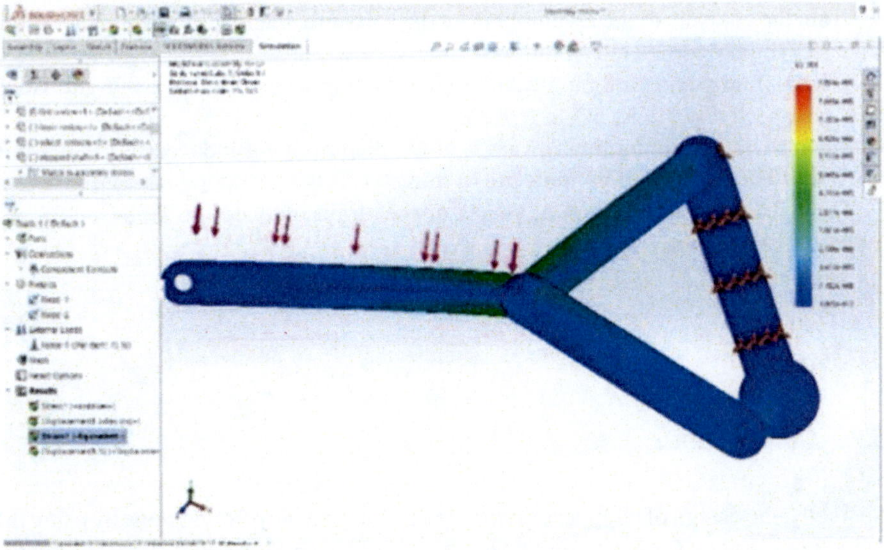

**Fig. 6** Principal of strain analysis

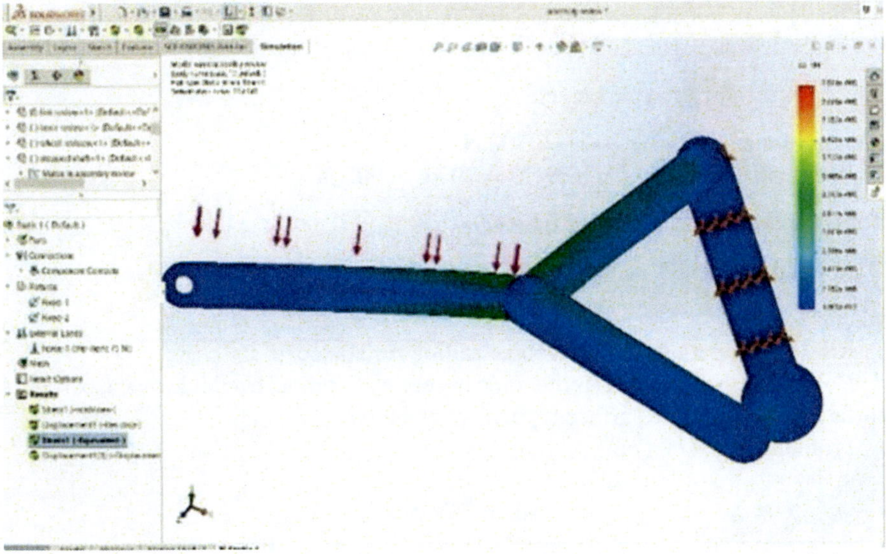

**Fig. 7** Displacement

From the analysis of the assembly from the software, the maximum and minimum values of the properties are as follows:

The maximum and minimum values of the stress induced is **8.584 e-005** and **1.095 e-012,** respectively, (units are in **N/m²**).

The maximum and minimum values of the displacement induced is **2.930 e-001** and **1.0 e-030,** respectively, (units are in **mm**).

The maximum and minimum values of the strain induced are **8.584 e-005** and **1.095 e-012,** respectively.

# 3 Results

## *3.1 Calculations*

So finally, the design of chainless drive mechanism for a bicycle was done by using the modeling software Catia and motion analysis was also performed by considering the various inputs obtained from the theoretical calculations. Now, it is time to analyze the mechanism components for stress and strain values for various loads. So initially the loads were considered and the design was evaluated and by using the software the analysis was performed to determine the safety of the design.

**Force required to movie bicycle**

Weight of the bicycle = 20 kg = 200 N.

Weight of the human on bicycle = 90 kg = 900 N.

**Torque required to propel the bicycle**

Total tractive force (TTE) = RR(N).

where

RR = Force required to overcome rolling resistance.

Rolling Resistance (RR) is the force necessary to propel a vehicle over a particular surface worst possible surface type, if we consider.

From the Eqs. (1)–(7).

$$\mathrm{RR} = \mathrm{GVW} * \mathrm{Crr}$$

where

GVW = Gross vehicle weight and

Crr = Surface friction between road and tyer, typical value is 0.055.

GVW = Man weight + cycle weight = 110 kg.

RR = 110 * 0.055 = 60.5 N.

**Torque required** $T_w = F * r = 60.5 * 0.3 = 18.15$ N-m

$$F = T_w/r = 18.15/0.05 = 363N$$

Bending Design Criteria for Lever.

The max bending moment on the lever applied is, $M_b = 297 * 125 = 45{,}375$ N-mm.

From bending stress $M/I = f/y$.

where

$M$ = bending moment.

$I$ = moment of inertia about neutral axis.

$f$ = bending stress.

$y$ = distance from outermost layer to neutral axis for aluminum value of yield strength, $S_{yt} = 207$Mpa considering F.S = 4.

bending stress, $f = (S_{yt}/ F.S)$.

= 51.74Mpa.

I = ((37,125*25)/51.75) = 20,719.17mm$^4$.

We considered the b and d values of the lever as, $b = 10$ mm and d = 50 mm.

Therefore I = 104,166.67mm$^4$.

Therefore our design against bending stress is safe.

### 3.2 *Stress and Strain Analysis of Components*

Since the most critical and failed front part of the system is lever arm, it is subjected to finite analysis using Solid works (Fig. 8 and 9).

From the analysis of the assembly from the software, the maximum and minimum values of the properties are as follows.

The maximum and minimum values of the stress induced is **9.584 e-005** and **2.095 e-012,** respectively, (units are in **N/m$^2$**).

The maximum and minimum values of the displacement induced is **3.930 e-001** and **1.0 e-030,** respectively, (units are in **mm**).

The maximum and minimum values of the strain induced are **9.584 e-005** and **2.095 e-012,** respectively.

Therefore induced stresses are within limits of allowable stresses. So our design is safe.

## 4 Conclusion

In the beginning, chainless bikes were less efficient than standard bikes, but changes in technology allowed to reach a level of efficiency that made them practical for day by day use. New design also improves the overall riding comfort and driving efficiency of bicycle. In many ways, this new chainless transmission system is superior to

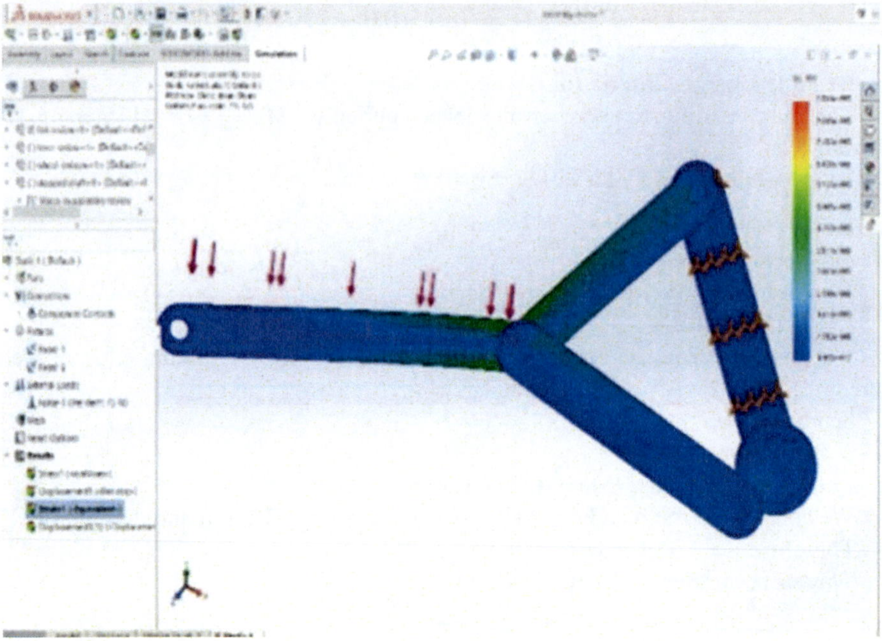

**Fig. 8** Principal of stress analysis

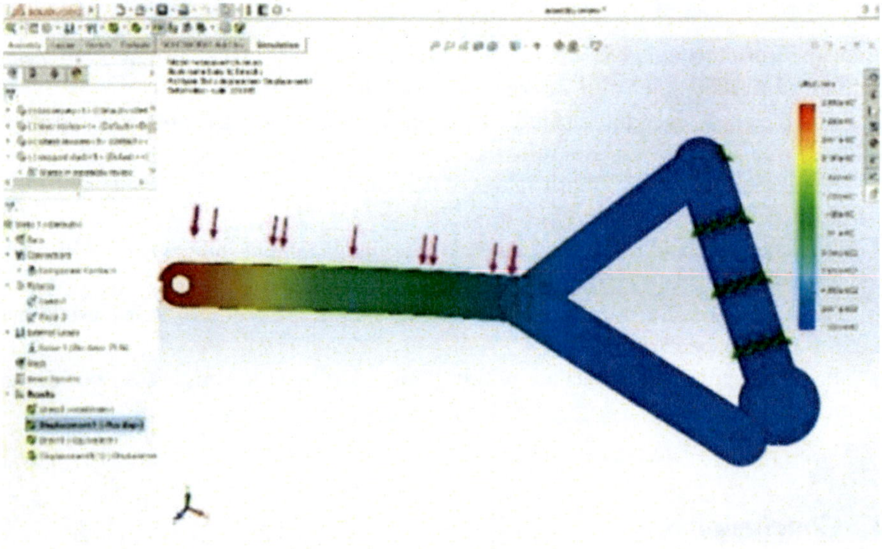

**Fig. 9** Principal of strain analysis

the traditional design. Effort can be reduced in a greater way. Gravitational effort can be utilized because of this reciprocating propulsion. Lubrication can be reduced compared to chain drive. Random disengagements of chain can be avoided. Life spam of the bicycle increased by using a basics four bar mechanism (oscillator motion to rotator motion). The results obtained from this work is an useful approximation to help in earlier stages of the development, saving development time and helping in the decision making process to optimize a design.

## References


1. Aakash P, Prajapati H, Patel V, Patel A (2017) Design and modelling of chainless bicycle with gear mechanism. Int J Sci Res Develop 5(05)
2. Reddy MRN, Prasad GH, Marurthi S, Ganapathi R, Janardhan M, Madhu Sudhan MP (2014) Design and febrication of shaft drive for bicycle. Int J Emerging Eng Res Technol 2(2)
3. Sharma Y, Banker P, Raikwar Y, Chauhan Y, Sharma M (2018) R&D on electric bike. IRJET Int Res J Eng Technol 05(02)
4. Zope SB, Linagariya ARM, Savsani D (2015) Dynamic chainless bicycle. IJAREST Int J Adv Res Eng Sci Technol 2(5)
5. Chandana MS, Kumar RS, Suhas S, Charan NS, Bhargav A, Design and analysis of shaft driven bicycle. IJRI-Int J Res Innovation
6. Babu MS, Mushtaq Quadri MD, Naga Prasad V, Kedarnath G (2017) Shaft driven bicycle. IJARIIE 3(02)

# A Review of Implantable Antenna for Biomedical Applications: Challenges and Solutions

**Pawan Kumar, Anil Sangwan, and Deepak Gangwar**

**Abstract** In this paper, an overview on the research progress of implantable antenna used for biomedical implantable devices is discussed. Biotelemetry is used for the transmission of physiological data from human body to a distance place by using the implantable medical devices (IMDs). These IMDs having many applications in biomedical communication mostly therapeutic and diagnostic application. Currently, patch antenna shows significant scientific interest in implantable medical devices due to their interesting feature like design flexibility, easy fabrication, conformability and its shape. The important factors for implantable medical devices need some requirement like its compact size, biocompatibility, patient safety, communication link ability, etc. This review paper focus to provide an overview on related research work of implantable antenna and challenges related to the designing and testing of implantable antenna and its solutions.

**Keywords** Implantable medical devices (IMDs) · Specific absorption rate (SAR) · External device (ED) · Wireless personal area networks (WPANs)

## 1 Introduction

Implantable medical devices (IMDs) are design to monitor physiological data inside human body and provide major contribution to improve the quality of life with disease prevention, therapy and diagnosis like drug delivery system, bone growth stimulators and treatment of various condition in medical profession [1]. With the exponential growth of implantable devices in medical profession body-centric wireless communication system (BWCS) becoming a focusing point. BWCS is a combined field of wireless body area networks (WBANs), wireless personal area networks (WPANs) and wireless sensor networks (WSNs) (Fig. 1).

P. Kumar · A. Sangwan (✉)
Department of ECE, UIET, Maharshi Dayanand University, Rohtak, India
e-mail: anilsangwan1979@gmail.com

D. Gangwar
Department of ECE, Bharati Vidyapeeth's College of Engineering, New Delhi, India

 
S. Yadav et al. (eds.), *Recent Advances in Metrology*, Lecture Notes in Electrical Engineering 906, https://doi.org/10.1007/978-981-19-2468-2_6

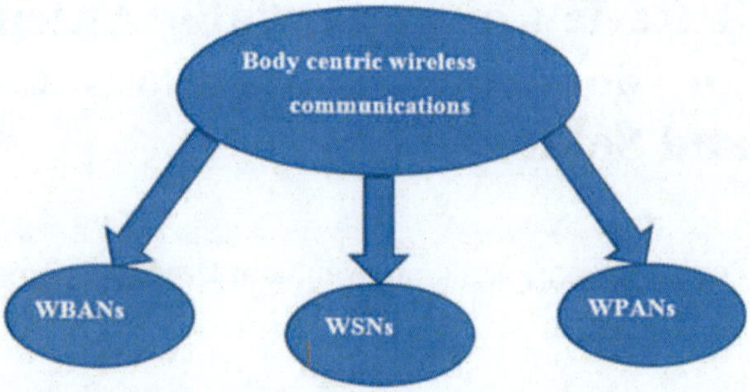


**Fig. 1** Different fields of body-centric wireless communications

BWCS consists of off-body, on body and in body communication. Communication with IMDs is an in body communication system. In body communication system means communication of IMDs with external device or communication of IMD with another IMDs. On body communication deals with communication between wearable device to another external devices [2]. Implantable medical devices facing more challenges as compare to wearable devices because of their complex and poor in body working environment (Fig. 2).

Different frequency bands are allocated to implantable medical devices like MICS, MedRadio, ISM bands. IMDs are capable to communicate with external devices by two types, either by radio frequency link or by inductive link [3]. Communication through inductive link is only possible when external device comes closer to implantable device, i.e., used only for short range communication. Communication through inductive link may also suffers from high bit error rate, low data rate and sensitivity increased due to inter coil positioning. While communication through radio frequency link have advantages on inductive link such as lower bit error rate, higher bandwidth, higher data rate, increases the range of the communication. So, focus is shifted toward RF linked implantable devices and these devices are most widely used in a number of health care applications.

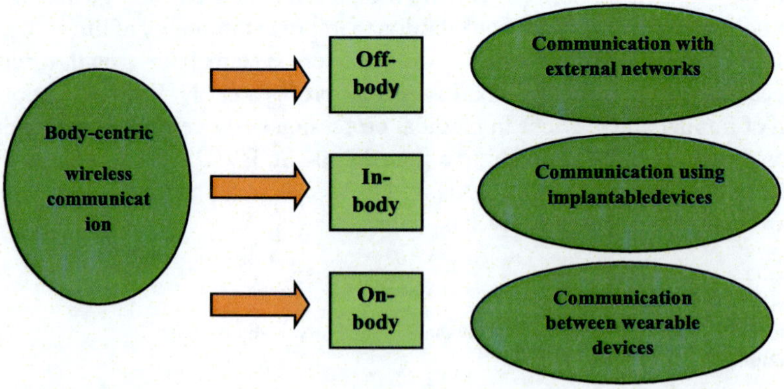


**Fig. 2** Types of BWCS [2]

This paper gives a review on challenges and solutions related to implantable antenna for IMDs. In Sect. 2, design requirements for implantable antenna like miniaturization, biocompatibility issues, patient safety consideration, and communication link ability are given, while antenna measurement and testing are summarized in Sect. 3, followed by conclusion remarks in Sect. 4.

## 2 Design Requirement

Unlike other transmitting antennas such as mobile antenna that operate in free space, implantable antenna which are embedded into human bodies consider some special kind of requirements. These requirements for implantable antennas are miniaturization, biocompatibility, safety consideration, communication link ability of implantable circuit.

### 2.1 *Miniaturization*

Due to implantable antennas are embedded into human bodies these antennas require an ultra-small design for IMDs. IMDs which are operating at very low frequency specially at Medical Implant Communications Systems (MICS) band ranging from 402 to 405 MHz or Medical Device Radio Communication Services (MedRadio) band ranging from 401 to 406 MHz, the size of implantable antenna at resonance frequency became very large which create problems during the embedded process in human tissue. Therefore the size of implantable antenna became very crucial and miniaturization techniques are used to solve these types of challenges. Miniaturization is done by changing the physical property of structure, changing the material characteristic or by adding additional components.

#### 2.1.1 High Permittivity Dielectric Substrate or Superstrate

One of the simplest miniaturization techniques of antennas, is by using a high permittivity substrate/superstrate which can reduce the effective implantable antenna length. Permittivity of substrate/superstrate material directly affect the resonant frequency, i.e., if a high permittivity material is used as a substrate/superstrate the resonant frequency moves toward the lower frequency because this shorten the effective wavelength. Some of the material which are commonly used for dielectric substrate of implantable antenna are alumina, ceramic with the $\varepsilon_r$ [4] and Roger 3210, RO 3010,6002 with the $\varepsilon_r$ 10.2 is utilized (Table 1).

A significant reduction in effective length is achieved by using $MgTa_{1.5}Nb_{0.5}O_6$ as a dielectric substrate which has dielectric constant $\varepsilon_r$ [7].

**Table 1** Different materials used as substrate

| References | Material | Relative dielectric constant (→$_r$) |
|---|---|---|
| [5] | Macor | 6.1 |
| [6] | Rogers 3210 | 10.2 |
| [7] | $MgTa_{1.5}$ $Nb_{0.5}$ $O_6$ | 28 |
| [8] | Rogers TMM10 | 9.1 |
| [9] | Rogers 3010 | 10.2 |

### 2.1.2 Lengthening the Current Flow Path of Antenna

By changing the physical properties of an antenna we can achieve the longer effective current flow path, which shifted the resonant frequency to lower resonant frequency and results in antenna size reduction. Several design techniques are used for this purpose, meandered, spiral, hook slotted [10], waffle type [11] and patch staking method [12] is also used.

### 2.1.3 Impedance Matching with Loading and Shorting Pin

When we design an implantable antenna it is not a perfectly resonant antenna, depending on its length impedance of antenna may be either capacitive or inductive. For a good impedance matching we should nullify the effect of reactance by using either capacitive or inductive loading, this will help in miniaturization of antenna. Shorting pin technique is also used for miniaturization, for this we have to insert shorting pins in between ground plane and patch plane [7, 13]. Shorting pin technique is work like a monopole antenna, in which the size of the antenna became half of the ordinary dipole antenna. Typically, a planner inverted-F antenna (PIFA) antenna which also has quarter wavelength size is also suggested in related work.

### 2.1.4 Higher Frequency Band

Different frequency bands are allotted by United State Federal communication commission (FCC) and European Radiocommunication committee (ERC)which are MICS, MedRadio, Industrial, Scientific and Medical (ISM) bands [12, 14]. By using higher operating frequency band we can shorten the wavelength of implantable antenna which cause reduction in the size of the antenna. Operating in higher frequency band provides wider bandwidth which is better for high data rate communication. However operating in higher frequency band can cause higher biological tissue attenuation in human body. So factors such as operating frequency, transmission distance, device dimensions needs carefully design of implantable antenna.

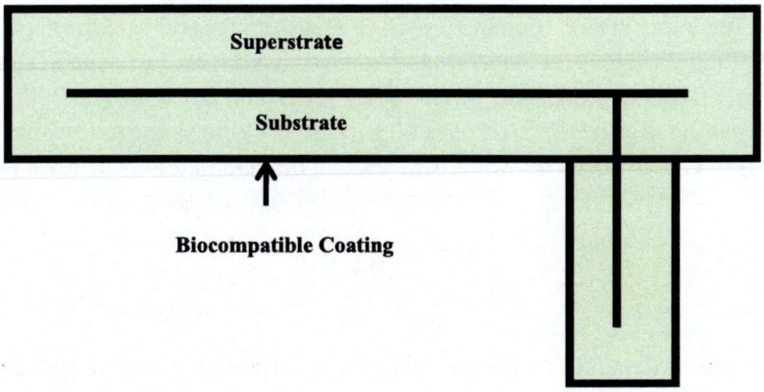


**Fig. 3** Biocompatible material coating on antenna

## 2.2 *Biocompatibility*

Implantable antennas are embedded into human bodies so they must be biocompatible nature in order to preserve patient safety. If we directly embedded the implantable antenna into human body, it get short circuited because human tissues are conductors. So prevention from undesired short circuited and biocompatibility becomes very crucial for long term implantation. Mainly two types of approaches are suggested for biocompatibility issues of implantable antenna. One of these approaches is using biocompatible substrate for antenna fabrication and the second approach is by covering the implantable antenna by a thin layer of low loss biocompatible material coating. The material which are proposed for antenna coating are PEEK [15], Zirconia [16], Silastic MDX-4210 Biomedical grade-based Elastomer[17], etc. and the material which are used as biomedical substrate are Teflon, MACOR, Dielectric loss tangent, Ceramic Alumina [5], etc. Thin layer of low loss biocompatible material coating increases the biocompatibility of implantable antenna, but the thickness of layer may affect the performance of implantable antenna, so carefully design of implantable antenna is needed [18]. To increase the biocompatibility of antenna Parylene C material coating on both side of antenna is suggested [19] (Fig. 3).

## 2.3 *Safety Consideration*

Implantable antenna are used to improve the quality of life and embedded into human body, so patient safety is a very important factor during the designing of antenna. When some power is incident on implantable antenna it can increase the temperature of the surrounding tissues, which may be harmful for human body. So the maximum power which is incident on implantable antenna is limited by factor known as Specific Absorption Rate (SAR). SAR determines the total amount of energy which is observed by nearest human body tissues. Two standards are used as preference for exposure of human tissues to electromagnetic radiations. The IEEE C95.1-1999 standard restrict the averaged value of SAR over 1g tissue in cubic shape to less than 1.6 W/kg, i.e., SAR $\leq$ 1.6 W/kg [20]. The IEEE C95.1-2005 standard restrict the averaged value of SAR over 10 g tissue in cubic shape to less than 2 W/kg, i.e., SAR $\leq$ 2 W/kg [20]. The value of SAR can be calculated as follows [21]:

$$\mathrm{SAR} = \frac{1}{2}\frac{\sigma}{\rho}|E|^2 \tag{1}$$

where

SAR variation depends on the permittivity of substrate material and the permittivity is depends on signal frequency [22]. Specific Absorption (SA) is another limitation for pulsed transmission which is given by

$$\mathrm{SA} = \mathrm{SAR} \times T_p \tag{2}$$

## 2.4 *Communication Link Ability*

In far field communication, the link power budget may be calculated as [9],

$$\text{Link margin (dB)} = \text{link}\frac{C}{N_o} - \text{Required}\frac{C}{N_0} \tag{3}$$

$$= P_\mathrm{t} - G_\mathrm{t} - L_\mathrm{t} - G_\mathrm{r} - N_o - \frac{E_b}{N_0} - 10\log_{10} B_\mathrm{r} + G_\mathrm{c} - G_\mathrm{d} \tag{4}$$

where $P_\mathrm{t}$ = transmitting power, $G_\mathrm{t}$ = transmitting antenna gain, $L_\mathrm{p}$ = path loss, $G_\mathrm{r}$ = receiving antenna gain, $N_o$ = noise power density (Fig. 4).

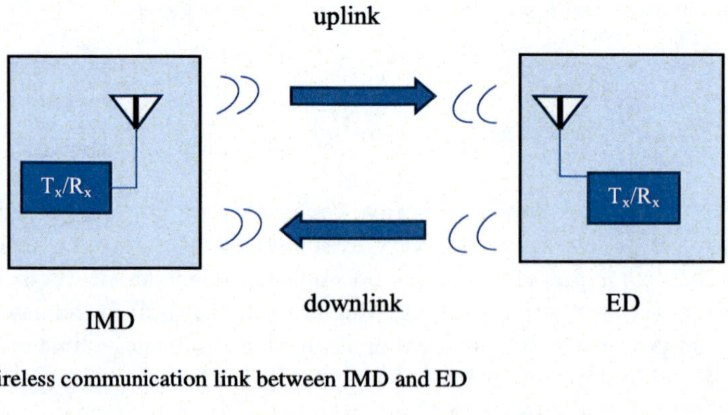


**Fig. 4** Wireless communication link between IMD and ED

In free space communications, with increase in distance between the transmitting and receiving antenna, the signal strength got reduces.

Path loss can be calculated as

$$L_f(\text{dB}) = 20 \log \frac{4\pi d}{\lambda} \tag{5}$$

where $d =$ distance between transmitting and receiving antenna.

Similarly, impedance matching losses is given by

$$L(\text{dB}) = -10 \log(1 - |\tau|^2) \tag{6}$$

$\tau =$ reflection coefficient.

## 3 Antenna Measurement and Testing

For designing and simulation of implantable antenna CST microwave suite, COMSOL, HFSS commercial simulation software's are suggested in related research work. In [23], spherical dyadic green's function (DGF) expansion and Finite-Difference Time-Domain (FDTD) codes are used to analyze the electromagnetic characteristics of antenna inside human body and head. When a thin layer of biocompatible material coating on implantable antenna, the current distribution on the patch surface of the antenna gets changed which caused to change in field pattern of the antenna. Implantable antennas which are embedded into human body and is surrounded by body tissues and fluids. Hence, human phantoms models are used to analyze the electromagnetic characteristics of the antenna.

One layer skin model is mostly used for testing of implantable antenna to make simulation more efficient. Beside this three layer tissue (skin, fat, muscle) model and 2/3 muscle model are also suggested in related research work. Testing of implantable antenna in animals tissues is also suggested in related research work. Different types of material are used for making these model to meet the requirement of equivalent environment human body tissue model having properties conductivity and permittivity. However in practical the efficiency of implantable antenna may decrease due to variation in factors one body to another body and these factors are frequency, age, sex, temperature, etc. In-vivo and In-vitro measurement for implantable antenna are suggested in related research work [24] (Table 2).

## 4 Conclusion

In this review paper, challenges and solution related to implantable antenna design are discussed. An overview on design requirement related to implantable antenna

**Table 2** Different $\sigma$ and $\varepsilon_r$ of different materials at 433 MHz (15)

| Material | Relative dielectric constant ($\varepsilon_r$) | Sigma ($\sigma$) $S/m$ |
|---|---|---|
| Fat | 5.03 | 0.045 |
| Bone | 17.36 | 0.1673 |
| Cartilage | 43.65 | 0.65 |
| Skin | 46.68 | 0.64 |
| Blood | 57.29 | 1.72 |
| Muscle | 47.81 | 0.6463 |

is presented. Different types of miniaturization techniques, biocompatibility issues, patient safety issues, communication link ability of implantable antenna, antenna measurement and testing methods are also presented. Meanwhile other parameter related to implantable antenna like operating frequency, radiation efficiency, bandwidth, power consumption, testing of antenna in different phantoms models also discussed. Both the quality of patient life and health care can be improved to a higher level with these implantable medical devices.

## References

1. Kiourti A, Nikita KS (2012) A review of implantable patch antennas for biomedical telemetry: challenges and solutions (wireless corner). IEEE Antennas Propag Mag 54(3):210–228
2. Liu C, Guo YX, Xiao S (2016) A review of implantable antennas for wireless biomedical devices. In: Forum for electromagnetic research methods and application technologies (FERMAT), vol 14(3)
3. Valanarasi A, Dhanasekaran R (2016) A review on design considerations of implantable antennas. In: 2016 international conference on advanced communication control and computing technologies (ICACCCT). IEEE, pp 207–211
4. Kiourti A, Christopoulou M, Nikita KS (2011) Performance of a novel miniature antenna implanted in the human head for wireless biotelemetry. In: IEEE international symposium on antennas and propagation, Spokane, Washington
5. Soontornpipit P, Furse CM, Chung YC (2004) Design of implantable microstrip antennas for communication with medical implants. IEEE Trans Microw Theory Tech 52(8):1944–1951
6. Karacolak T, Hood AZ, Topsakal E (2008) Design of a dual-band implantable antenna and development of skin mimicking gels for continuous glucose monitoring. IEEE Trans Microw Theory Tech 56:1001–1008
7. Chien TF, Cheng C-M, Yang H-C, Jiang J-W, Luo C-H (2010) Development of nonsuperstrate implantable low profile CPW- fed ceramic antennas'. IEEE Antennas Wirel Propag Lett 9:599–602
8. Merli F, Bolomey L, Zürcher J-F, Corradini G, Meurville E, Skrivervik AK (2011) Design, realization and measurements of a miniature antenna for implantable wireless communication systems. IEEE Trans Antennas Propag 59(10):3544–3555
9. Liu C, Guo YX, Xiao S (2014) Capacitively loaded circularly polarized implantable patch antenna for ISM-band biomedical applications. IEEE Trans Antennas Propag 62(5):2407–2417
10. Kiourti A, Nikita KS (2011) Meandered versus spiral novel miniature PIFAs implanted in the human head: tuning and performance. In: Proceedings of the 2nd ICST international conference on wireless mobile communication and healthcare (MobiHealth 2012), Kos Island, Greece

11. Soontornpipit P, Furse CM, Chung YC (2005) Miniaturized biocompatible microstrip antenna using genetic algorithm. IEEE Trans Antennas Propag AP-53 6:1939–1945
12. Kiourti A, Nikita KS, Miniature scalp-implantable antennas for telemetry in the MICS and ISM bands: design, safety considerations and link budget analysis. IEEE Trans Antennas Propag (to appear)
13. Li H, Guo YX, Liu C, Xiao S, Li L (2015) A miniature implantable antenna for MedRadio-band biomedical telemetry. IEEE Antennas Wirel Propag Lett 14:1510–1513
14. Yilmaz T, Karacolak T, Topsakal E (2008) Characterization and testing of a skin mimicking material for implantable antennas operation at ISM band (2.4 GHz-2,48 GHz). IEEE Antennas Wirel Propag Lett 7:418–420
15. Abadia J, Merli F, Zurcher JF, Mosig JR, Skrivervik AK (2009) 3D spiral small antenna design and realization for biomedical telemetry in the MICS band. Radioengineering 18(4):359–367
16. Skrivervik AK, Merli F (2011) Design strategies for implantable antennas. In: Proceedings of the antennas and propagation conference, Loughborough, UK, Nov 2011
17. Karacolak T, Cooper R, Butler J, Fisher S, Topsakal E (2010) In vivo verification of implantable antennas using rats as model animals. IEEE Antennas Wirel Propag Lett 9:334–337
18. Merli F, Fuchs B, Mosig JR, Skrivervik AK (2011) The effect of insulating layers on the performance of implanted antennas. IEEE Trans Antennas Propag 59(1):21–31
19. Duan Z, Guo YX, Xue RF, Je M, Kwong DL (2014) Design and in vitro test of a differentially fed dual-band implantable antenna operating at MICS and ISM bands. IEEE Trans Antennas Propag 62(5):2430–2439
20. IEEE standard for safety levels with respect to human exposure to radiofrequency electromagnetic fields, 3 kHz to 300GHz. IEEE Standard C95.1 (1999)
21. Lin H-Y, Takahashi M, Saita K, Ito K (2013) Characteristics of electric field and radiation pattern on different locations of the human body for in body wireless communication. Electron Lett 61(10):5350–5354
22. Thotahewa KMS, Redouté J-M, Yuce MR (2013) SAR, SA, and temperature variation in the human head caused by IRUWB implants operating at 4 GHz. IEEE Trans Microw Theory Tech 61(5):2161–2169
23. Kim J, Rahmat-Samii Y (2004) Implanted antennas inside a human body: simulations, designs, and characterizations. IEEE Trans Microw Theory Tech 52(8):1934–1943
24. Kawoos U, Tofighi M-R, Warty R, Kralick FA, Rosen A (2008) In-vitro and in-vivo trans-scalp evaluation of an intracranial pressure implant at 2.4 GHz. IEEE Trans Microw Theory Tech 56:2356–2365

# Stability Evaluation of the Reference Standard PMU at CSIR-NPL

**Swati Kumari, Sunidhi Luthra, Archana Sahu, Saood Ahmad, and Anurag Gupta**

**Abstract** Phasor measurement units (PMU) has becoming an emerging and promising technology in monitoring the safety and security of the power grids. Synchro phasor measurements using PMU facilitate dynamic real-time measurements round the clock that will help in realization of smart grid. As PMUs are increasingly being deployed in power grids, it is critical that their results must be traceable, reliable, and accurate. Recently, established PMU calibration system traceable to the primary standards of CSIR-NPL is used to carry out comprehensive PMU calibration as per IEEE Std. C37.118.1a-2014. In this paper, a reference standard PMU, which is being calibrated using PMU calibration system, has been studied for its stability over the period of time that is significant for analyzing reference standard PMU performance. Reference standard PMU has been calibrated for *M* class at 50 frame rate. Static and dynamic tests were performed at particular voltage and current level to study the stability. Measured parameters are defined in terms of total vector error (TVE), frequency error (FE), and rate of change of frequency error (ROCOF). As per IEEE protocol, acceptable limit defined for TVE is 1.00%. The CSIR-NPL reference standard PMU has maximum estimated TVE of 0.6%. The measured result shows that reference standard PMU has very good stability and measurements taken so far is well within the acceptable IEEE limits.

**Keywords** Steady state · Stability · Phasor measurement unit · PMU calibration system · Traceability

## 1 Introduction

Monitoring of power grid is the key factor to enable safe and efficient power distribution, as well as it reduces overall operation and maintenance cost. Earlier SCADA/EMS [1] systems were used to monitor power grid system that has the capability to provide only steady state view of the power system. These systems

S. Kumari (✉) · S. Luthra · A. Sahu · S. Ahmad · A. Gupta
CSIR-National Physical Laboratory, New Delhi, India
e-mail: kumaris1@nplindia.org

 
S. Yadav et al. (eds.), *Recent Advances in Metrology*, Lecture Notes in Electrical Engineering 906, https://doi.org/10.1007/978-981-19-2468-2_7

take a minute to measure status of a system whose characteristics are changing very fast [2]. But with PMU system, measurement can be done with the accuracy of few milliseconds. Synchro phasor measurements using PMU over wide area facilitate dynamic real-time measurements and visualization of power system which are useful in monitoring safety and security of the grid as well enable in taking control/corrective actions in the new regime of grid management. Due to technological advancements, Indian grid has become smart having modern sensors, communication links, and computational power are used to improve efficiency, stability, and flexibility of the system. With PMUs dispersed throughout the system, the power transfer capacity of lines can also be increased.

CSIR-NPL being an National Measurement Institute (NMI) of the country by an act of parliament has the responsibility to establish national standards and provide apex level calibration and testing services to the nation. PMU being the key measurement device in modern smart grids, it is necessary that device must be able to measure accurate data. In view of that CSIR-NPL is now providing calibration of phasor measurement units (PMUs) for effective monitoring of power grid system so that their data will be consistent, accurate, and credible [3]. In this paper, we present the stability evaluation of our reference standard PMU which is being tested at particular configuration over a period of time.

## 2 PMU Calibration/Test Setup

PMU calibration system is an integrated system that completes number of automated tests and meets the entire compliance testing as per IEEE C37.118.1a-2014 standard. Calibration time is shortened to five to six hours instead of a week and handles all the steady state, dynamic state, and reporting latency tests for both *M* class and *P* class at all required frame rates. PMU calibration system is capable of generating required voltage and current signals in similar conditions that occurs in power grid. Figure 1 shows the PMU calibration/test setup used for reference PMU stability evaluation.

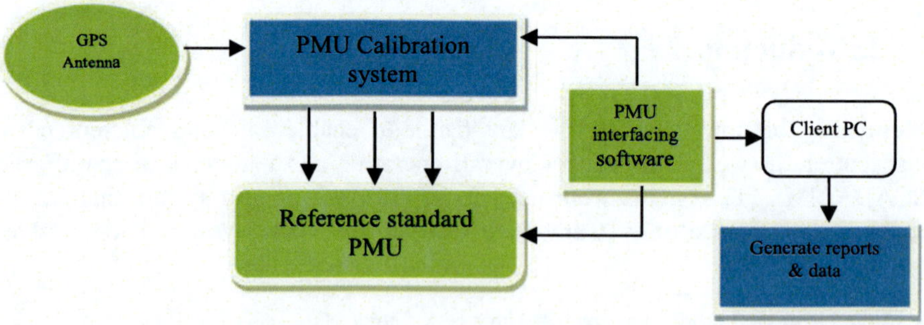


**Fig. 1** PMU calibration test setup

The reference standard PMU is being tested using three phase voltage and current signals generated using PMU calibration system [3]. The reference standard PMU then report estimates of synchro phasor, frequency, and rate of change of frequency with time stamping [4]. Calibration system is fully automated LabVIEW-based software with full compliance as per IEEE C37.118.1a-2014 standard.

## 3 Measurement Parameters

Reference standard PMU is evaluated for TVE, FE, and ROCOF. The determination of stability for the PMU is based on the graphical analysis of TVE, FE, and ROCOF.

1. *Total vector error*: Total vector error is an expression of the difference between a 'perfect' sample of a theoretical synchro phasor and the estimate given by the unit under test at the same instant of time [4]. The TVE is expressed as

$$\mathrm{TVE}(n) = \sqrt{\frac{\left(\hat{X}_r(n) - X_r(n)\right)^2 + \left(\hat{X}_i(n) - X_i(n)\right)^2}{(X_r(n))^2 + (X_i(n))^2}}$$

   where $\hat{X}_r(n)$ and $\hat{X}_i(n)$ are the sequences of estimates given by the unit under test, and $X_r(n)$ and $X_i(n)$ are the sequences of theoretical values of the input signal at the instants of time ($n$) assigned by the unit to those values.
2. *Frequency and ROCOF errors:* are the absolute value of the difference between the theoretical values and the estimated values given in Hz and Hz/s, respectively [4].

$$\mathrm{FE} = |f_{\mathrm{true}} - f_{\mathrm{measured}}| = |\Delta f_{\mathrm{true}} - \Delta f_{\mathrm{measured}}|$$

$$\mathrm{RFE} = \left| \left(\frac{df}{dt}\right)_{\mathrm{true}} - \left(\frac{df}{dt}\right)_{\mathrm{measured}} \right|$$

The measured and true values are for the same instant of time, which will be given by the time tag of the estimated values.

## 4 Test Results

Calibration was performed at 50 Hz. Basic configuration used for stability analysis is at 70 V, 1 A, and at 50 frame rate for M class. Tests were repeated over a period of time to study the stability of reference standard PMU. Test results are shown in Tables 1 and 2.

**Table 1** Steady state frequency response test results

| Steady state frequency response test | $V_a$ | $V_b$ | $V_c$ | $I_a$ | $I_b$ | $I_c$ |
|---|---|---|---|---|---|---|
| TVE | 0.002 | 0.003 | 0.003 | 0.005 | 0.003 | 0.002 |
| FE | 0.00002 | 0.00002 | 0.00002 | 0.00002 | 0.00002 | 0.00002 |
| RFE | 0.0032 | 0.0032 | 0.0032 | 0.0032 | 0.0032 | 0.0032 |

**Table 2** Dynamic state frequency ramp test results

| Dynamic state frequency ramp test | $V_a$ | $V_b$ | $V_c$ | $I_a$ | $I_b$ | $I_c$ |
|---|---|---|---|---|---|---|
| TVE | 0.003 | 0.002 | 0.003 | 0.004 | 0.006 | 0.004 |
| FE | 0.00005 | 0.00005 | 0.00005 | 0.00005 | 0.00005 | 0.00005 |
| RFE | 0.0030 | 0.0030 | 0.0030 | 0.0030 | 0.0030 | 0.0030 |

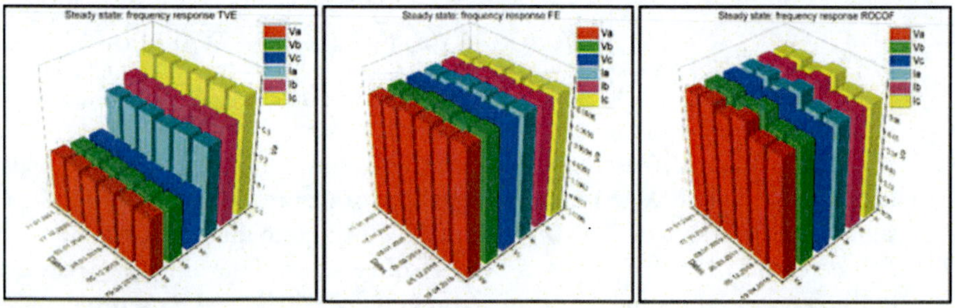


a) TVE b) FE c) ROCOF

**Fig. 2** Steady state frequency response test results

Figure 2a shows TVE analysis for steady state frequency response tests. TVE limits defined as per IEEE protocol is 1% [4]. Graph shows results are highly stable for all the three phases of voltage and current over the period of time as well as within the TVE limits defined as per IEEE limits. Similarly, Fig. 2b, c shows stability of frequency error and rate of change of frequency error. Figures 3 and 4 show tests results for harmonic distortion and frequency ramp test.

## 5 Conclusion

Synchronized phasor measurements plays an important role in wide area measurement systems used in power grid. PMU is an encouraging device for smart grids to provide protection, monitoring, and control. Due to the recent development of the smart Grid, PMU calibration, its stability, and measurement traceability are

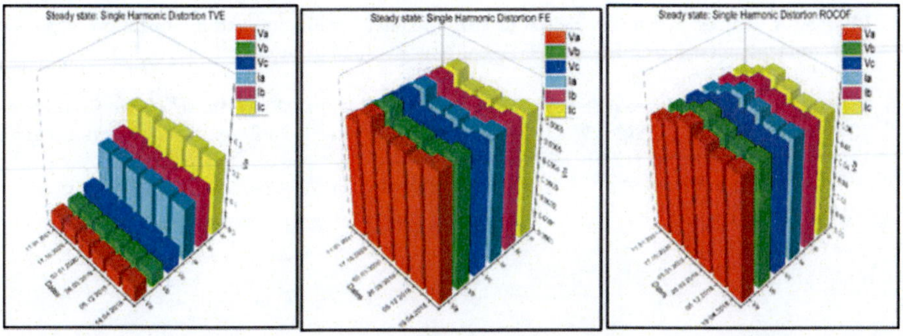


a) TVE b) FE c) ROCOF

**Fig. 3** Steady single harmonics distortion test results

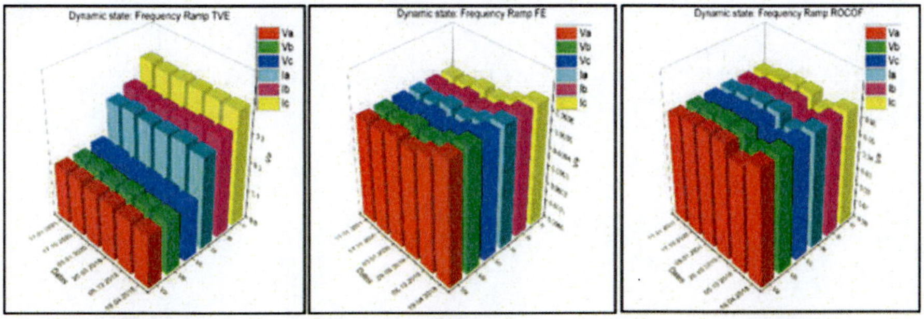


a) TVE b) FE c) ROCOF

**Fig. 4** Dynamic state frequency ramp test results

significant. This paper presented a brief review on phasor measurement unit and its related detailed stability analysis over a period of time. The measurement results have confirmed that the CSIR-NPL reference standard PMU's is highly stable and measurement results are well within the IEEE standard limits. A stable PMU measurement will enable to determine the exact cause of events which may have led to the failure of the power system.

## References

1. https://www.electricalindia.in/phasor-measurement-units-for-power-systems/
2. Rihan M, Ahmad M, Beg M (2011) Developing smart grid in India: background and progress. https://doi.org/10.1109/ISGT-MidEast.2011.6220788

3. Bhargav A, Ahmad S, Kumari S et al (2020) Technical evaluation and optimization of phasor measurement unit using CSIR-NPL PMU calibrator system to ensure reliability. Mapan 35:117–124
4. IEEE standard for synchro phasor measurements for power systems. In: IEEE Std C37.118.1-2011 (Revision of IEEE Std C37.118-2005) vol., no., pp 1–61, 28 Dec 2011. https://doi.org/10.1109/IEEESTD.2011.6111219

# Optimization of Electron Glass at Zero Disorder Using Rejection-Free Kinetic Monte Carlo Algorithm

**Vikas Malik**

**Abstract** In this paper, we try to find an electron glass system's ground state at zero disorder. We have used rejection-free kinetic Monte Carlo simulation for optimization. We use Kawasaki dynamics since the particle number is always conserved. We only allow nearest neighbor hops in our simulation. The optimization is done by quenching the system from an infinite temperature to a temperature below the transition temperature. Keeping the same run time for all temperatures considered, we check the algorithm's efficiency for different temperatures. One sees that as the temperature decreases, the optimization becomes more and more inefficient. This inefficiency is explained here in terms of the domain pictures.

**Keywords** Coulomb glass · Random field Ising model · Kinetic Monte Carlo · Optimization

## 1 Introduction

Various systems exhibit glassy behavior because of competition between disorder and interactions. Such systems (having short-range interactions) are modeled by the spin glass and random field Ising model (RFIM). Amorphous and compensated semiconductors showed a glassy phase (termed as Coulomb glass or electron glass) at very low temperatures due to competition between long-ranged Coulomb interaction and disorder [1–3]. It is essential to understand that if the Coulomb interactions get screened, one will get random field Ising model results.

The Coulomb Glass (CG) model represents a disordered insulator with unscreened Coulomb interactions whose electronic states are localized close to the Fermi level. Examples of Coulomb Glass systems are amorphous and compensated semiconductors. In these systems, at low temperatures, conductivity is governed by hopping among impurity states. Recently Coulomb glass system has been applied to nanocomposites as well. Efros [4] has shown that the long-range Coulomb interaction creates

V. Malik (✉)
JIIT, Noida, Uttar Pradesh, India
e-mail: vikas.malik@jiit.ac.in

 
S. Yadav et al. (eds.), *Recent Advances in Metrology*, Lecture Notes in Electrical Engineering 906, https://doi.org/10.1007/978-981-19-2468-2_8

a gap in the density of states so that it tends to zero as $g(\epsilon) \sim \epsilon^{d-1}$ where $d$ is the dimension of the system. This leads to Efros-Shklovskii's $T^{1/2}$ law in conductivity for CG systems. According to the Efros model, electrons can occupy a simple periodical lattice's sites. The total energy of the system consists of the random initial energies of the sites and the electrostatic interaction energy of the charges. The Hamiltonian for CG system on a lattice [5, 6] is,

$$H = \frac{1}{2} \sum_{i \neq j}^{N} J_{ij} S_i S_j - \sum_{i=1}^{N} \phi_i S_i \tag{1}$$

where $\phi_i$ is the on-site disorder term which is chosen from a box-distribution of width $W$ and $J_{ij} = e^2/r_{ij}$ is the unscreened Coulomb interactions between electrons on sites $i$ and $j$. $W$ represents the strength of disorder. For a long-range Coulomb interaction system with site energy disorder, one cannot find the ground state analytically. Baranovskii [7] tried to find the ground state using a minimization procedure. They found many minimum energy states close in energy to each other. This set of states were called pseudo ground states. Davies et al. [5, 6] performed a similar study but with periodic boundary conditions to study the characteristics of the density of state close to the Fermi level. The density of states showed a gap for 2-d and 3-d lattices as predicted by Efros.

A lot of work has been done on the random field Ising model (RFIM) at zero temperature ($T$) to determine the possibility of ferromagnetic ordering below a critical disorder and the nature of phase transition there. The Hamiltonian for this model is,

$$H = -\sum_{ij}^{N} J_{ij} S_i S_j - \sum_{i=1}^{N} h_i S_i \tag{2}$$

where $h_i$ is the random field at the $i$th site. $J_{ij}$ is equal to $J$ when $i, j$ are nearest neighbors and zero otherwise. In the random field Ising model [8], the ordered phase in 3-d is ferromagnetic, which changes into a paramagnetic phase at $T > T_c$ and $h > h_c$. At $T = 0$, Imry and Ma [9] showed that ferromagnetic ordering becomes unstable when $d < 2$ and $d = 2$ is the critical dimension. It was shown using roughening of domain walls [10] that there can be no ferromagnetic ordering at finite disorders for 2-d RFIM at zero temperature. For zero random field, the RFIM reduces to the simple Ising model in zero magnetic fields.

The 2-d CG system undergoes a first-order phase transition from a charge-ordered phase (COP) to a paramagnetic phase as the disorder is increased [11]. A single domain in the charge-ordered phase for $W < W_c$ breaks into two large domains for a disorder just above the critical disorder ($W_c$) [12]. This leads to discontinuity in staggered magnetization, which was the reason for the first-order transition. The domain formed just above the critical disorder was compact and became fractal at high disorders. Notably, there is no such phase transition in 2-d RFIM.

At zero disorder in two and three dimensions, for the Ising system, there is a transition from paramagnetic to ferromagnetic phase as the temperature becomes less than the transition temperature. This transition is of second-order in which the magnetization changes continuously from zero to one. For a CG system, a similar second-order transition was observed [13–16] in two and three dimensions. Since CG is like an antiferromagnet with long-range interactions, the ground state is the charge-ordered phase. The order parameter is staggered magnetization, which changes from zero to one as the system goes from a paramagnetic to antiferromagnetic phase. The critical exponents obtained for the CG system are the same as those for the Ising model. For finite disorders ($W \ll W_c$), the CG model in two and three dimensions undergoes a second-order phase transition from paramagnetic to charge-ordered phase. In three dimensions, Goethe et al. [14] claim that the critical exponents are the same as RFIM and do not change with the disorder. In contrast, Preeti et al. [15, 16] have found that the critical exponents do not match the RFIM exponents and are a function of disorder in two and three dimensions.

## 2 Rejection-Free Kinetic Monte Carlo Algorithm

The kinetic algorithms are used mainly to study the dynamics of the system. There are two types of kinetic Monte Carlo algorithms (KMC). One would choose two nearest neighbor sites randomly from the system in the first type. If the sites have different spins that imply one is an electron, and another is a hole, one will consider a possible spin-exchange. The spin-exchange is done according to the Metropolis algorithm [17]. If the spin-exchange leads to lowering the system's energy, then one does the spin-exchange. If the spin-exchange leads to an increase in the system energy by, let us say $\Delta E$, then the system at temperature $T$ does the exchange with the probability $\exp(-\beta \ \Delta E)$ where $\beta = 1/T$. So the spin-exchange probability for $\Delta E > 0$ decreases with the lowering of temperature. This implies that some moves will be rejected. In the second type of Monte Carlo, called rejection-free Monte Carlo [18], one always moves or exchanges at each Monte Carlo step. In this, the system's transition rates for all possible nearest neighbor hops are calculated first. Then one adds all the transition rates to find tr_total. Then one makes an $i$th transition with probability given by tr(i)/tr_total. Here tr($i$) is the transition rate for the $i$th transition. The rejection-free kinetic Monte Carlo is more time-consuming than the first one because one calculates transition rates for all the possible transitions at each step. The phase ordering Kinetics is mostly done with the first type of Monte Carlo and was applied to the CG system by Preeti et al. [19].

**Fig. 1** A plot of $\sigma$ versus the number of KMC steps for two states (black filled circles) indicating success in reaching the ground state and (red filled squares) indicating unsuccessful in getting the ground state

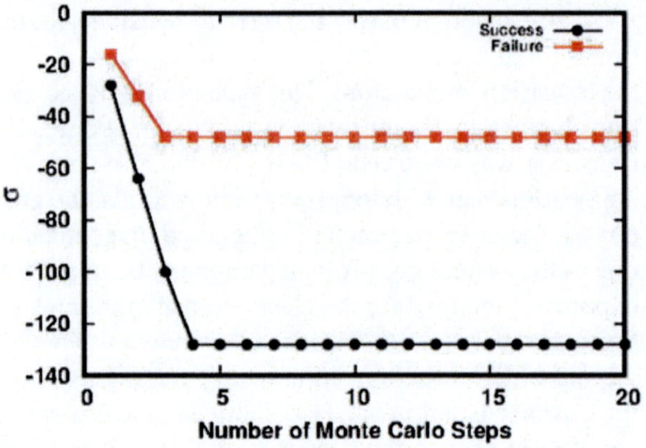


## 3 Results

The system used for the optimization is a two-dimensional (2-d) square lattice. The disorder is kept zero in the system. The size of the system is $16 \times 16$. To take the range nature of Coulomb interaction, we have used the Ewald Summation technique [20]. Since the system has antiferromagnetic ordering below the transition temperature, the order parameter is defined as sigma ($\sigma$) given by $\sigma = \Sigma\, S_i\, (-1)^i$.

Since the value of $S_i$ can take values $+1/2$ and $-1/2$, the saturation staggered magnetization ($\sigma$) for the system is $+128$ or $-128$. Initially, the optimization is done at different temperatures for 20 KMC steps. Defining $\beta$ as the inverse of temperature $T$, the different beta used are 10, 30, 50, 75, 100. Figure 1 shows the change of $\sigma$ with the number of KMC steps for two different configurations. In one configuration, the system achieves the ground state, and in the other, it fails to do so. One also sees that the final state is achieved very fast, and after that, the system remains in that state. We next increased the number of steps to 50 and again checked if the system achieves the ground state or not. The results for two states in which the system did not reach the ground state are shown in Fig. 2. One found no change in success rate with the simulation's increased time. This implies that the system gets stuck in a metastable state out of which it cannot get out. In Fig. 3, we show such a metastable state. One can see that the system contains two domains. The size of the domains for different metastable states but the topology remains the same. Finally, we show the frequency of success with temperature in Table 1.

## 4 Conclusion

In this paper, we have studied the efficiency of rejection-free algorithms in finding the ground state of a 2-d CG model on a square lattice. Once the system gets into a

**Fig. 2** A plot of $\sigma$ versus the number of KMC steps for two states (black filled circles) and (red filled squares) unable to reach the ground state for the extended time to 50 steps

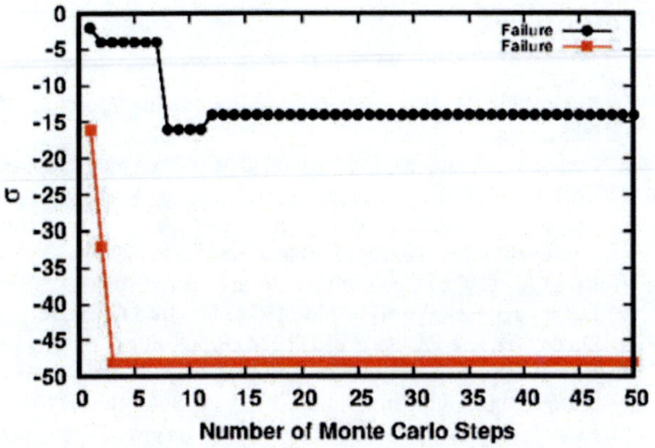


**Fig. 3** Example of a metastable state in which the system gets trapped

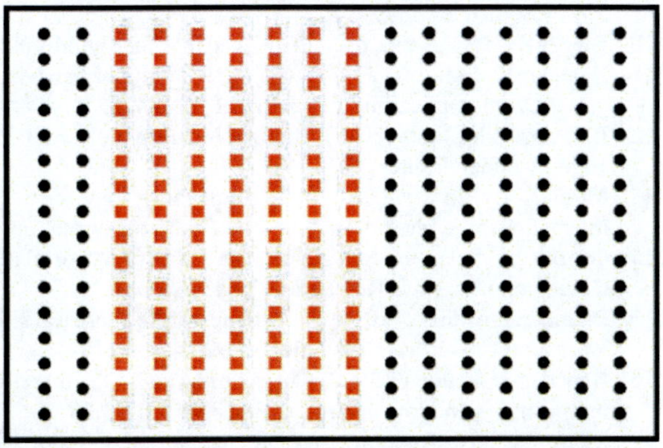

**Table 1** Percentage of success in getting the ground state

| Temperature ($T$) | Beta ($\beta = 1/T$) | Percentage of success |
|---|---|---|
| 0.1 | 10 | 92 |
| 0.033 | 30 | 62 |
| 0.02 | 50 | 67 |
| 0.0133 | 75 | 20 |
| 0.01 | 100 | 8 |

metastable state of topology shown in Fig. 3, it gets stuck. The ground state represented by antiferromagnetic ordering is not achieved. Increasing the simulation time does not help the system get out of this two-domain metastable case in most cases. One should note that it is easier for the system to make transitions for which energy of the system increases at a higher temperature. This means that the system can get out of the metastable by making higher energy transitions. So, the system can sample the ground state with increasing probability. This is clearly seen from the variation of success probability with temperature in Table 1.

## References


1. Shklovskii BI, Efros AL (1984) Electronic properties of doped semiconductors. Springer, Heidelberg
2. Pollak M, Ortuno M, Frydman A (2013) The electron glass. Cambridge University Press, New York
3. Pollak M (1970) Effect of carrier-carrier interactions on some transport properties in disordered semiconductors. Discuss Faraday Soc 50:13–19
4. Efros AL (1976) Coulomb gap in disordered systems. J Phys C 9:2021–2030
5. Davies JH, Lee PA, Rice TM (1978) Electron Glass. Phys Rev Lett 49:758–761
6. Davies JH, Lee PA, Rice TM (1984) Properties of the electron glass. Phys Rev B 29:4260–4271
7. Baranovskii SD, Efros AL, Gelmont BL, Shklovskii BI (1979) Coulomb gap in disordered systems: computer simulation. J Phys C Solid State Phys 12:1023–1034
8. Esser J, Nowak U, Usadel KD (1975) Exact ground-state properties of disordered Ising systems. Phys Rev B 55:5866–5872
9. Imry Y, Ma SK (1975) Random-field instability of the ordered state of continuous symmetry. Phys Rev Lett 35:1399–1401
10. Binder K (1983) Random-field induced interface widths in Ising systems. Z Phys B 50:343–352
11. Bhandari P, Malik V, Ahmad SR (2017) Critical behavior of the two-dimensional Coulomb glass at zero temperature. Phys Rev B 95:184203
12. Bhandari P, Malik V (2017) Effect of increasing disorder on domains of the 2d Coulomb glass. J Phys Conden Matter 29:485402
13. Möbius A, Rössler UK (2009) Short-range type critical behavior in spite of long-range interactions: the phase transition of a Coulomb system on a lattice. Phys Rev B 79:174206
14. Goethe M, Palassini M (2009) Phase diagram, correlation gap, and critical properties of Coulomb glass. Phys Rev Lett 103:045702
15. Bhandari P, Malik V (2019) Finite temperature phase transition in the two dimensional Coulomb glass at low disorders. Eur Phys J B 92:147
16. Bhandari P, Malik V (2020) Charge ordering in the three-dimensional Coulomb glass at finite temperatures and low disorders. Eur Phys J B 93:66
17. Metropolish N, Rosenbluth AW, Rosenbluth MN, Teller AH, Teller E (1953) Equation of state calculations by fast computing machines. J Chem Phys 21:1087
18. Newman MEJ, Barkema GT (1999) Monte Carlo methods in statistical physics. Oxford University Press, Oxford
19. Bhandari P, Malik V, Puri S (2019) Logarithmic coarsening in the Coulomb glass. Phys Rev E Phys Rev E 99:052113
20. Ewald PP (1921) Die Berechnung optischer und elektrostatischer Gitterpotentiale. Ann Phys 369:253–287

# Development of Narrow Size Distribution Silver Nanoparticles as Standard Reference Material/Bhartiya Nideshak Dravay for TEM/HRTEM and Particle Size Analyzer Instruments

**Santosh Singh Golia, D. K. Singh, and Manju Arora**

**Abstract** Silver nanoparticles (Ag NPs) with narrow size distribution were derived chemically by sol–gel process and characterized by XRD, and high resolution transmission electron microscopy (HRTEM) techniques to confirm their formation, morphology, particle size and distribution properties. The variation in size of as prepared Ag NPs were monitored periodically every month by HRTEM for the last six months to check the stability of nanoparticles with time to assign the shelf life of Bhartiya Nideshak Dravay (BND) and estimated the overall expanded uncertainty in size of nanoparticles. The traceability of the developed silver nanoparticles will be derived from the NIST standard reference material or from other NMI which established facility for assigning particle size. Such Ag NPs BNDs of different particle sizes can be used for calibration of instruments like TEM/HRTEM and particle size analyzers for research and find usage in many industrial applications with global acceptance.

**Keywords** Silver nanoparticles · Sol–gel · Uncertainty measurement · HRTEM · Particle size analyzer

## 1 Introduction

The vast emergence in the field nanoscience and nanotechnology has encouraged development of variety of nanomaterials for application in many existing and new advanced technologies in all walks of life. As we know that nanomaterials have one dimension in nanometer range and acquire different morphological forms like nanoparticles, nanorods, nanowire, nanotube, flower/ring/spring or in quantum dot form, etc. These nanoparticles are prepared by physical/chemical/green chemistry route. The size and morphology of nanomaterials are controlled by the selection of appropriate synthesis route and operating parameters as they are very important for their device application. The structure, shape, size and distribution of nanoparticles

S. S. Golia (✉) · D. K. Singh · M. Arora
CSIR-National Physical Laboratory, Dr. K.S. Krishnan Marg, New Delhi, India
e-mail: santoshs@nplindia.org

 
S. Yadav et al. (eds.), *Recent Advances in Metrology*, Lecture Notes in Electrical Engineering 906, https://doi.org/10.1007/978-981-19-2468-2_9

prepared by different routes is very precisely and accurately measured by transmission electron microscope (TEM)/HRTEM techniques. While the other indirect measurement technique, e.g., particle size analyzer gives information of particles size only and do not provide information about the shape, and structure of nanoparticles. That's why TEM/HRTEM are the most preferred techniques for the nanoparticles size and size distribution estimation. For this and global acceptance of data, the instruments in use should be calibrated and the record for its periodical calibration as recommended by the company should also be maintained to keep the performance of instrument at par. The sample used for calibration of TEM should have the capability of image resolution calibration for recording images, the camera constant calibration for indexing diffraction patterns, and the image/diffraction pattern rotation calibration for crystal directions viewed in the image [1–3]. The different magnification standards have been proposed for different magnification ranges. For example in HRTEM, magnification exceeds 300,000 times when thin cross-sections of single crystals and single-crystalline gold islands are used. In single crystals, e.g., gold and silver, the magnification calibration is done via lattice-plane spacing. The optics aberration correction and image reconstruction methods have markedly improved the point resolution of TEM to ~ 0.05 nm.

The magnification and resolution are interrelated to each other and reveals the efficiency of the microscope. At present polystyrene spheres are used but they are damaged either by the radiation or the grating. The resolution of TEM and HRTEM is evaluated by lattice imaging of the standard Au and Ag nanoparticles. This problem encouraged researchers for the development of new robust materials which can tolerate the variations in temperature or radiation [2]. The noble gold nanoparticles are considered to bear such adverse conditions for the calibration of TEM instrument. In this work, the use of economic Ag NPs derived by sol–gel method under optimized conditions is proposed as a secondary standard reference material/BND for the calibration of TEM/HRTEM or particle size analyzer instruments. The uncertainty in as prepared silver nanoparticles size from recorded TEM/HRTEM image is evaluated as per standard GUM and Eurachem guidelines for the measurement of overall uncertainty.

## 2 Experimental Measurements

### 2.1 *Synthesis of Ag Nps*

The monodispersed silver nanoparticles were prepared by a sol–gel method, the aqueous solution of $AgNO_3$ as silver precursor was prepared and stored in dark brown colored bottle. Trisodium citrate (TSC) solution in water was heated at 90 °C having Teflon coated magnetic stirrer for 30 min to undergo reduction reaction. To this hot solution $AgNO_3$ was added dropwise in dark with continuous stirring and pH of solution was maintained at 10.5 by adding 0.1M NaOH aqueous solution.

On further heating for 20 min, the reaction completed and solution color changed to yellow color. In this reaction, TSC plays dual role (i) as reductant and (ii) as stabilizing agent and performed in dark to restrict oxidation of $AgNO_3$ on exposure to air. Then suspension was cooled to room temperature homogenous solution and centrifuged at 12,000 rpm for 15 min to separate out Ag NPs and washed several time with DI water to remove excessive unreacted salts. The as obtained Ag NPs were again dispersed in DI water to obtain stable aqueous suspension for storage. These suspensions are kept in refrigerator for further use.

## 2.2 *Measurements*

The formation of Ag NPs is confirmed from XRD pattern recorded on Bruker d8 Advance X-ray diffractometer, using CuKα radiation ($\lambda = 1.5406$ Å), 40 kV–30 mA in $2\theta$ range of 30°–90°. The recorded XRD pattern is presented in Fig. 1.

The five peaks pertaining to silver at $2\theta$ values of 38.2209°, 44.5253°, 64.9748°, 77.7057° and 81.8620° of (111), (200), (220), (311) and (222) planes, respectively, are observed in diffraction pattern. These peaks matches well as reported in the standard powder diffraction card of JCPDS, silver file No. 04-0783.

HRTEM images of Ag NPs suspension were recorded in different intervals on M/S Technai G2 F30 STWIN HRTEM to reveal the particle size, distribution and uncertainty measurement.

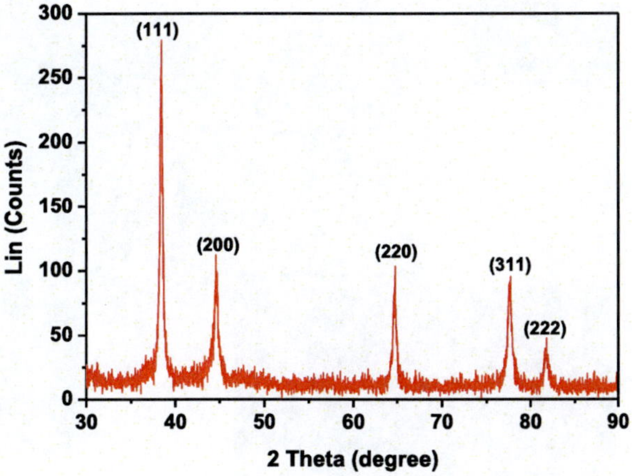


**Fig. 1** XRD pattern of Ag NPs

# 3 Results and Discussion

The aim of present work is to develop different size Ag NPs BNDs for the calibration of TEM/HRTEM instruments available in different laboratories/institute/universities in India as well as outside. These are extensively used for the characterization of nano-materials. For the global acceptance of data as per ISO 17025 standard guidelines, all the assigned values to any parameter should be accompanied by the uncertainty value and its traceability to some standard issued by apex body or NMI of any country where such type of work is going on. The size of any type of procured or developed nanoparticle is generally measured by TEM/HRTEM instruments. To check the accuracy of data obtained from these instruments need calibration these instruments. For this purpose, CSIR-NPL, New Delhi initiated project on the development of BNDs in lab pertaining to different activities in lab and in India to aware materials research community for the need of these standards and to earn revenue to become self-sustainable laboratory in India.

HRTEM images of the as prepared Ag NPs were recorded under different resolution along with specific area electron diffraction pattern at (20 ± 2) °C temperature and (45 ± 5) % relative humidity are presented in Fig. 2 (a–e). The tentative planes of silver are mentioned pertaining to different rings. The images exhibit the spherical

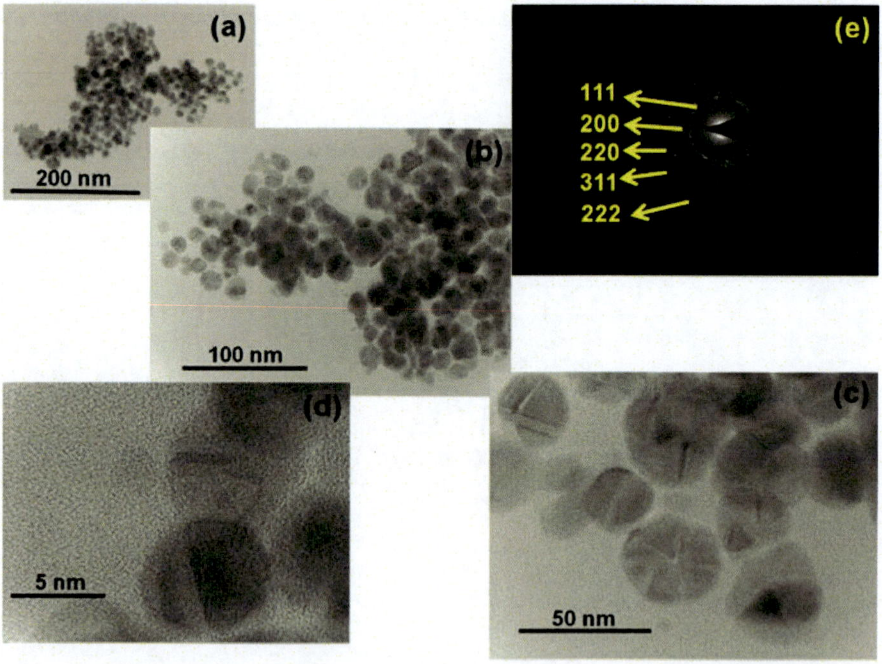


**Fig. 2** TEM images (**a–d**) of as prepared Ag NPs under different resolutions and (**e**) SAEDP of crystalline ring structure

shape nanoparticles with size in the range of 10–30 nm range ($10 \pm 2$ nm, $18 \pm 2$ nm and $25 \pm 2$ nm) with the dominance of $10 \pm 2$ nm size particles.

As per EAL, ISO, GUM or NIST guidelines [4–15] for evaluation of the overall uncertainty, this can be divided into two sections: Type A (random errors) and Type B (systematic errors). Uncertainty evaluated from experimental data statistically, repeated number of times under substantially similar conditions comes under Type A category. This comprised of small independent random variables like measuring process, environmental conditions, inherent instability of the instrument, personal judgment of the operator, etc. The random component of uncertainty, generally for infinite number of observations gives information about the population of results. But practically, a finite number of measurements were carried out to evaluate a particular parameter as defined in standard procedure [6–9, 11–15]. Type B uncertainty was evaluated from the contribution of three main sources (i) measuring instrument, (ii) operating procedure and (iii) characteristics of the sample under calibration. Uncertainty value for these components were generally taken from the calibration certificate provided by the manufacturer/ literature available. The variations in uncertainty from systematic errors generally follows normal, rectangular or triangular probability distribution and to calculate their uncertainty values, divide their uncertainty value by 2, $\sqrt{3}$ and $\sqrt{6}$, respectively. Sometimes contribution to the systematic uncertainty do not follow rectangular distribution and in such cases the standard deviation should be determined separately for each contribution and assign values. Then Type A (random errors) and Type B (systematic errors) components of uncertainty were combined for the estimation of single value of uncertainty, i.e., overall uncertainty. The final result was expressed as overall uncertainty at 95% confidence level.

To estimate the overall uncertainty in particle size measurement, all the electron microscopic images are recorded under same environmental conditions and instrumental parameter settings to nullify/minimize their effects on measurements and contribution in uncertainty. For uncertainty calculations, the following Ag NPs picture (Fig. 3) at 100 n and 50 nm resolution is selected. This figure clearly depicts three different sized nanoparticles labeled as 1, 2, 3 with the dominance of smaller size nanoparticles, i.e., 1. While number 3 labeled are least.

The images are scanned at the same position and repeated ten times to observe the variation in image. Then in the image similar size nanoparticles are marked and through statistical standard deviation calculations Type A uncertainty ($u_A$) value is obtained. The instrumental parameters, environmental conditions, human effects comes under Type B uncertainty. Since all the measurements were carried out by the same operator under same environmental conditions, that's why their contribution in uncertainty measurement is ignored and considered only instrumental parameters contribution. These are given as:

(i) Uncertainty in layer thickness ($u_1$) of reference standard provided by the supplier, i.e., 0.02 nm. Used triangular distribution $u_1(\delta r1) = 0.02\sqrt{6} = 0.008165$ nm with degree of freedom $(\nu_1) = \infty$.
(ii) Uncertainty in TEM Point Resolution: TEM point resolution value of is 0.205 nm for HRTEM as provided by the supplier for thickness in 50 nm

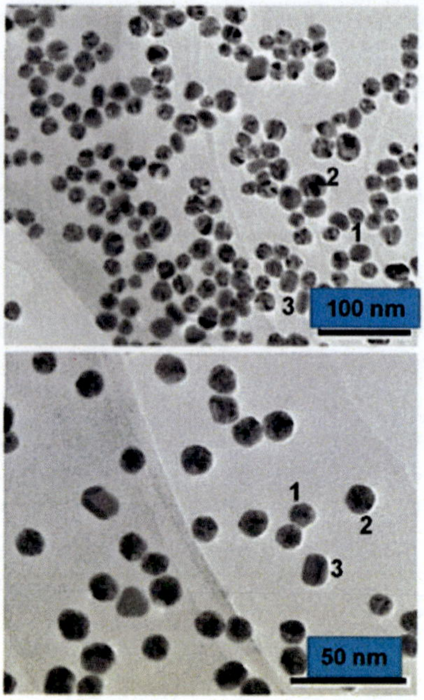


**Fig. 3** TEM image of Ag NPs under 100 and 50 nm resolution with particles marked 1, 2 and 3 for uncertainty measurements

to150 nm. By using normal distribution, standard uncertainty in the value of measurement of particle size by this resolution: $u_2(\delta r_2) = 0.205/2 = 0.1025$ nm with degree of freedom $(\nu_2) = \infty$.

(iii) Uncertainty in TEM Line Resolution: TEM Line resolution value of is 0.144 nm, i.e., $0.144 \times 10^{-9}$ m for thickness in the range of 50–150 nm for HRTEM as provided by the supplier. By assuming normal distribution, standard uncertainty in the value of measurement of particle size by this resolution: $u_3(\delta r_3) = 0.144/2 = 0.072$ nm with degree of freedom $(\nu_3) = \infty$.

(iv) Uncertainty in STEM resolution: STEM resolution value of is 0.17 nm, i.e., $0.17 \times 10^{-9}$ m for HRTEM as provided by the supplier. By assuming normal distribution, standard uncertainty in the value of measurement of particle size by this resolution: $u_4(\delta r_4) = 0.17/2 = 0.085$ nm with degree of freedom $(\nu_3) = \infty$.

By estimating the values of Type A and Type B uncertainty components then combined uncertainty value is obtained by using the following equation:

$$\text{Combined uncertainty}(u_c) = [(u_A(x))^2 + (u_1(r_1))^2 + (u_2(r_2))^2 + (u_3(r_3))^2 + (u_4(r_4))^2]1/2$$

Overall Expanded uncertainty is $U = 2^*u_c$ at 95% confidence level, and Coverage factor $(K)$: 2

**Table 1** Particle labeled in Fig. 3, size and overall uncertainty of prepared Ag NPs

| Particles label | Size (nm) | Overall uncertainty (nm) |
|---|---|---|
| 1 | 8.458 | 0.0368 |
| 2 | 17.162 | 0.0419 |
| 3 | 26.249 | 0.0467 |

In the present case of sol–gel derived Ag NPs, the size of three different size nanoparticles along with overall uncertainty estimated values at 95% confidence level are summarized in Table 1.

## 4 Conclusion

These preliminary investigations report the synthesis of Ag NPs with optimized parameters to get desired size nanoparticles with narrow size distribution. These nanoparticles are characterized by XRD and HRTEM techniques to confirm their formation and particle size and shape evaluation. The uncertainty in particle size is estimated and the stability of nanoparticles in suspension is derived by recording their HRTEM images in regular interval of time which in turn helps in assigning their shelf life an important parameter required for releasing BND.

## References

1. Orji NG, Dixson RG, Garcia-Gutierrez DI, Bunday BD, Bishop M, Cresswell MW, Allen RA, Allgair JA (2007) TEM calibration methods for critical dimension standards. In: Proceedings of SPIE 6518, metrology, inspection, and process control for microlithography XXI, 651810
2. Filippov MN, Gavrilenko VP, Kovalchuk MV, Mityukhlyaev VB, Ozerin YV, Rakov AV, Roddatis VV, Todua PA, Vasiliev AL (2011) Reference material for transmission electron microscope calibration. Meas Sci Technol 22(094014):5
3. Reimer L (1997) Transmission electron microscopy: image formation and microanalysis, 4th edn. Springer, Berlin
4. Dietrich CF (1973) Uncertainty, calibration & probability. Adam Hilger, London
5. Glossary of terms used in metrology. Commonwealth Science, Council document, CSC 80, MS—8.
6. Guide to the expression of uncertainty in measurement. International Organization for Standardization (ISO) (1995) 1993
7. Expression of the uncertainty of measurement in calibration. European Cooperation for Accreditation of Laboratories (EAL), EAL-RL (1997)
8. Ojha VN (1998) Evaluation & expression of uncertainty of measurement, Mapan. J Metrol Soc India 13:71–84
9. EURACHEM/CITAC guide on quantifying uncertainty in analysis measurement, 2nd edn. (2000)
10. ISO Guide 35, Certification of reference materials—general & statistical principles, 2nd edn. (1989)

11. Guidelines for estimation and expression of uncertainty in measurement: NABL—141, National Accreditation Board for Testing & Calibration Laboratories (2000)
12. http://www.citac.ws
13. http://www.ilac.org
14. http://www.eurachem.org
15. http://physics.nist.gov

# Establishment of Metrological Traceability of AC Resistance Using Electrical Equivalent Circuit at CSIR-NPL

**Satish, Priyanka Jain, Poonam Bist, Sachin Kumar, R. P. Aloysius, and J. C. Biswas**

**Abstract** The traceability of AC resistance is established from audio frequencies to MHz band using electrical equivalent circuit (EEC) of resistance standard. The EEC consists of resistance and its parasitic components in the form of either capacitance or inductance or both depending upon the construction and range of the resistance standard. The procedure for the establishment of metrological traceability involves the measurement of reference resistance at DC and parasitic components at AC frequencies. The uncertainty in measurement of AC resistance is also computed as per the GUM document.

**Keywords** Electrical equivalent circuit · AC resistance · Impedance · Parasitic components · Metrological traceability · Uncertainty in measurement

## 1 Introduction

CSIR-National Physical Laboratory (CSIR-NPL) is the "National Measurement Institute" (NMI) by the Act of Parliament and is the custodian of "National Standards" with a responsibility of the dissemination of measurements to the needs of country. It serves nation's R&D institutions and laboratories like ISRO, DRDO, BEL, ERTL, ETDC and STQC by providing them apex-level calibration services. It also ensures the comparability with the national standards of other countries for the purpose of international trade.

The proposed work is performed at LF, HF impedance and DC metrology section of CSIR-NPL, which is mainly responsible for the establishment, maintenance and advancement of measurement capabilities of impedance [1, 2], voltage, current, charge and resistance at DC [3, 4]. The work performed is the part of establishment of metrological traceability for impedance standards. The AC resistance standards ranging from 1 Ω to 1 MΩ are characterized from audio frequencies to MHz band.

Satish (✉) · P. Jain · P. Bist · S. Kumar · R. P. Aloysius · J. C. Biswas
LF, HF Impedance and DC Metrology, CSIR-National Physical Laboratory, Dr. K. S. Krishnan Marg, New Delhi, India
e-mail: singhsp3@nplindia.org

 
S. Yadav et al. (eds.), *Recent Advances in Metrology*, Lecture Notes in Electrical Engineering 906, https://doi.org/10.1007/978-981-19-2468-2_10

# 2 AC Resistance

AC resistance differs from the DC resistance as former is effected by the skin effect as the frequency progress to MHz band. AC resistance has the contribution from the residual or parasitic components, which are the function of geometry of resistance standard. The impedance can be defined by Eq. (1) given below:

$$Z = \left(R_f + jX\right) \tag{1}$$

where $R_f$ is real part, also known as AC resistance, and $X$ is the imaginary part, also known as reactance, which may be due to capacitance, $C$, or inductance, $L$, or both as shown in EEC of AC resistance [5] in Fig. 1.

## 2.1 High-Value AC Resistance

The AC resistance of the high-value resistance standard may be computed from the EEC shown in Fig. 1b and is given by Eq. (2). It may be observed that it is the function of parasitic capacitance.

$$R_f = \frac{R}{1 + (\omega RC)^2} \tag{2}$$

where $\omega$ is angular frequency, $R$ is reference resistance at DC and $C$ is parasitic capacitance.

**Fig. 1** **a** Electrical equivalent circuit for AC Resistance where 1 and 2 are the termination ports. **b** Circuit for high-value resistance standard reduces to parallel combination of resistance and parasitic capacitance. **c** Circuit for low value of resistance standard reduces into the parallel combination of conductance and susceptance

The uncertainty in measurement of AC resistance $R_f$ is computed using uncertainty propagation method by employing equation (3).

$$\Delta R_f = \sqrt{\left(\left(\frac{\partial R_f}{\partial R}\Delta R\right)^2 + \left(\frac{\partial R_f}{\partial C}\Delta C\right)^2\right)} \tag{3}$$

where $\Delta R$ and $\Delta C$ are the standard uncertainties of reference resistance and parasitic capacitance, respectively.

### 2.2 *Low-Value AC Resistance*

AC resistance of the low-value resistance may be computed from the circuit as shown in Fig. 1c from Eq. (4).

$$R_f = R\left(1 + \left(\frac{\omega L}{R}\right)^2\right) \tag{4}$$

$$\Delta R_f = \sqrt{\left(\left(\frac{\partial R_f}{\partial R}\Delta R\right)^2 + \left(\frac{\partial R_f}{\partial L}\Delta L\right)^2\right)} \tag{5}$$

where $\Delta L$ is the standard uncertainty of parasitic inductance.

## 3 Measurement Setup and Observations

The 8.5 digit digital multimeter (Fluke 8508A) or direct current comparator bridge is used to calibrate resistance standards at DC, while precision LCR meter (Keysight E4980A) is used to calibrate parasitic capacitance and inductance at 1 kHz. The traceability chart of AC resistance standard for the proposed work is shown in Fig. 2.

### 3.1 *Measurement Procedure*

The measurement procedure to establish the metrological traceability of ac resistance involves the following steps:

1. Measurement of reference resistance at DC using 8.5 digit digital multimeter/direct current comparator bridge.

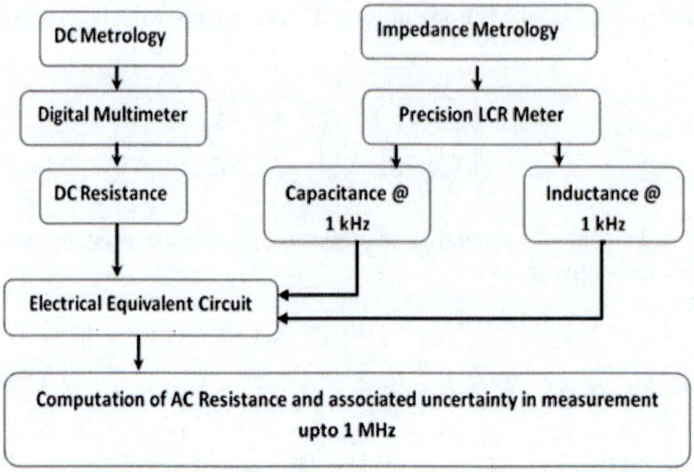


**Fig. 2** Traceability chart for AC resistance standards

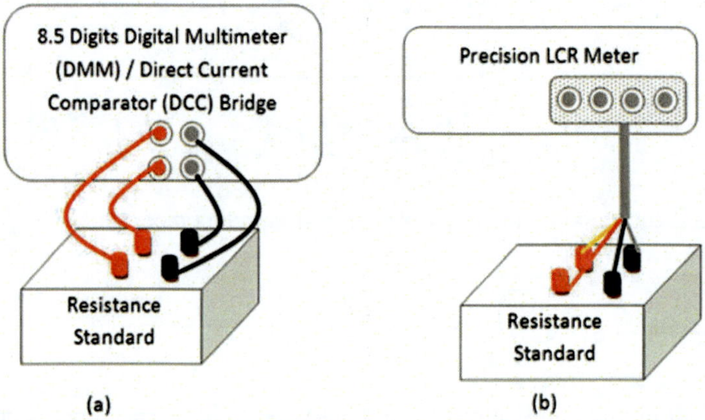


**Fig. 3** **a** Measurement setup for calibration of reference resistance at DC. **b** Measurement setup for calibration of parasitic components at 1 kHz

2. Measurement of parasitic component (capacitance/inductance) at 1 kHz using precision LCR meter as defined in electrical equivalent circuit.
3. Computation of AC resistance for high values using Eq. (2) and for low values using Eq. (4) and
4. Computation of respective uncertainty in measurements using Eqs. (3) and (5).

### 3.2 Measurement Results

AC resistances of 1 Ω to 1 MΩ are computed from 1 kHz to 1 MHz frequency range. Table 1 shows the resistance at DC and computed low-value AC resistances, while Table 2 shows that resistance at DC and computed high-value AC resistance along with uncertainty in measurement [6] for ~95% confidence level at $k = 2$, respectively.

**Table 1** Measurement results for low-value AC resistance

| Resistance standard (Ω) | $R \pm u_{e,}$ (Ω) | | Freq. (kHz) | $R_f \pm u_e$ (Ω) | |
|---|---|---|---|---|---|
| 1 | 1.000000 | 0.00000025 | 0.1 | 1.000000 | 0.000040 |
| | | | 1 | 1.000000 | 0.000008 |
| | | | 10 | 1.000023 | 0.000400 |
| | | | 100 | 1.002296 | 0.000400 |
| | | | 1000 | 1.229553 | 0.250000 |
| 10 | 9.999990 | 0.0000025 | 0.1 | 9.99999 | 0.00060 |
| | | | 1 | 9.99999 | 0.00030 |
| | | | 10 | 10.00000 | 0.00030 |
| | | | 100 | 10.00099 | 0.00400 |
| | | | 1000 | 10.09978 | 0.30000 |
| 100 | 99.999960 | 0.0000050 | 0.1 | 100.0000 | 0.0060 |
| | | | 1 | 100.0000 | 0.0020 |
| | | | 10 | 100.0000 | 0.0050 |
| | | | 100 | 100.0025 | 0.0050 |
| | | | 1000 | 100.2535 | 0.2299 |

**Table 2** Measurement results for high-value AC resistance

| Resistance standard (kΩ) | $R \pm u_e$ (kΩ) | | Frequency (kHz) | $R_f \pm u_e$ (kΩ) | |
|---|---|---|---|---|---|
| 10 | 9.999992 | 0.000006 | 1 | 9.99999 | 0.00005 |
| | | | 10 | 9.99995 | 0.00010 |
| | | | 100 | 9.99542 | 0.00300 |
| 100 | 100.000120 | 0.00004 | 1 | 99.9993 | 0.0050 |
| | | | 10 | 99.9212 | 0.0500 |
| | | | 100 | 92.6797 | 5.0000 |
| 1000 | 1000.00242 | 0.004 | 1 | 999.292 | 1.000 |
| | | | 10 | 933.638 | 20.000 |

## 4 Conclusion and Summary

The metrological traceability of AC resistance standards is successfully established using the electrical equivalent circuit. The resistance defined at DC is used as a reference to define the resistance at AC frequencies up to 1 MHz. The frequency characteristics of resistance standards from 1 Ω to 1 MΩ established with uncertainty in measurement. The uncertainty in measurement is computed using uncertainty propagation method as defined in GUM document. The parasitic components of resistance standards such as capacitance and inductance are measured with precision LCR meter. The measurement of parasitic capacitance and inductance plays critical role in defining the uncertainty of the resistances. The proposed method may be utilized for the resistance values ranging from 1 mΩ to 100 mΩ in the future.

**Acknowledgements** Authors would like to thank Director, CSIR-NPL, for his constant support and encouragement to carry out proposed work.

## References

1. Satish, Babita, Khurana B, Kumar S, Saxena AK (2015) Evaluation of four-terminal-pair capacitance standards using electrical equivalent circuit model. Measurement 73:121–126
2. Singh S, Kumar S, Babita, John T (2017) Realization of four-terminal-pair capacitors as reference standards of impedance at high frequency using impedance-matrix method. IEEE Trans Instrum Meas 66(08):2129–2135
3. Babita, Satish, Singh A, Saxena AK (2013) Determination and validation of average value of national standard of DC voltage using different methods at CSIR-NPLI. Mapan-JMSI 28(2):99–104
4. Babita, Sharma DK, Satish, Ansari MA, Saxena AK (2014) A versatile automation program using LabVIEW for low dc current measurement. JSIR 73(2):91–94
5. Agilent (2009) Agilent impedance measurement handbook
6. Joint Committee for Guides in Metrology (2008) Evaluation of measurement data—guide to the expression of uncertainty in measurement. Int Organ Stand Geneva 50:134. ISBN

# Calibration of Pentaprism and Its Uncertainty Evaluation at NPL, India

**M. Arif Sanjid and R. Sharma**

**Abstract** In the field of angle metrology, pentaprism is used to steer optical beams exactly 90° irrespective of the angle of incidence of the light. The maximum angular deviation may ranges from few arc second to an arc minute. Traditional analysis omits the residues of fitting the measurements to the analytical model. The calibration of pentaprism is set up at NPL, length and dimension and nanometrology division of NPL, India. Authors attempted to device a new analytical method contrary to the conventional method. The necessary experiments are performed to validate the results.

**Keywords** Uncertainty of measurement · Error · Calibration · Traceability · Angle

## 1 Introduction

The profilometers based on the deflectometry technique use pentaprism to measure the precision form of optical surfaces by deflecting the collimated beam of the autocollimator [1]. Pentaprism is a constant deviation prism having five faces. Pentaprism or similar optical configuration is invariantly used for the purpose of bending an optical beam exactly 90° irrespective of the angle of incidence of the light. The basic configuration of active facets of pentaprism is shown in Fig. 1. By construction, its five faces need to be perpendicular to each other.

Pentaprism is extensively used for deflectometric scanning of flat surfaces, establishing optical datum useful for inner diameter calibration and straightness calibrations of guide ways, etc. [2–5]. When the optical beam of the autocollimator is passed to pentaprism, it deflects it by 90° plus the deflection error. The deviation is due to the optical properties of the pentaprism.

M. Arif Sanjid (✉) · R. Sharma
Length, Dimension and Nanometrology, National Physical Laboratory, New Delhi, India
e-mail: sanjid@nplindia.org

R. Sharma
e-mail: rina@nplindia.org

 
S. Yadav et al. (eds.), *Recent Advances in Metrology*, Lecture Notes in Electrical Engineering 906, https://doi.org/10.1007/978-981-19-2468-2_11

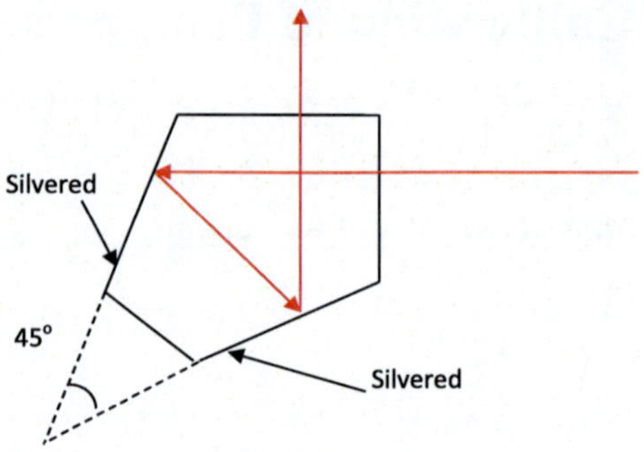


**Fig. 1** Scheme of construction of pentaprism

By construction, pentaprism is glued inside a square metal housing. The legs/base pads of the metal housing are lapped to achieve desirable optical square functionality. The faces of the pentaprism are set with an angular offset in the mounted frame to avoid ghost reflections. The concurrent engagement of these optical axes results in errors due to the crosstalk. Despite of many parameters of pentaprism, the maximum angular deviation is mostly used in the metrological applications. Traditional analysis neglected the residues of fitting the measurements in the analytical model [6]. Authors devised a new analytical method contrary to the conventional method.

## 2 Experimental Setup

The measurement setup needs a surface plate, two autocollimators, and two high-quality optical reflectors. The scheme of setting up calibration setup is given in Fig. 2. The mirrors mounted are perpendicular on a flat surface plate. The autocollimators are aligned to shine its collimated optical beam approximately 90° (i.e., normal) though the pentaprism. The autocollimator will read the angular offset of the mirror concerning 90°. This configuration is identified as the autocollimator at position A on to mirror 1. Similarly, the mirror 2 is place opposite to mirror 2. The pentaprism is rotated by 90° to read the angular orientation of the mirror 2 using the autocollimator A. The autocollimator B is fixed opposite to the autocollimator A. Now, the pentaprism is rotated 90° to read the reflected beam of autocollimator B from mirror 2. Finally, autocollimator B is adjusted to receive the reflected beam from mirror 1 through the rotated pentaprism. One needs to adjust the orientations of the autocollimators, mirror to obtain readings for all four orientations of the pentaprism.

## 3 Measurement

Generally, determination of the deviation error of optical beam about 90° is considered as an important parameter for the calibration. However, it is equally important

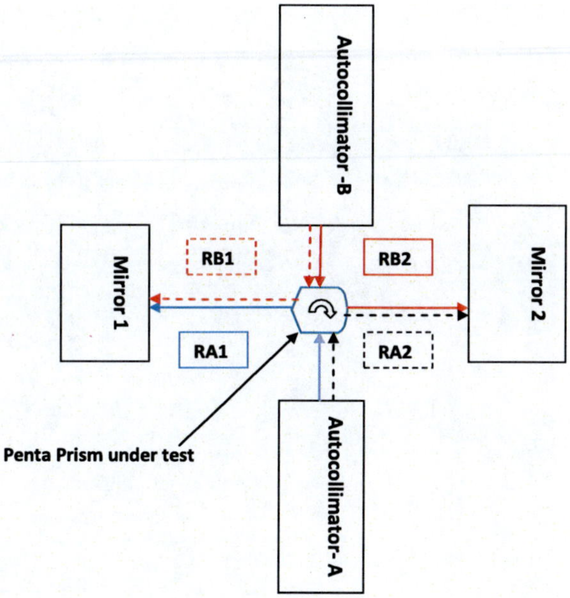


**Fig. 2** Schematic diagram of angle deviation measurement

to assess the squatness of pentaprism faces. Nevertheless, one can verify the tilt of pentaprism by reading the vertical tilt. When there is no change in the horizontal angle, one can ensure that the autocollimators aligned within the range of squareness tolerance of the pentaprism. After all adjustments, the experimental setup is temperature stabilization. The arrangement of autocollimators and mirrors on the surface plate at the NPL, India, is shown in Fig. 3.

# 4 Analysis of Results

The orientations of the optical components relative to the measuring instruments are considered to develop the four simultaneous equations. The measurement acquired by the autocollimator A concerning the mirror 1(2) is represented by Eqs. (1, 2), respectively. Figure 4 shows the orientation of autocollimators and reflectors. The deviation of pentaprism from the 90° will have opposite sign for opponent mirrors. Let the mirror 1(2) is slightly off by angle $\beta_1(\beta_2)$ from the normal or the beam of incident of autocollimator.

$$\boldsymbol{m}_{A1} = \boldsymbol{\alpha} + \boldsymbol{\beta}_1 \tag{1}$$

$$\boldsymbol{m}_{A2} = -\boldsymbol{\alpha} + \boldsymbol{\beta}_2 \tag{2}$$

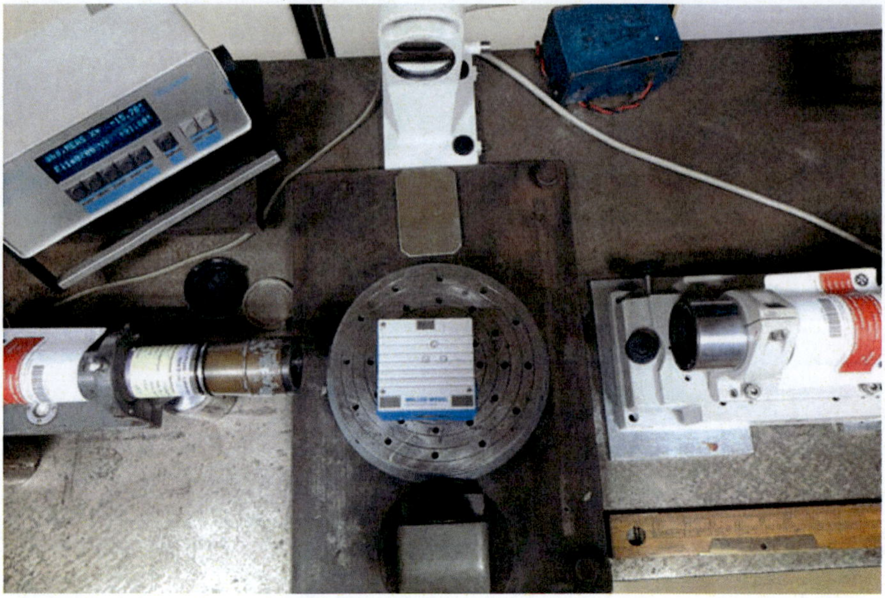

**Fig. 3** Experimental setup at NPL, India

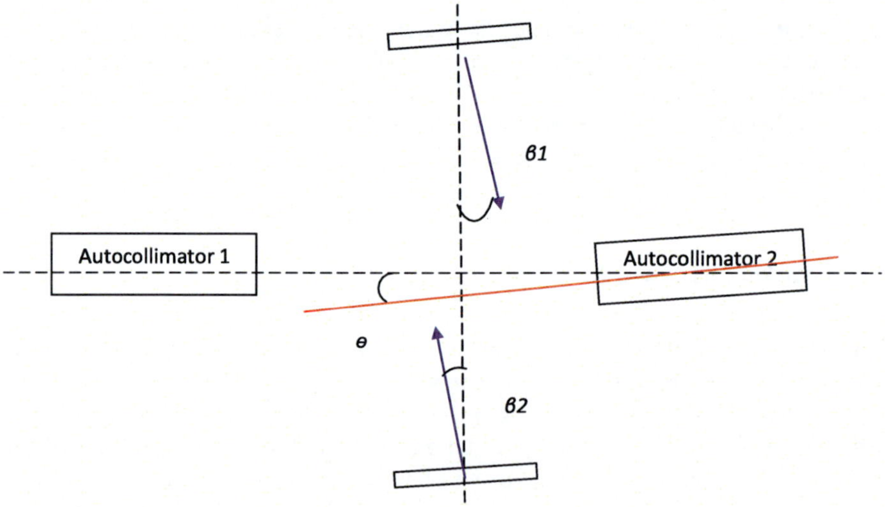


**Fig. 4** Orientation of optical components

**Table 1** Measured result

| Reading identification | Beam trajectory | Autocollimator reading | | | | | | |
|---|---|---|---|---|---|---|---|---|
| | | Set 1 | Set 2 | Set 3 | Set 4 | Set 5 | Average | $u_A$ |
| $m_{A1}$ | Autocollimator A—Mirror 1 | 7.1 | 7.1 | 6.7 | 6.8 | 6.9 | 6.9 | 0.08 |
| $m_{A2}$ | Autocollimator A—Mirror 2 | 7.5 | 7.5 | 7.4 | 7.5 | 7.5 | 7.5 | 0.02 |
| $m_{B1}$ | Autocollimator B—Mirror 1 | 2.3 | 2.2 | 2.3 | 2.3 | 2.2 | 2.3 | 0.02 |
| $m_{B2}$ | Autocollimator B—Mirror 2 | 8.8 | 8.6 | 8.5 | 8.7 | 8.7 | 8.7 | 0.05 |

Similarly, Eqs. (3, 4) represent the measurements acquired with the autocollimator B. Let the autocollimator A(B) is not parallel and deviate by an angle θ.

$$\boldsymbol{m}_{B1} = -\boldsymbol{\alpha} + \boldsymbol{\beta}_1 + \boldsymbol{\theta} \tag{3}$$

$$\boldsymbol{m}_{B2} = \boldsymbol{\alpha} + \boldsymbol{\beta}_2 + \boldsymbol{\theta} \tag{4}$$

The relation among these orientations and their signs is carefully examined in separate experiments. The matrix form of the four simultaneous equations is summarized as Eq. (5).

$$\begin{bmatrix} \boldsymbol{m}_{A1} \\ \boldsymbol{m}_{A2} \\ \boldsymbol{m}_{B1} \\ \boldsymbol{m}_{B2} \end{bmatrix} = \begin{bmatrix} +1\,1\,0\,0 \\ -1\,0\,1\,0 \\ -1\,1\,0\,1 \\ +1\,0\,1\,1 \end{bmatrix} \begin{bmatrix} \boldsymbol{\alpha} \\ \boldsymbol{\beta}_1 \\ \boldsymbol{\beta}_2 \\ \boldsymbol{\theta} \end{bmatrix} \tag{5}$$

The readings are given in Table 1. The repeated measurements and their average value repeatability are calculated.

Using the Eq. (6), the maximum angular deviation is result calculated as 1.5 arcs second.

$$\boldsymbol{\alpha} = \frac{1}{4}([\boldsymbol{m}_{A1} - \boldsymbol{m}_{A2}] - [[\boldsymbol{m}_{B1} - \boldsymbol{m}_{B2}]]) \tag{6}$$

## 5 Uncertainty Estimation

The uncertainty of measurement is estimated according to the uncertainty propagation Eq. 7 [7]. The repeatability of the four readings is reduced by four. The calibration of autocollimator and flatness deviations of the reflector contributes the measurement uncertainty. The uncertainty component $u^2(I)$ is due to squareness of the mounting frame of pentaprism.

$$u^2(\alpha) = \frac{1}{4}\left[u^2(m_{A1}) + u^2(m_{A2}) + u^2(m_{B1}) + u^2(m_{B2})\right] + u^2(AC\_A) + +u^2(AC\_B) + u^2(\beta 1) + +u^2(\beta 2) + u^2(I) \tag{7}$$

Table 2 gives the uncertainty of measurement of the pentaprism. The expanded uncertainty of measurement is 1.2 arc second. The flatness tolerance of reflector is taken as quarter a wavelength.

**Table 2** Uncertainty budget

Uncertainty of measurement

| Source of uncertainty of measurement | Limits | Type/Probability distribution | Standard uncertainty | Sensitivity coefficient | Uncertainty contribution" | Degree of freedom |
|---|---|---|---|---|---|---|
| $u_A$ Repeatability | | Type-A Normal | 0.08" | 1/4 | 0.02 | 3 |
| $u_A$ Repeatability | | Type-A Normal | 0.02" | 1/4 | 0.01 | 3 |
| $u_A$ Repeatability | | Type-A Normal | 0.02" | 1/4 | 0.01 | 3 |
| $u_A$ Repeatability | | Type-A Normal | 0.05" | 1/4 | 0.01 | 3 |
| $u_1$ Calibration Autocollimator 1 | | Type-B Normal | 0.2" | 1 | 0.2 | Infinite |
| $u_2$ Calibration autocollimator 2 | | Type-B Normal | 0.5" | 1 | 0.5 | Infinite |
| $u_3$ Flatness variation of mirror 1 | $\lambda/4 =$ 0.12 μm | Type-B Rectangular | 0.06 μm | 0.35″/μm | 0.02 | Infinite |
| $u_4$ Flatness variation of mirror 2 | $\lambda/4 =$ 0.12 μm | Type-B Rectangular | 0.06 μm | 0.35″/μm | 0.02 | Infinite |
| *Misalignment* | 4" | Type-B Rectangular | 2.3 | 0.11 | 0.25 | Infinite |
| Combined uncertainty of measurement of measurement $(k = 1)u_c$ | | | | 0.6" | | 3888 |
| Expanded uncertainty of measurement of measurement $(k = 2)U_e$ | | | | 1.2" | | |

Effective degrees of freedom,

$$v_{eff} = u_c^4 / \left( \sum \frac{u_{type-A}^4}{n-1} + \sum \frac{u_j^4}{\infty} \right)$$

$$v_{eff} = \frac{(0.6)^4}{\left( \frac{(0.1)^4}{4} + \frac{(0.01)^4}{\infty} + \ldots \right)} = 3888 \text{ (Here , } 3888 > 30)$$

There from, referring to student-$t$ table, the coverage factor $k = 2$ is chosen to expand the combined uncertainty to 95% confidence level for the normal distribution.

## 6 Conclusion

The instruments that use deflectometric technique scan the surface using autocollimator by moving the calibrated pentaprism [8]. Some workers used mirror-type pentaprism to reduce the systematic errors that are due to the non-homogeneous optics and construction faults of the pentaprism [9].

The offset of the angular orientation between the faces of the pentaprism causes the deflection error. The benefit of simultaneous equation is exploited to improve the uncertainty of measurement [10]. The novelty of the proposed method involves the formation of simultaneous equations. The value of determinant corresponds to the simultaneous equations. Consequently, the measurement uncertainty of the proposed reduces by 25% to that of the uncertainty achievable by the conventional method.

## References

1. Geckeler RD, Kranz O, Just A, Krause M (2012) A novel approach for extending autocollimator calibration from plane to spatial angles. Adv Opt Technol 1(6):427–439
2. Ehret G, Schulz M, Stavridis M, Elster C (2012) Deflectometric systems for absolute flatness measurements at PTB. Meas Sci Technol 23(9):094007
3. Mahammad SA, Chaudhary KP, Yadav S, Sen M, Ghoshal SK (2020) An accurate inner diameter measurement. Rev Sci Instrum 91(6):065112
4. Ehret G, Schulz M, Baier M, Fitzenreiter A (2013) Optical measurement of absolute flatness with the deflectometric measurement systems at PTB. J Phys Conf Ser 425(15):152016
5. Bitou Y, Kondo Y (2016) Scanning deflectometric profiler for measurement of transparent parallel plates. Appl Opt 55(32):9282–9287
6. Reeve CP, Reeve CP, Veale RC (1976) The calibration of a pentaprism. US Department of Commerce, National Bureau of Standards
7. BIPM I, IFCC I, ISO I, IUPAP O (2008) Evaluation of measurement data—guide to the expression of uncertainty in measurement, JCGM 100: 2008 GUM 1995 with minor corrections. Joint Committee for Guides in Metrology.

8. Barber SK, Morrison GY, Yashchuk VV, Gubarev MV, Geckeler RD, Buchheim J, Siewert F, Zeschke T (2011) Developmental long-trace profiler using optimally aligned mirror-based pentaprism. Opt Eng 50(5):053601
9. Barber SK, Yashchuk VV, Geckeler RD, Gubarev MV, Buchheim J, Siewert F, Zeschke T (2011) Optimal alignment of mirror-based pentaprisms for scanning deflectometric devices. Opt Eng 50(7):073602
10. Sanjid MA (2013) Improved direct comparison calibration of small angle blocks. Measurement 46(1):646–653

# Fuzzy-Based MPPT Controlled 3Z Boost Converter for PV Applications

**G. D. Anbarasi Jebaselvi and S. Paramasivam**

**Abstract** Since the efficiency of conventional boost converters are limited by the presence of parasitic parameters of the components used and remarkable power loss, it is necessary to cascade impedance networks to achieve high voltage boost in the output. To obtain the desired voltage, boost converters can be connected in series which is, however, complicated due to additional switches and control units. The inclusion of additional switches and control units degrades the reliability of the system. A high-frequency isolation DC-DC converter with a high transformer turns ratio was applied to solve these problems, but due to weight of transformer, the whole circuit becomes bulky which increases the cost. In this proposed system, a fuzzy-based MPPT control is employed to track the maximum power output from the PV panel, and a microcontroller is employed to produce 100 kHz PWM pulses such that when the duty cycle increases, the voltage also increases.

**Keywords** 3Z boost converter · Fuzzy logic · MPPT · Pulse width modulation · Photovoltaic

## 1 Introduction

The voltgae gain of boost converters can maximally reach five to six times than the input voltage supplied which is far below compared to industrial standards. But here in this 3Z boost converter, the voltage gain increases ten times than that of the conventional boost convertor and even the current gain by 4 times. Formerly, a high-frequency isolation DC-DC converter with a high transformer turns ratio has been used to solve these problems, but later cases, a single MOSFET switch has been employed to create shoot-through by switching ON and OFF at very high frequency (100 kHz). Shoot-through is creating current impulses by short circuit, which results

G. D. Anbarasi Jebaselvi (✉)
Sathyabama Institute of Science and Technology, Chemmencherry, Chennai, IndiaJeppiaar Nagar,
e-mail: anbarasi.enc@sathyabama.ac.in

S. Paramasivam
Engineering Services Division, ESAB, Chennai, India



S. Yadav et al. (eds.), *Recent Advances in Metrology*, Lecture Notes in Electrical Engineering 906, https://doi.org/10.1007/978-981-19-2468-2_12

in charging and discharging of energy in LCR filter so quickly with many voltage surges leading to hazards and havoc. Hence, here, a microcontroller is employed to produce 100 kHz pulse width modulation (PWM) pulses with variable duty cycles (i.e., variable ON time) such that when the duty cycle increases, the voltage boost also increases. In this proposed system, a fuzzy-based MPPT control is employed to track the maximum power output from the PV panel. Fuzzy logic is an intelligent decision-making algorithm that can decide the best duty cycle to achieve maximum power output from it.

### *1.1 PV Cell Characteristics—An Overview*

Since solar modules are rated according to standard test conditions which do not give evidence of the performance under real outdoor conditions and the most important properties like temperature coefficient, spectral response, and mechanical construction have a great influence on the performance of photovoltaic modules under different weather conditions, it is necessary to review all these parameters to get higher conversion efficiency of the solar panel [1]. A solar PV cell basically is a P–N semiconductor junction diode, when exposed to light, generating DC current. The generated current varies linearly with the solar irradiance. The equivalent electrical circuit of an ideal solar PV cell can be treated as a current source parallel with a single diode as shown in Fig. 1. When light incident on the surface of a PV cell, some portion of the solar energy is absorbed by the semiconductor material. If the absorbed photon energy is greater than the band gap energy of semiconductor, the electrons bounce from valence band to the conduction band creating hole-electron pairs throughout the illuminated region of the semiconductor. These electrons and hole pairs flow in opposite directions across the junction thereby creating dc current flow.

I-V characteristics of an equivalent solar cell circuit can be determined by the equations given below. The output current of an ideal PV cell is given by

$$I = I_{sc} - I_d. \tag{1}$$

where

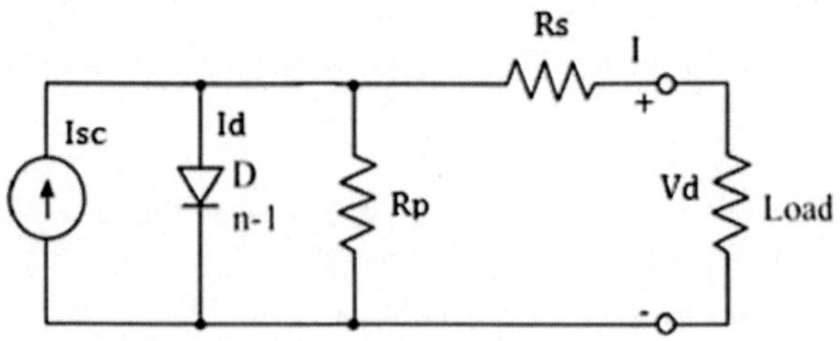


**Fig. 1** Equivalent electrical circuit of PV cell with a single diode

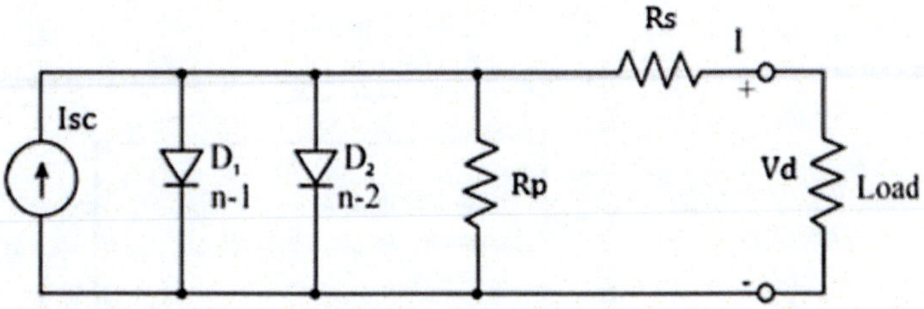


**Fig. 2** Accurate model of PV cell with double diodes

$I$: output current in amps.
$I_{sc}$: short circuit current in amps.
$I_d$: shunted current through the diode in amps.
$R_s$ and $R_p$: series and parallel resistances.
Equation (2) is used to model PV cell, based on the theory of semiconductors and is given by

$$I_d = I_0\left[e^{\frac{qV_d}{\beta kT}} - 1\right]. \tag{2}$$

Combining Eqs. (1) and (2), the current I through the load is found out and shown by Eq. (3).

$$I = I_{sc} - I_0\left[e^{\frac{qV_d}{\beta kT}} - 1\right]. \tag{3}$$

The most accurate model is shown in Fig. 2 which includes an extra diode $D_2$ along with series and parallel resistances.

The two diodes $D_1$ and $D_2$ are combined and become Eq. (4).

$$I = I_{sc} - I_0\left[e\frac{q(V_d + IR_s)}{nkT} - 1\right] - \left[\frac{V_d + IR_s}{R_P}\right]. \tag{4}$$

where $n$ is the ideality factor lies between 0 and 1.

Multiple solar PV cells are connected in series and enclosed in a common framework to form a photovoltaic panel or module. PV modules are cascaded first and then paralleled to meet the power need. By connecting several cells in series, the combined voltage of the PV module is increased, and the conduction losses in the cables get minimized. The typical range of power capacity of a PV string lies between few hundred watts to 5 kW.

Practically, PV panels possess nonlinear voltage–current characteristics as in Fig. 3, with a distinct maximum power point (MPP), which depends mainly on environmental factors like temperature and irradiance. In order to continuously harvest maximum power from the solar panels, they have to be operated at their MPP despite the random availability of sun energy. The intersection of load line with its V-I

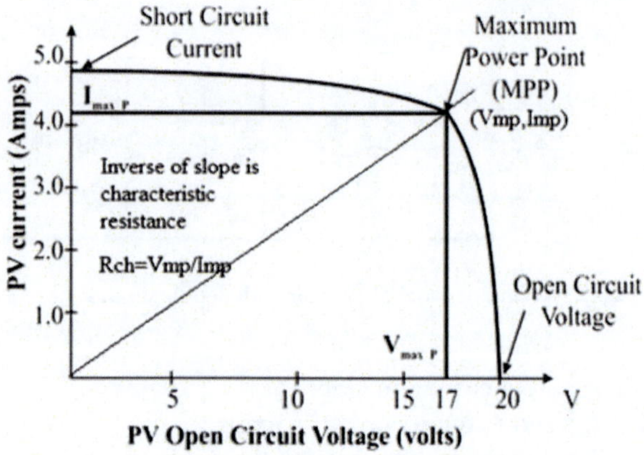


**Fig. 3** V-I characteristics of an ideal PV cell

characteristics is the operating point of PV cell with its characteristic resistance designated by Rch as shown in Fig. 3. The three basic parameters involved in the photovoltaic technology are short circuit current Isc, open-circuit voltage Voc, and the maximum power point Imp, Vmp. The power delivered by the PV cell attains maximum value only at the points Imp, Vmp and reaches zero at points Isc and Voc. PV cells and modules made from different semiconductors and for subcategories within each semiconductor grouping namely crystalline, polycrystalline and thin film would have different operation and conversion efficiencies [2]. Polyimide (PI) has proved to be a promising roll-to-roll compatible substrate yielding very high efficiency devices for Cu(In,Ga)Se2 and that too done with low temperature processing method [3].

Fill factor is another important parameter in PV characteristics which indicates the ability of semiconductor junction in the cell and measures how efficiently the solar cell is able to collect the carriers generated by light. It is defined by Eq. (5) as

$$\text{FF} = \frac{V_{\text{mp}} I_{\text{mp}}}{V_{\text{oc}} I_{\text{sc}}}. \tag{5}$$

It can be re-written in Eq. (6) as below.

$$V_{\text{oc}} I_{\text{sc}} * FF = V_{\text{mp}} I_{\text{mp}} = P_{\text{max}}. \tag{6}$$

It is obvious that the fill factor is always less than 1 and varies with respect to the type of semiconductor material used to produce the PV cell. The closer the values of fill factor to unity, the better are the operation of PV cell. The nonlinear current–voltage characteristics of PV cells are characterized with a unique MPP, which is highly dependent on weather and load conditions. An MPPT algorithm is an analog or digital-based technique allows the PV cell to operate at the MPP at any given environmental conditions. MPPT controllers or algorithms are integrated with solar power conditioning systems to maximize the output power extracted from PV

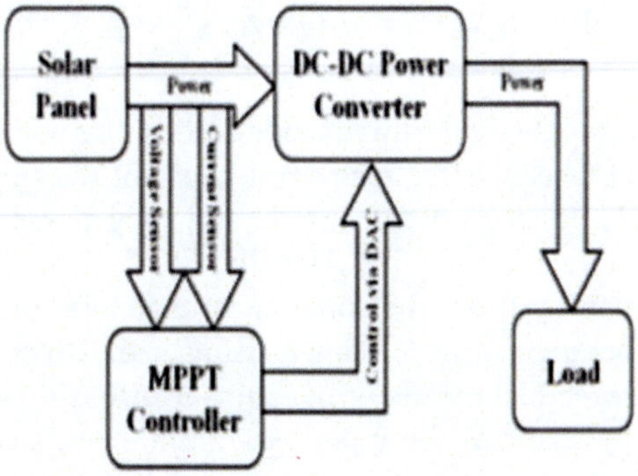


**Fig. 4** Block diagram of MPPT controller with PV cell

generator. Various MPPT techniques have been proposed including perturbation and observation (P&O), incremental conductance, fractional open-circuit voltage, fractional short circuit current, fuzzy logic controller, neural network, ripple correlation control, and DC link capacitor droop control.

The perturbation and observation method is the most commonly implemented technique among other algorithms although oscillations around the MPP may occur. In this technique, the controller adjusts the output voltage of the PV cell based on its instantaneous output power. The incremental conductance algorithm uses the slope of the power–voltage curve of PV cell to spot the voltage reference. The derivative of the cell output power with respect to the cell output voltage at the MPP is zero. This method requires more computations compared to P&O but may reach the MPP faster.

A MPPT is used for extracting maximum power from the solar PV module and transferring that power to the load. A DC-DC converter (steps up/step down) serves the purpose of transferring maximum power from the solar PV module to the load. A DC/DC converter acts as an interface between the load and the modules shown in Fig. 4. By changing the duty cycle, load impedance is varied and matched at the point of peak power with the source so as to transfer maximum power. MPPT techniques are needed to maintain the PV array's operating at its MPPT. At few times, a modulated hysteretic current control (MHCC) technique is tried to improve the transient response of a DC-DC boost converter, when it suffers from low bandwidth [4]. A novel indirect solar cooker with outdoor elliptical cross section, wickless heat pipes, flat-plate solar collector and integrated with indoor PCM thermal storage is best suited to improve the overall efficiency of the solar cell [5].

Periodically, the voltage and the current values are taken from the solar panel and given to MPPT controller, wherefrom the control signals were given to DC-DC converter. This DC-DC converter boosts the voltage, and the corresponding power drives the load.

## 1.2 DC-DC Converters

A single switch three diode dc-dc pulse width modulated (PWM) converters operating at constant frequency and constant duty cycle are different from the conventional dc-dc step-up converters, and they posses higher voltage gain with small output voltage ripples [6]. Usually, DC-DC converters are used to change DC electrical power efficiently from one voltage level to another, as unlike AC, DC cannot simply be stepped up or down using a transformer. In many ways, a DC-DC converter resembles a transformer and performs the conversion with highest possible efficiency by changing the input into output energy with different impedance level. The basic power flow in a converter can therefore be represented with this Eq. (7)

$$P_{in} = P_{out} + P_{losses}. \quad (7)$$

where $P_{in}$ is the power fed into the converter, $P_{out}$ is the output power, and $P_{losses}$ is the power wasted inside the converter. The converter efficiency,

$$\text{Efficiency}(\%) = \frac{P_{out}}{P_{in}}. \quad (8)$$

Converters achieve an overall efficiency of 90% by utilizing the latest version of electronic components and circuit techniques. There are many types of DC-DC converter, each of which tends to be more suitable for some types of application than the others. Among them, the boost converter converts an input voltage to a higher output voltage, called as a step-up converter. Boost converters are used in battery-powered devices, where the electronic circuit requires higher operating voltage than the battery could supply, e.g., mobile phones and camera flashes.

The function of the boost converter in Fig. 5 can be described in terms of energy balance. While during the on-time of the switch, the inductance is charged with energy, and during off time, this energy is transferred from the inductor through the diode to the output capacitor. Output voltage is maintained constant by virtue of large value of capacitor C. When switch S is closed, diode is reversed isolating the output. The input supplies energy to the inductor, i.e., the voltage across L is equal to Vin, and the current $I_L$ increases linearly. When switch S is opened, the current $I_L$ flows through the diode and charges the output capacitor C. Thus, the output stage receives

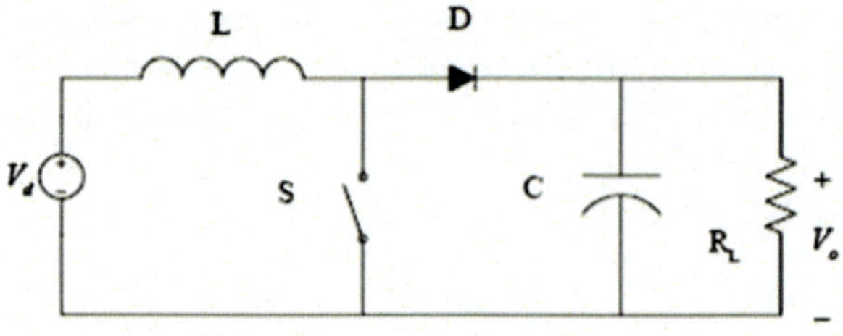


**Fig. 5** Basic boost converter circuit

energy from the input as well as from the inductor, and hence, the output across the load is large.

### 1.3 Interleaved Step-Up Converter with Voltage Multiplier Cell

Experimental results for a 40V to 760V converter verify the significant improvements in efficiency by using this proposed ZVT converters [7].High voltage gain can be achieved by adjusting the turn's ratio of two same coupled inductors without affecting the duty cycles. This interleaved topology utilizes current sharing technique at the input allowing the use of smaller inductors and lower power-rated switches. The voltage multiplier cell is composed of two diodes and the capacitor. The secondary windings of coupled inductors are inserted in conventional interleaved boost converter. Lower Rds means higher input capacitance, but also higher efficiency and less heat at a given current in the case of MOSFET, and hence, low Rds switches can be used to improve the converter performance. An effective way to prevent system instability is defining impedance specifications for modules/subsystems [8, 9]. Presented here is the circuit which works in turn-on zero current switching (ZCS) mode which greatly reduces switching losses and EMI noise as well.

## 2 Simulation Results

The solar PV module/system is simulated with the case of maximum solar radiation on a sunny day [10]. The power demand from the PV module/ network is periodically monitored and control signals adjust the angle and magnitude of the Inverter's voltage [11].With the help of Simulink/MATLAB, the complete block diagram has been constructed for the entire model with PV module and 3Z boost converter as shown in Fig. 6. A fuzzy logic controller is designed with the proposed MPPT algorithm.

The subsystem of an interleaved boost converter showing PV reference panel along with the working panel is shown in Fig. 7.

The subsystem made with fuzzy logic controller is shown in Fig. 8.

Simulation has been carried out, and the simulated outputs have been presented here in the following graphs. Figure 9 shows the generation of PWM signals after the control signal has been superimposed over the high-frequency carrier triangular wave. The first scope in this graph is the generation of gate pulse, second being the control signal; third scope is triangular wave, and the last one is the overlapping of control signal over the triangular wave.

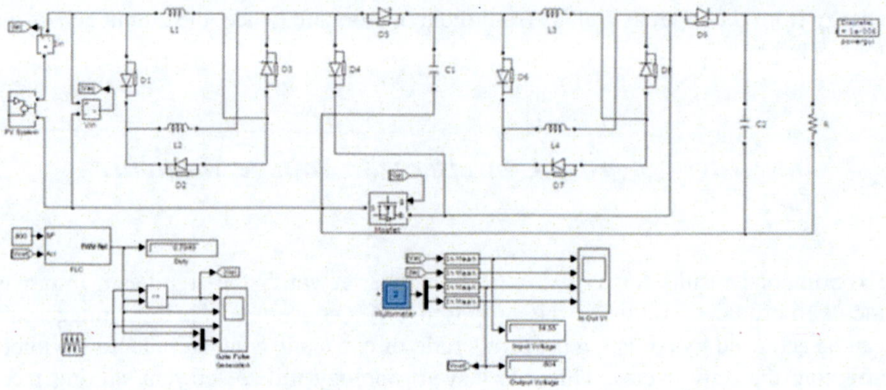


Fig. 6 Model of an entire PV system in Simulink

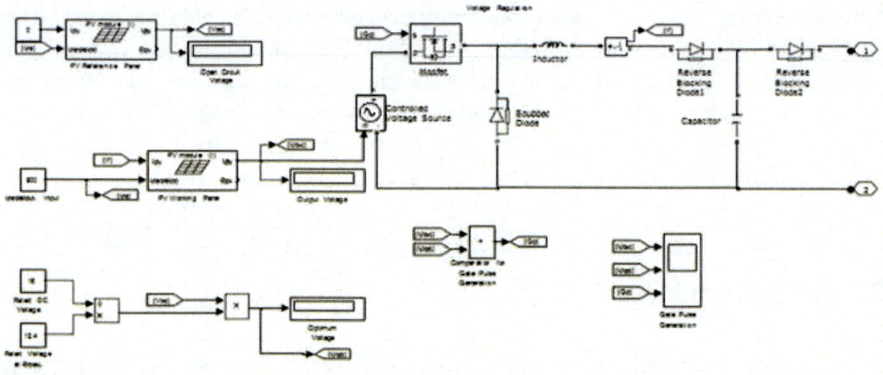


Fig. 7 Subsystem of an interleaved converter using MOSFET

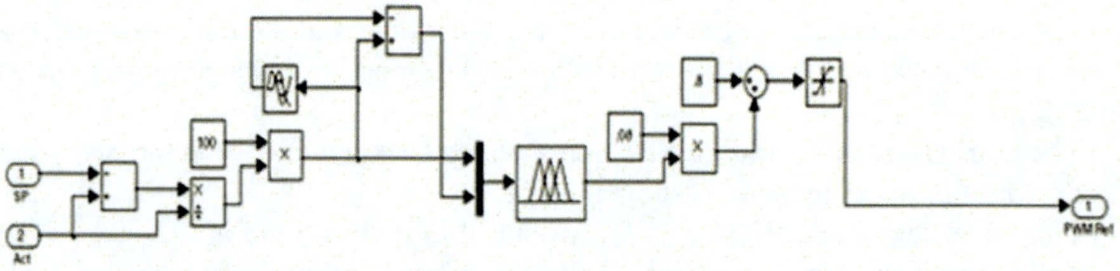


Fig. 8 Subsystem showing the implementation of fuzzy logic controller

The boosted voltage of the converter is shown in Fig. 10 along with the input current and voltage waveforms. The output voltage is found to be increased from 15 V DC to 750 V DC.

The observations in the simulation have been given in Table 1.

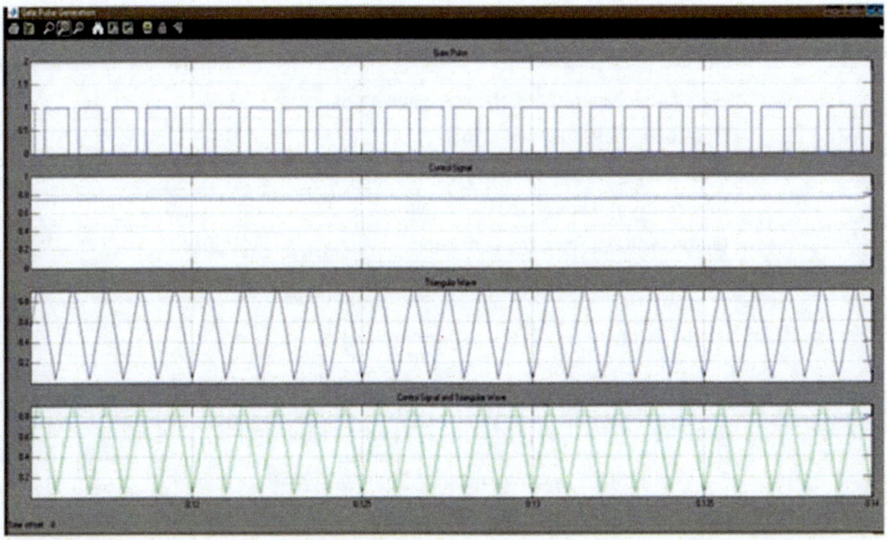

**Fig. 9** PWM output of the proposed converter

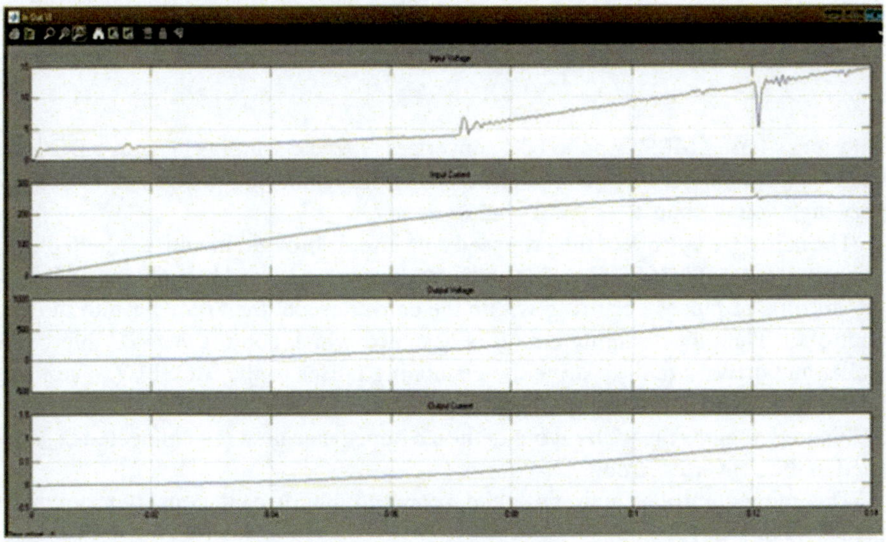

**Fig. 10** Input and output current and voltage waveforms

# 3 Hardware Description

A family of ZC-ZVS converters with wide input voltage range and remarkable voltage gain is achieved [12].The complete block diagram showing solar PV panel associated

**Table 1** Results obtained in simulation

| S. No. | Related parameters | Simulated values |
|---|---|---|
| 1 | Gate pulse generated in the pulse generator | 1 V |
| 2 | Control signal/reference signal | 0.8 V |
| 3 | Input voltage to the converter | 15 V |
| 4 | Input current through the converter | 250 mA |
| 5 | Output voltage from the converter | 750 V |
| 6 | Output current through the converter | 1A |

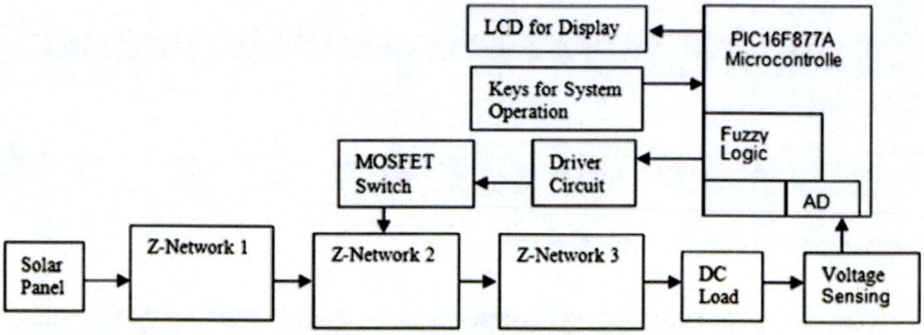


**Fig. 11** Block diagram of the complete hardware setup

with fuzzy-based MPPT applied 3Z converter is depicted in Fig. 11 with a DC load. This system consists of three impedance network connected in cascade to achieve very high voltage boost.

The hardware system of the proposed converter is brought in with the help of PIC microcontroller PIC16F877A. The embedded software is used to write the coding of on–off time of pulses occurring within the converter and then transfer into the PIC controller. The power supply circuit is designed which control the PIC microcontroller and driver circuit to drive the generating pulses to the MOSFET. The firing circuit of MOSFET is shown in Fig. 12.

Figure 13 shows the PCB layout of the 3Z impedance network connected in series used in the converter circuit.

The microcontroller unit with the proposed interleaved converter circuit is presented in Fig. 14.

In this hardware setup, a step-down transformer of 230 V/12–0–12 V is used to perform the step-down operation by which 230 V AC appears as 12 V AC across the secondary winding. A bridge rectifier of four diodes (4*1N4007) is used to obtain full-wave rectification. Filter circuit, in which capacitor acts as a surge arrester, follows the rectifier unit. The bypassing capacitor is used not only to short the ripple with frequency of 130 Hz to ground but also to leave the frequency of the DC to appear at the output.

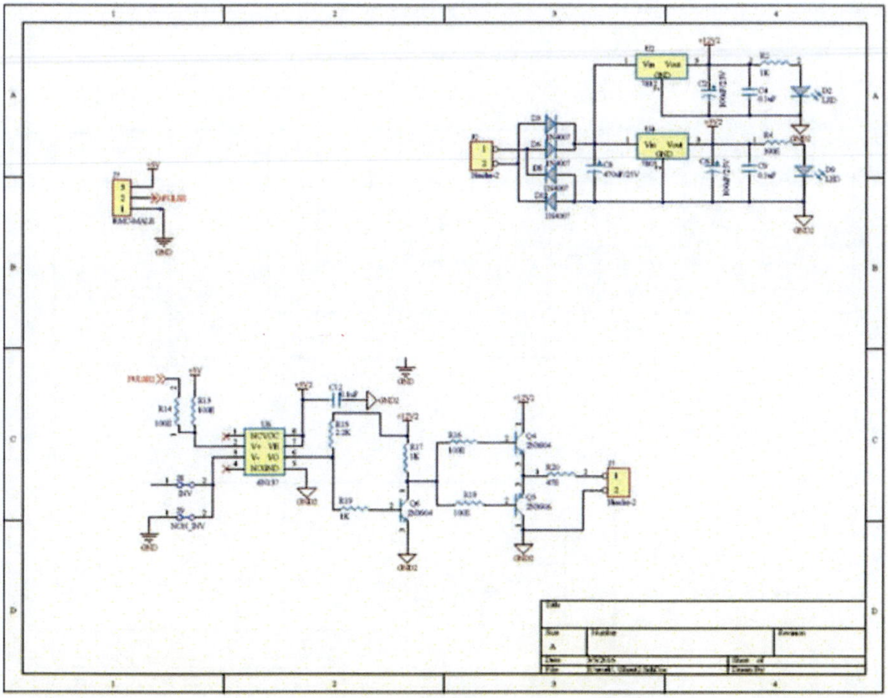

**Fig. 12** PCB layout of the firing circuit using MOSFET

With a voltage regulator connected to the DC output, the voltage is maintained within a close tolerant region to the desired output. The timing and control unit will generate necessary control signals for internal and external operation of the microcontroller. The switching pulses required for inverter operation are generated using PIC16F877A microcontroller, thus reducing the overall system cost and complexity. This microcontroller generates PWM pulses of particular frequency and switching pulses required to trigger MOSFET switches ON. PIC 16F877A op codes are 14-bit wide enabling a single-word instruction. A two-stage pipeline overlaps fetch and execution cycles of instructions consequently by which all instructions are carried over by a single cycle except for program branches and conditions test. The program memory and data memory block have separate buses so that concurrent access can occur. The inverter is built with two MOSFET switches which are used to convert the DC voltage into AC. When the switch S1 is closed (S2 is open), a positive voltage will be applied across the load. By opening S1 switch and closing S2 switch, this voltage is reversed allowing reverse operation.

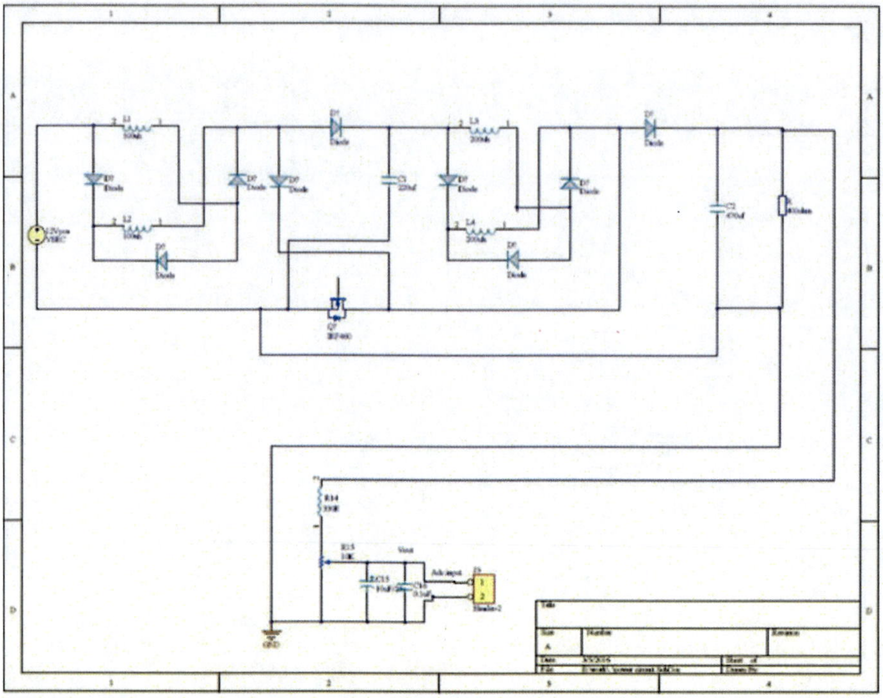

**Fig. 13** PCB layout of the 3Z boost converter connected in series

# 4 Conclusion

A 3Z boost converter has been designed and implemented in PV applications both in simulation and real time. From the simulated results, it is found that by using this interleaved converter with MOSFET, the output voltage has been increased from 15 V DC to 750 V DC. An innovative effort has been tried with two stages which overlaps fetch and execution cycles of instructions consequently by which all instructions are carried out in a single cycle except for program branches and conditional tests. Since the program memory and data memory blocks have separate buses to transmit the data, concurrent access of the data can be done which in turn increases the response time of the entire boost converter. The inverter is built with two MOSFET switches which are used to convert the DC voltage into AC. When the switch S1 is closed (S2 is open), a positive voltage will be applied across the load. By opening S1 switch and closing S2 switch, this voltage is reversed allowing reverse operation. The entire system can be used to meet the load demand of multiple loads. A serial-parallel converter can be used instead of this 3Z converter to obtain higher current values in future.

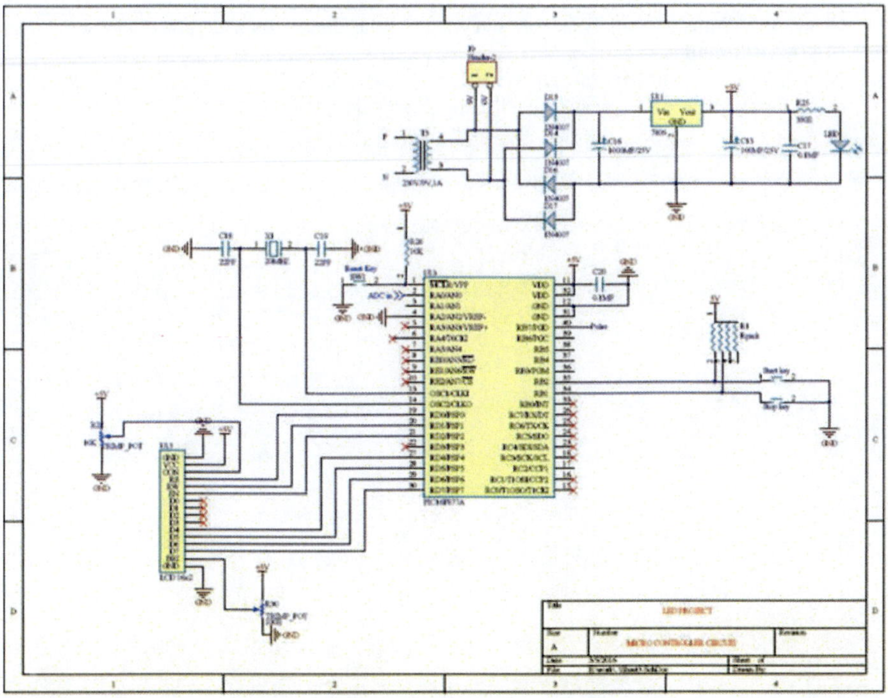

**Fig. 14** PCB layout of the interleaved converter with microcontroller unit

# References


1. Bogenrieder J et al (2017) Technology specific yield analysis of various photovoltaic module technologies under specific real weather conditions, pp 74–85
2. Martin A. Green et al (2017) Solar cell efficiency tables. Version 51:3–12
3. Romero IB et al (2017) CZTSe solar cells developed on polymer substrates: effects of low temperature processing, pp 55–68
4. Tsai JC et al (2011) Modified hysteretic current control (MHCC) for improving transient response of boost converter. IEEE Trans. Circuit Syst I Reg Papers 58(8):1967–1979
5. Muthusivagami RM, Velraj R, Sethumadhavan R (2010) Solar cookers with and without thermal storage-a review. Renew Sustain Energy Rev 14(2)
6. Ismail EH, Al-Saffar MA, Sabzali AJ, Fardoun AA (2008) A family of single-switches PWM converters with high step-up conversion ratio. IEEE Trans Circuit Syst I 55(4):1159–1171
7. Li W, Liu J, Wu J, He X (2007) Design and analysis of isolated ZVT boost converters for high-efficiency and high step-up applications. IEEE Trans Power Electron 22(6):2263–2374
8. Feng XG, Liu JJ, Lee FC (2002) Impedance specifications for stable DC distributed power systems. IEEE Trans Power Electron 17(2):157–162
9. Wildrick CM, Lee FC, Cho BH, Choi B (1995) A method of defining the load impedance specification for a stable distributed power system. IEEE Trans Power Electron 5(3):280–285
10. Tyson Denherder (2014) Design and simulation of photovoltaic super system using Simulink. Project report submitted to California Polytechnic State University

11. Chan TT Transient analysis of integrated solar/diesel hybrid power system using MATLAB/Simulink
12. Al-Saffar MA, Ismail EH, Sabzali AJ (2013) Family of ZC-ZVS converters with wide voltage range for renewable energy systems. Renew Energy 56:32–43

# Real-Time Realization and Validation of LabVIEW-Based Automation Program for AC–DC Voltage and Current Measurements of AC Calibrators

**Sunidhi Luthra, Vishnu Parammal, S. Kumari, B. Pal, S. Ahmad, and A. Gupta**

**Abstract** In measurement technology and metrology, the voltage- or current-generating instruments are of extreme importance. They are used to calibrate high-precision multimeters, voltmeters, ammeters, measurement standard, etc. It is therefore necessary to ensure that these devices remain accurate within the uncertainty limits. Obtaining accurate and precise value of voltage or current requires accurate AC–DC transfer difference measurement of such devices. Manual measurement of AC–DC transfer difference at low frequency using null or Budvosky method is tedious and time-consuming as it involves collecting large number of datasets for varying frequencies, voltage and current values and performing statistical operations on them to obtain uncertainty. Hence, the system has been automated using NI LabVIEW. The benefit of automation is that it saves time, energy, labor and minimizes the human involvement, improving the measurement accuracy. The datasets are obtained using data acquisition tools and passed on to excel sheet for calculating uncertainty. The software developed prepares the final report in excel in a ready to print format.

**Keywords** AC calibrator · AC–DC transfer difference · LabVIEW · Software validation

## 1 Introduction

AC voltage- and current-generating instruments alias calibrators are worldwide used for measurements in low-frequency range for calibration of various electronic

---

S. Luthra (✉) · S. Kumari · S. Ahmad · A. Gupta
CSIR-National Physical Laboratory, New Delhi, India
e-mail: madaansunidhi@nplindia.org

V. Parammal
Veermatabai Jijabai Technological Institute, Matunga, Mumbai, India

B. Pal
CSIR-Humar Resource and Development, Ghaziabad, India



S. Yadav et al. (eds.), *Recent Advances in Metrology*, Lecture Notes in Electrical Engineering 906, https://doi.org/10.1007/978-981-19-2468-2_13

measurement devices. The usage of these sourcing instruments in the long-run might degrade their performance. To provide confidence in the accuracy of calibration results, the measurement must have demonstrable *traceability.* It is therefore necessary to re-calibrate these instruments as per standards. The method used to perform this calibration is discussed in this paper. The entire process of calibration is tedious, time-consuming and error-prone if done manually, as it involves recording a large number of observations in fixed time intervals and performing complex mathematical and statistical analysis. Later, a summary of this big data pool has to be obtained in the form of a report. A solution is therefore needed for automating the process that will overcome all above-mentioned problems and also remove any human errors involved in the process.

## 2 Calibration of Alternating Voltage/Current Sources

The alternating voltage/current sources (AC calibrators) are devised for calibration of equal or lower resolution digital multimeter, oscilloscope that are widely used in various laboratories across the country. Being an NMI of the country, it is the prime responsibility of CSIR-NPLI to provide accurate and precise measurement of such instruments. Therefore, it becomes important to precisely evaluate the AC–DC transfer difference of AC voltage- and current-generating sources [3]. The Budvosky method has been used for evaluation of AC–DC transfer difference.

The AC–DC transfer difference is defined in terms of $\delta$ as

$$\frac{S_{\mathrm{ac}} - S_{\mathrm{dc}}}{n_s S_{\mathrm{dc}}} - \frac{X_{\mathrm{ac}} - X_{\mathrm{dc}}}{n_s X_{\mathrm{dc}}} \tag{1}$$

where $S_{\mathrm{ac}}$ is the mean emf output of the standard when AC is applied.

$S_{\mathrm{dc}}$ is the mean emf output of the standard, when DC with reverse polarity is applied.

$X_{\mathrm{ac}}$ is the mean emf output of the DUC, when AC is applied.

$X_{\mathrm{dc}}$ is the mean emf output of the DUC, when DC with reverse polarity is applied.

## 3 Measurement Procedure

The precision AC calibrator is compared against a thermal voltage converter (TVC) or AC/DC transfer standard (TTS). The calibrator to be calibrated in AC voltage/current must be calibrated first in DC voltage. The desired voltage level with frequency is selected at calibrator and is applied to the calibrated TVC/TTS. The emf output generated by the TVC/TTS is recorded in the nanovoltmeter. Then, the same rated DC+, DC− and AC voltage is applied, and their response is recorded, respectively.

The average $V_{\mathrm{dc}}$ is calculated using Eq. (2).

**Fig. 1** Measurement setup

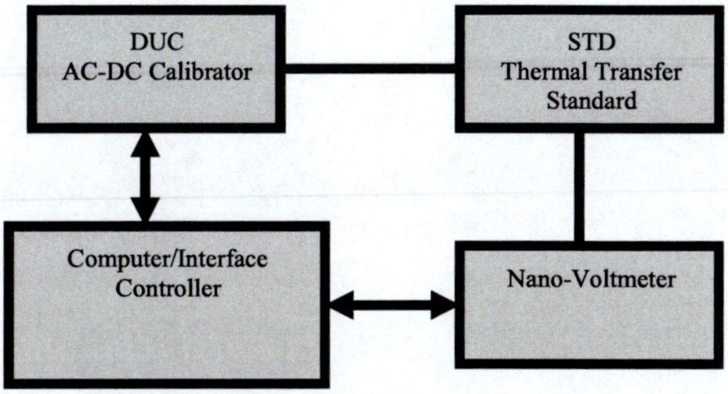


$$V_{dc} = (V_{dc+} + V_{dc-})/2 \quad (2)$$

Using $V_{dc}$ and AC–DC transfer difference ($\delta$) of thermal converter, $V_{ac}$ is calculated as per Eq. (1). This calculated $V_{ac}$ is actual AC voltage of calibrator at that frequency. For AC current, calibration voltage is replaced by current signal of a particular frequency and the steps are repeated as in case of ac voltage. The AC current $I_{ac}$ is calculated as per Eq. (2). A set of five or ten measurements is taken at each voltage/current and frequency point.

The measurement can be carried out manually or through automation. This paper emphasizes automatic evaluation of AC–DC transfer difference of AC calibrator with its validation by comparing the results with manual calibration (Fig. 1).

## 4 Traceability

The AC calibrator is traceable to reference standard, viz. thermal converter or thermal transfer standard, which in turn are traceable to the primary standard (multijunction thermal converters) of low frequency AC voltage and current.

The traceability chain is as shown in Fig. 2. As a part of Asia Pacific Metrology Program (APMP), CSIR-NPL provides calibration to regional laboratories like ERTLs, ETDCs, defence laboratories like DRDO, 13 BRD, BEL.

## 5 Programming Software

LabVIEW is a graphical programming platform that provides an easy-to-use application environment to its users for designing virtual instruments by creating a graphical user interface) (GUI) on computer screen [2]. This platform has three basic components: front panel and back panel as shown in Figs. 3 and 4. The LabVIEW programming environment is simple and easy that a non-programmer can also build programs by just dragging and dropping tools.

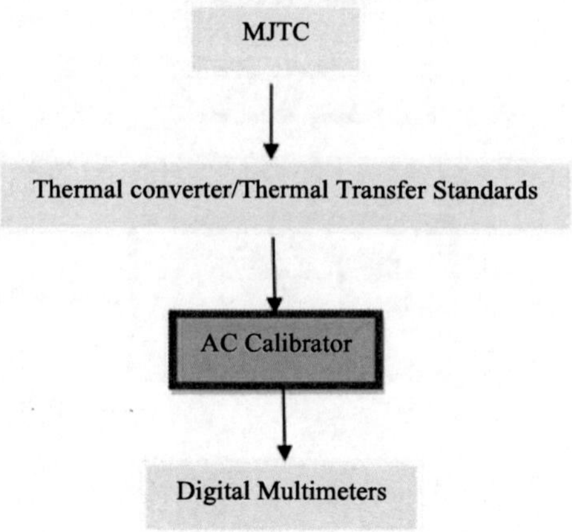


**Fig. 2** Traceability chart

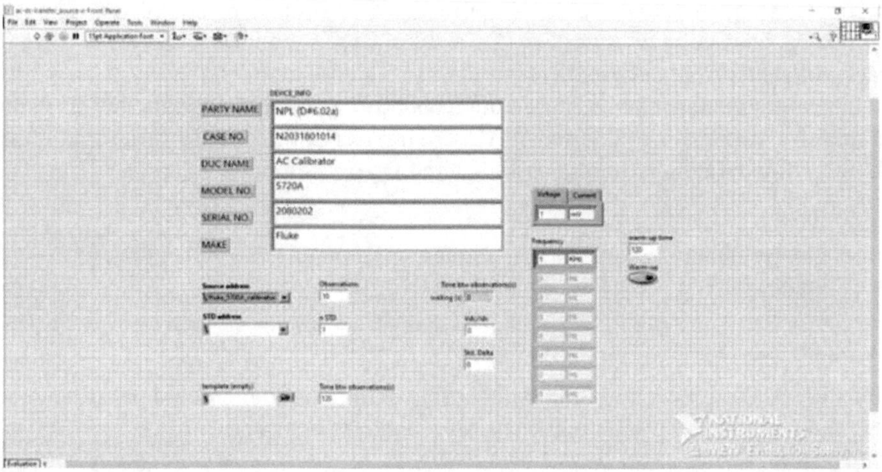


**Fig. 3** Front panel

The attractiveness of the software is that it itself acquires, analyzes and presents the data in Microsoft Excel or word file, as per the need of the user.

## 6 Instrument Connection/Control

The instruments used in the measurement setup are connected to the computer using National Instruments GPIB-USB-HS. It is an IEEE 488 controller device for computers with USB slots [1]. The device includes a license for the NI-488.2

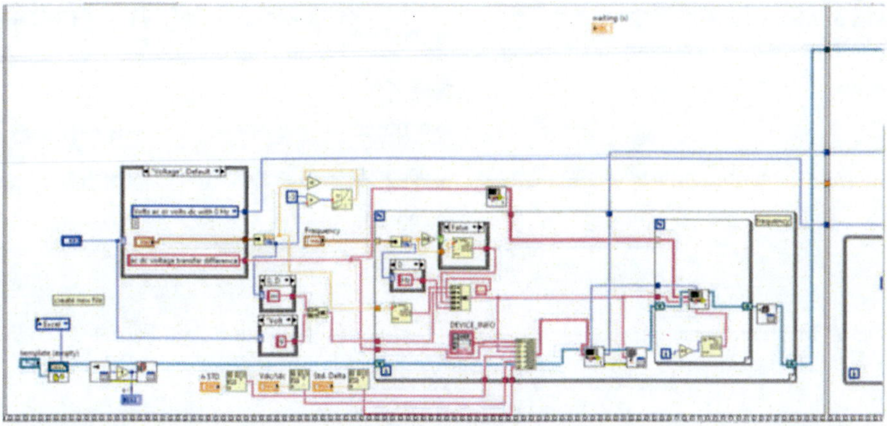

**Fig. 4** Block diagram

driver software, providing maximum reliability for connecting to third-party instruments with GPIB. For each instrument to be controlled, instrument drivers (available online) have been used. The programming interface between hardware and development environment like LabVIEW is done using VISA. It is a standard for configuring, programming and troubleshooting instrumentation systems comprising GPIB (Fig. 5).

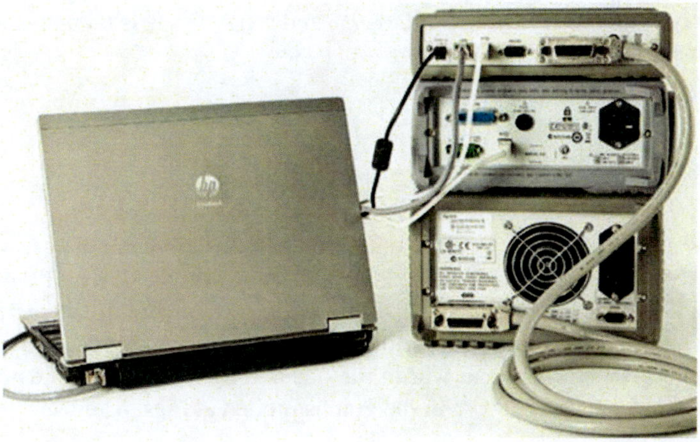

**Fig. 5** Instrument connection/control

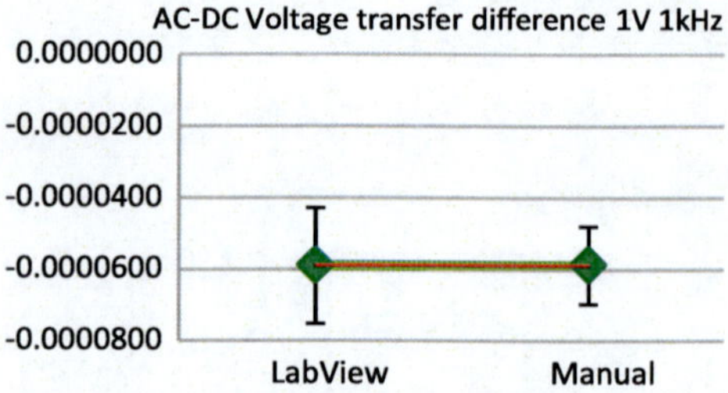


**Fig. 6** Measurement at 1 V 1 kHz

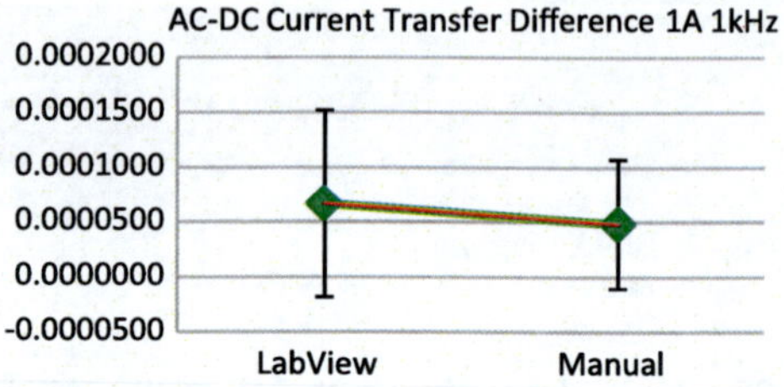


**Fig. 7** Measurement at 1A 1kHz

## 7 Results with Uncertainty Calculation

The results for calibration of AC–DC transfer difference of source (AC calibrator) for manual and automated calibration are as shown in Figs. 6 and 7. The reported result at 1 V 1 kHz for manual and automated measurement is 58 ppm and 59 ppm, respectively, with an uncertainty of 0.009% and at 1A 1 kHz is 66 ppm and 50 ppm with an uncertainty of 0.002%.

## 8 Conclusion

The software for AC–DC transfer measurements of AC calibrator (source) was successfully designed in LabVIEW platform. The results for 1 V and 1 A at 1 kHz are shown above. The results are well within the uncertainty limits as specified in our calibration and measurement capabilities. The results obtained through automated calibration are satisfactory and within the uncertainty limits and validate with the manual measurements as shown in the graphs plotted at Figs. 6 and 7.

## References

1. Ponce-Cruz P, Molina-Gutierrez A (2010) LabVIEW for intelligent control research and education. In: 2010 4th IEEE International conference on e-learning in industrial electronics

2. Elliott C, Vijayakumar V, Zink W (2007) National instruments LabVIEW: a programming environment for laboratory automation and measurement. Published February 1, 2007 https://doi.org/10.1016/j.jala.2006.07.012
3. Govil AK, Ahmad S, Bijendra Pal, Kothari PC (2008) Development of automated AC-DC transfer measurement system for voltage and current at low frequencies. CPEM

# Biomedical Signal Processing: ECG Signal Analysis Using Machine Learning in MATLAB

**Monu Malik, Tanya Dua, and Snigdha**

**Abstract** An arrhythmia is an abnormality in the heart rhythm or heartbeat pattern. ECG beats can be classified into different arrhythmias beat types (bigeminy, trigeminy, ventricular tachycardia (VT)). Early and accurate detection of arrhythmia types is important in detecting heart diseases and choosing appropriate treatment for a patient. We have used MIT-BIH arrhythmia database for data collection and prepared different data sets. Features, such as amplitude, RR interval, heart rate (Speed), gender, age, are used for the analysis. In classification learner application, the extracted features are used as inputs to different classifiers: support vector machines (SVM) and Naïve Bayes. Some other techniques that also have been employed for arrhythmia classification are decision trees and ensemble learning. Results show high classification accuracy of over 99.3% with either of these classifiers. The performance comparison of these classifiers is carried out using accuracy. Each classifier can show the confusion matrix, which summarizes the accuracy for each true label class.

**Keywords** MATLAB classification learner application · Arrhythmia detection · ECG signal analysis · Machine learning · MATLAB signal processing

## 1 Introduction

An electrocardiogram (ECG) is a diagnostic tool that measures and records the electrical activity of the heart of a patient. ECG is useful in finding the cause of chest pain and detecting abnormal heart rhythm or cardiac abnormalities. Normal healthy hearts have a cardiac ECG shape. Any irregularity in the heart rhythm can change the shape of ECG signal. Classification of ECG beats plays an important role in detecting various heart diseases. Early and accurate detection of ECG arrhythmia helps doctors to detect various heart diseases. ECG beat classification is a challenging problem due to the following reasons [1]. Variability in normal ECG waveform

M. Malik (✉) · T. Dua · Snigdha
Department of Electronics and Communication, Maharaja Surajmal Institute of Technology, New Delhi 110075, India
e-mail: msmalikmonu@msit.in

 
S. Yadav et al. (eds.), *Recent Advances in Metrology*, Lecture Notes in Electrical Engineering 906, https://doi.org/10.1007/978-981-19-2468-2_14

of each person, dissimilar signs for one disease on different patients' ECG waveform, and two distinct diseases may have approximately similar effects on different patients' ECG waveform, lack of standardization of ECG features, and non-existence of optimal classification rules for ECG beat classification [2].

Researchers have used different preprocessing techniques, feature extraction techniques, and classification techniques for ECG beat classification. For preprocessing (noise removal), various filters are used such as low-pass linear phase filter, linear phase high-pass filter, and median filter.

For feature extraction, techniques such as discrete wavelet transform (DWT), and Pan–Tompkins algorithm are used. The following classifiers are used by research community for ECG classification: artificial neural network (ANN), support vector machine (SVM), fuzzy C-Means clustering (FCM), and ID3 decision tree [3].

This paper has support vector machines (SVM) and Naïve Bayes Model for ECG beat classification. We have used MIT-BIH arrhythmia database for collecting raw ECG data. Pan–Tompkins algorithm is applied on raw ECG data for noise removal and features extraction. Extracted features are used as inputs to these classifiers. Principal component analysis is used for feature selection. The remainder of this paper is organized as follows: Sect. 2 gives introduction to ECG waveform and different types of arrhythmia. The background knowledge is presented in Sect. 2.3. Section 2.3 shows methodology. The results are presented in Sect. 3. The conclusions are given in Sect. 4.

## 2 Methodology

### *2.1 Contextual Information*

An ECG signal consists of P, Q, R, S, T, and U peaks as shown in Fig. 1 PR, RR, QRS, ST, and QT intervals and PR and ST segments. R-peak, RR interval, and QRS complex are the basic features of ECG signal, where R-peak is the second most upward movement of the ECG tracing; RR interval is the interval between two

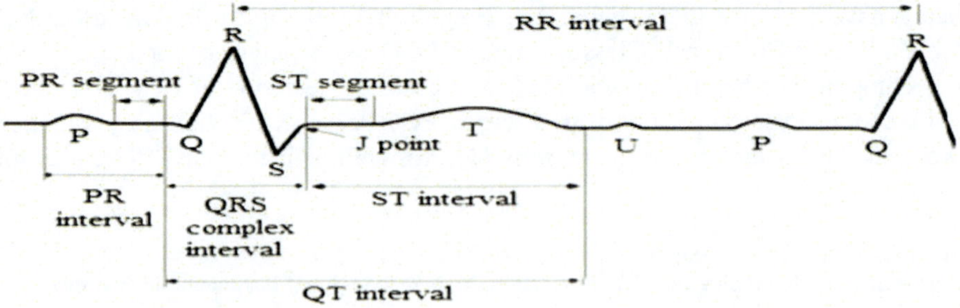


**Fig. 1** ECG signal peaks

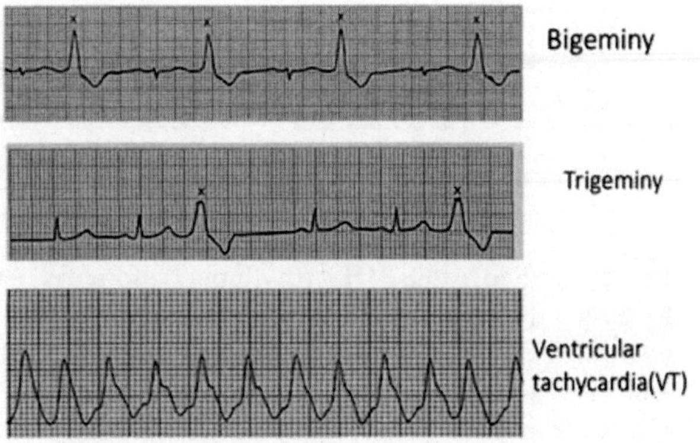


**Fig. 2** Different types of Arrhythmia

consecutive R-peaks; and QRS complex normally begins with a downward deflection Q, a larger upward deflection R, and ends with a downward S wave.

Any irregularity in the rhythm of heart can change the shape of ECG signal. There can be many types of abnormalities present in the ECG signal. The heartbeat can be too slow, too fast, have extra beats, or skip a beat.

## 2.2 *Types of Arrhythmia*

There are three different types of Arrhythmia as depicted in Fig. 2.

1. Bigeminy—If heart is skipping a beat or having rapid beats.
2. Trigeminy—These are of two kinds:
   a. Two normal heartbeats followed by one extra beat
   b. One normal beat followed by two extra beats
3. Ventricular tachycardia (VT)—It is a fast, abnormal heart rate. It starts in your heart's lower chambers called the ventricles.

Features of ECG signal like R-peak, QRS interval, RR interval, RR ratio, and RR mean are useful to classify ECG beats into these arrhythmia types.

## 2.3 *About Data Set*

The MIT-BIH database was the first set of standard test material that generally available to evaluate arrhythmia detection. Since 1980, this database has been used for the basic research for cardiac dynamics. The MIT-BIH Arrhythmia database contains 48 half-hour excerpts of two-channel ambulatory ECG recordings, obtained from 47

**Fig. 3** Block diagram of Arrhythmia detection using machine learning techniques

subjects studied by the BIH Arrhythmia Laboratory between 1975 and 1979. Twenty-three recordings were chosen at random from a set of 4000 24-h ambulatory ECG recordings collected from a mixed population of inpatients (about 60%) and outpatients (about 40%) at Boston's Beth Israel Hospital; the remaining 25 recordings were selected from the same set to include less common but clinically significant arrhythmias that would not be well-represented in a small random sample. The recordings were digitized at 360 samples per second per channel with 11-bit resolution over a 10-mV range. Figure 3 shows the steps involved during Arrhythmia detection using machine learning techniques.

## 3 Results and Discussion

### *3.1 Feature Collection*

A study presents a comparative study between the use of principal component analysis (PCA) and independent component analysis (ICA) to reduce the noise and artifacts of the ECG signal and showed that PCA is a better technique to reduce noise, while ICA is better one to extract features. The ICA technique enables statistically separate individual sources from a mixing signal whereas The PCA technique separates the sources according to the energy contribution to the signal.

In this paper, we have used PCA technique which is a dimensionality reduction method that transforms a large set of variables into a smaller one that still contains most of the information in the large set. Reducing the number of variables of a data set naturally comes at the expense of accuracy, but the trick in dimensionality reduction is to trade a little accuracy for simplicity. Smaller data sets are easier to explore and visualize and make the process of data analysis much easier and faster for machine learning algorithms without extraneous variables to process. Figure 4 shows test and training data sets for the considered system.

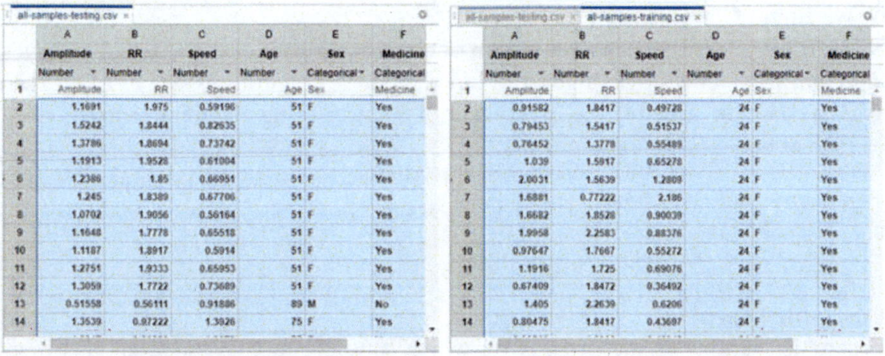

all-samples-testing.csv

| | A | B | C | D | E | F |
|---|---|---|---|---|---|---|
| | Amplitude | RR | Speed | Age | Sex | Medicine |
| | Number | Number | Number | Number | Categorical | Categorical |
| 1 | Amplitude | RR | Speed | Age | Sex | Medicine |
| 2 | 1.1691 | 1.975 | 0.59196 | 51 | F | Yes |
| 3 | 1.5242 | 1.8444 | 0.82635 | 51 | F | Yes |
| 4 | 1.3786 | 1.8694 | 0.73742 | 51 | F | Yes |
| 5 | 1.1913 | 1.9528 | 0.61004 | 51 | F | Yes |
| 6 | 1.2386 | 1.85 | 0.66951 | 51 | F | Yes |
| 7 | 1.245 | 1.8389 | 0.67706 | 51 | F | Yes |
| 8 | 1.0702 | 1.9056 | 0.56164 | 51 | F | Yes |
| 9 | 1.1648 | 1.7778 | 0.65518 | 51 | F | Yes |
| 10 | 1.1187 | 1.8917 | 0.5914 | 51 | F | Yes |
| 11 | 1.2751 | 1.9333 | 0.65953 | 51 | F | Yes |
| 12 | 1.3059 | 1.7722 | 0.73689 | 51 | F | Yes |
| 13 | 0.51558 | 0.56111 | 0.91886 | 89 | M | No |
| 14 | 1.3639 | 0.97222 | 1.3926 | 75 | F | Yes |

all-samples-training.csv

| | A | B | C | D | E | F |
|---|---|---|---|---|---|---|
| | Amplitude | RR | Speed | Age | Sex | Medicine |
| | Number | Number | Number | Number | Categorical | Categorical |
| 1 | Amplitude | RR | Speed | Age | Sex | Medicine |
| 2 | 0.91582 | 1.8417 | 0.49728 | 24 | F | Yes |
| 3 | 0.79453 | 1.5417 | 0.51537 | 24 | F | Yes |
| 4 | 0.76452 | 1.3778 | 0.55489 | 24 | F | Yes |
| 5 | 1.039 | 1.5917 | 0.65278 | 24 | F | Yes |
| 6 | 2.0031 | 1.5639 | 1.2809 | 24 | F | Yes |
| 7 | 1.6881 | 0.77222 | 2.186 | 24 | F | Yes |
| 8 | 1.6682 | 1.8528 | 0.90039 | 24 | F | Yes |
| 9 | 1.9958 | 2.2583 | 0.88376 | 24 | F | Yes |
| 10 | 0.97647 | 1.7667 | 0.55272 | 24 | F | Yes |
| 11 | 1.1916 | 1.725 | 0.69076 | 24 | F | Yes |
| 12 | 0.67409 | 1.8472 | 0.36492 | 24 | F | Yes |
| 13 | 1.405 | 2.2639 | 0.6206 | 24 | F | Yes |
| 14 | 0.80475 | 1.8417 | 0.43697 | 24 | F | Yes |

**Fig. 4** Preprocessing training data

## 3.2 *Learning Algorithms*

Once the set of features has been selected, models can be trained using the in-built classification learner application in MATLAB. The main algorithms studied for this task are support vector machines (SVM), Naïve Bayes, and ensemble learning (Fig. 5).

**Support Vector Machines (SVM)**
Support vector machine (SVM) is one of the most popular machine learning classifiers. It falls under the category of supervised learning algorithms and uses the concept of margin/hyperplanes to classify between classes. In most situations, it gives better accuracy than KNN, decision trees, and Naive Bayes classifier and hence is quite useful.

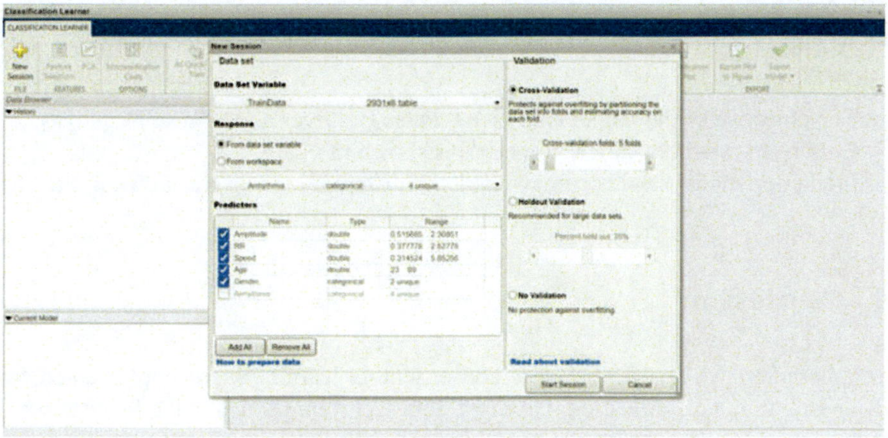


**Fig. 5** Classification learner application configuration

**Naïve Bayes Classification**

Naive Bayes classifiers are a collection of classification algorithms based on Bayes' theorem. It is not a single algorithm but a family of algorithms where all of them share a common principle; i.e., every pair of features being classified is independent of each other.

a. Gaussian Naïve Bayes
b. Kernel Naïve Bayes.

**Ensemble Learning**

Ensemble learning is the process by which multiple models, such as classifiers, are strategically generated and combined to solve a particular computational intelligence problem. Ensemble learning is primarily used to improve the (classification, prediction, function approximation, etc.) performance of a model, or reduce the likelihood of an unfortunate selection of a poor one.

### *3.3 Evaluation Metric: Confusion Matrix*

The confusion matrix is one of the most intuitive and easiest metrics used for finding the correctness and accuracy of the model. It is only used for classification problem where the output can be of two or more types of classes.

1. True Positives (TP): It is the case where we predicted Yes and the real output was also yes.
2. True Negatives (TN): It is the case where we predicted No and the real output was also No.
3. False Positives (FP): It is the case where we predicted Yes but it was actually No.
4. False Negatives (FN): It is the case where we predicted No but it was actually Yes.

The objective of the model is to give 0 False Positives and 0 False Negatives. Confusion matrix of different models is shown in Fig. 6.
Prediction models and accuracy results of different models are shown in Fig. 7.

## 4 Conclusion

The classifiers SVM, Naïve Bayes and ensemble learning were implemented for ECG arrhythmia classification. Different data sets were used, and the features were fed as inputs to the classifiers. The trained classifiers were applied on test data, and obtained results were compared.

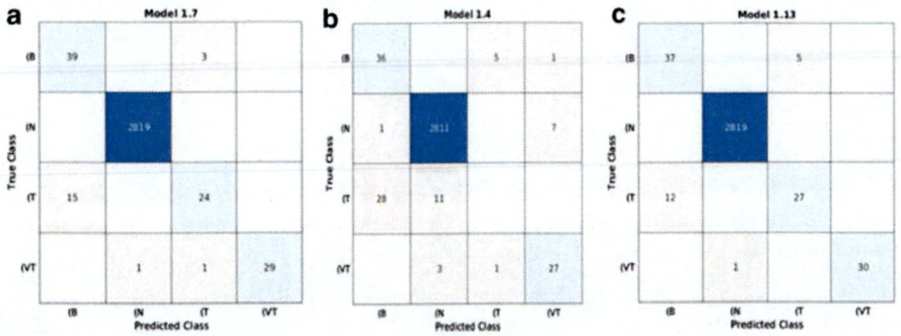


**Fig. 6** **a** SVM model; **b** Naïve model; **c** Ensemble learning

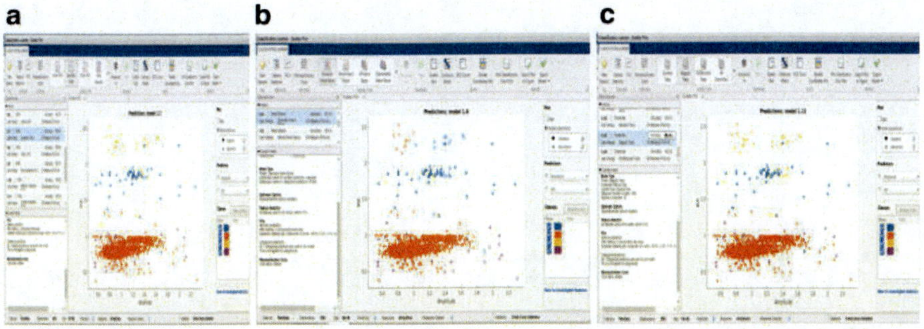


**Fig. 7** **a** SVM model accuracy 99.3%; **b** Naive model accuracy 98.1%; **c** Ensemble learning model accuracy 99.4%

With the evaluation using accuracy results and confusion matrix for each MATLAB model for the ECG arrhythmia extraction and analysis, the best results came from ensemble learning algorithms and the Gaussian SVM, while Naïve Bayes algorithms' results were the worst ones.

## References

1. Jambukia SH, Dabhi VK, Prajapati HB (2018) ECG beat classification using machine learning techniques. Int J Biomed Eng Technol 26(1):32–53
2. Jambukia SH, Dabhi VK, Prajapati HB (2015) Classification of ECG signals using machine learning techniques: a survey. In: Computer engineering and applications (ICACEA), 2015 International conference on advances in IEEE
3. Alickovic E, Subasi A (2016) Medical decision support system for diagnosis of heart arrhythmia using DWT and random forests classifier. J Med Syst

# Study of DC Sputtered Undoped NiO Thin Films

**Aasim Hussain, A. M. Siddiqui, Anju Dhillon, Shafaque Rahman, Navjyoti Boora, and A. K. Hafiz**

**Abstract** Thin films of nickel oxide were fabricated on glass substrate using the DC sputtering method at ambient temperature under 4 bar base pressure and 100-W power for a period of 15 min in the Argon atmosphere with subsequent post-annealing for 2 h at 500 °C. Using ultrahigh-resolution electron scanning microscopy (SEM), the surface morphology of these films was investigated. For the study of its electrical properties, the resistance of grown films was measured using the conventional four-probe method at 300 K with 1 T applied magnetic field and at zero magnetic fields. The optical characterization of NiO film was performed with UV–Vis, and the bandgap comes out to be 3.6 eV. Nickel oxide thin films can be used in electrochromic display applications, functional layer material for chemical sensors, and as a hole transport layer and window layer in inverted perovskite solar cells.

**Keywords** DC sputtering · SEM · Thin film

## 1 Introduction

In recent era, nickel oxide (NiO) has attracted the scientific world due to its numerous applications in diverse fields. Low cost, electrochemical stability, non-toxicity, large spin optical density, one of the promising ion storage materials and possibility of fabrication by variety of techniques are the key features of attraction of NiO thin films [1]. Nickel oxide is the principal oxide of nickel [2], and there are many other oxidation states, namely nickel peroxide ($NiO_4$), nickelosic oxide ($Ni_3O_4$), nickel sesquioxide ($Ni_2O_3$), and nickel dioxide ($NiO_2$) have been reported [1].

A. Hussain (✉) · A. M. Siddiqui
Department of Physics, Jamia Millia Islamia (Central University), New Delhi, India
e-mail: aasim169255@st.jmi.ac.in

S. Rahman · N. Boora · A. K. Hafiz
Centre for Nanoscience and Nanotechnology, Jamia Millia Islamia (Central University), New Delhi, India

A. Dhillon
Maharaja Surajmal Institute of Technology, New Delhi, India

 
S. Yadav et al. (eds.), *Recent Advances in Metrology*, Lecture Notes in Electrical Engineering 906, https://doi.org/10.1007/978-981-19-2468-2_15

Crystal structure of NiO below the Néel temperature (523 K) is rhombohedral, whereas above the Néel temperature, the crystals become ordered and acquire cubic structure [3]. NiO shows p-type semiconducting behavior grown at room temperature with wide range (3.5–4.0 eV) of band gap energy [4]. NiO thin films can be manufactured by several techniques like sol–gel, plasma-enhanced chemical vapor deposition, spray pyrolysis, chemical deposition, pulse laser deposition, vacuum evaporation, and electron beam evaporation [5–8]. Among these techniques, DC magnetron sputtering permits relatively high deposition rate and control over film's atomic ratio [5]. NiO thin films are deposited by DC reactive magnetron sputtering from a nickel target. Parameters like substrate bias voltage, oxygen and argon gas flow mixture, sputtering power, sputtering pressure, the deposition time and rate portray important role to fabricate good quality thin films [9]. NiO has numerous applications such as oxygen evolution, p–n junction electrodes in dye-sensitized solar cells, organic light-emitting diodes, solar thermal absorber, photo-electrolysis, anode in oxygen fuel cells, lightweight aerospace component, gas or temperature sensors, electrochromic coating, and catalyst. Due to its magnetic nature, it is also used as antiferromagnetic layers [10, 11].

In this present work, NiO thin film is deposited by DC sputtering with optimized parameters. Deposited conductive thin film is semi-transparent and graphite black in color. It becomes transparent after annealing at 500 °C for 2 h. Deposition conditions and annealing are some of the influential factors for surface morphology and structural properties of thin-film deposition. Further study is done by different characterizations such as optical properties carried out by UV–Vis spectroscopy, and surface morphology by scanning electron microscopy (SEM), electrical and magnetic behavior is studied via four-probe method.

## 2 Experimental Section

Thin films of nickel oxide were grown on glass substrate using NiO target(99.99% purity) via DC sputtering deposition technique at ambient temperature and under 4 bar base pressure and 2 bar argon pressure at 100 W power for a time span of 15 min with subsequent post-annealing at 500 °C for 2 h. The surface morphology of the as deposited films was analyzed using scanning electron microscopy (Zeiss and Nova Nano SEM 450, FEI). The electrical and magnetic measurements were done using conventional four-probe method with (1 T) or without magnetic field. Absorbance spectra of these films were examined using UV–Vis spectrophotometer (Varian Cary 500 UV–vis-NIR spectrometer) (Fig. 1).

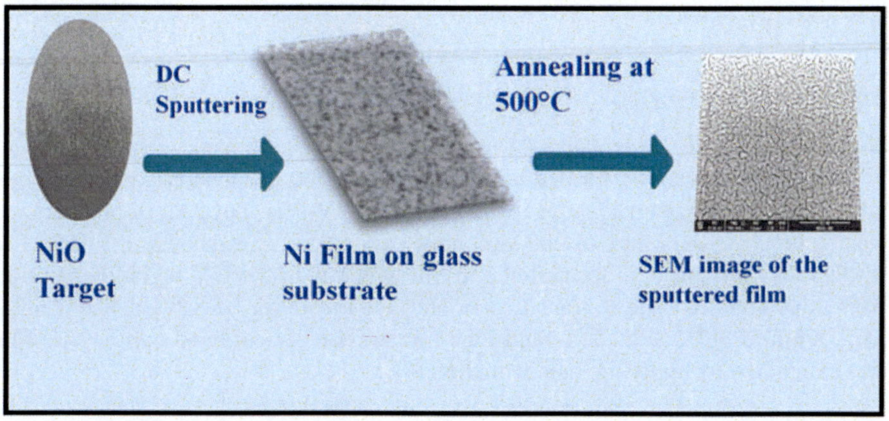


**Fig. 1** Schematic of NiO thin deposition by DC sputtering

# 3 Results

## *3.1 Morphological Studies*

The ultrahigh scanning electron microscopy revealed the homogeneity of as grown films on glass substrate. SEM micro-graphs showed the films were polycrystalline in nature and are of long-range order. It is evident from the images that grain boundaries are well-connected with no discontinuities. SEM images were taken at different resolutions (Fig. 2).

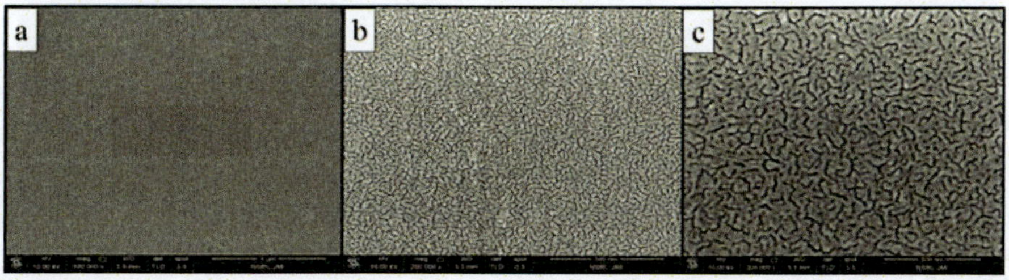


**Fig. 2** SEM micro-graphs (**a**), (**b**), (**c**) were taken at different resolutions of 1 μm, 500 nm, 300 nm, respectively

# 4 UV–Visible

## 4.1 Optical Studies

UV–Vis spectra were performed by a Varian Cary 500 UV–Vis-NIR spectrometer and energy band gap analyzed using Tauc's plot. UV–Vis spectrum of Ni sputtered was recorded in the range of 200–800 nm (6.20–1.55 eV) as shown in Fig. 3. The result shows that peak is located at 346 nm which is in the UV region because of band edge emission due to transition of $Ni^{2+}$ ions from top of $3d^8$ valence band to the bottom of $3d^8$ conduction band Fig. 4 shows the optical band gap of NiO film was calculated by applying Tauc's method [9].

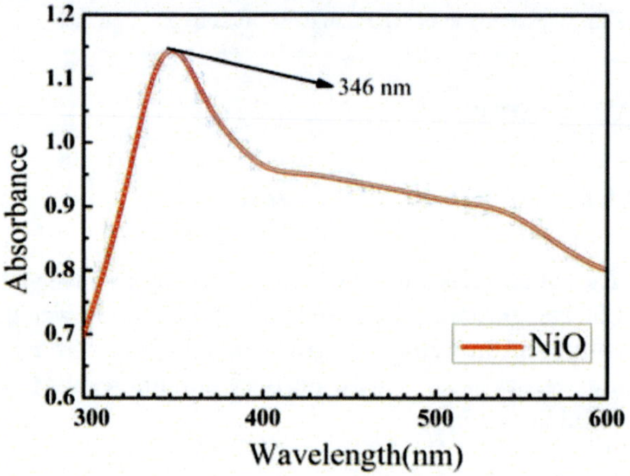


**Fig. 3** UV–Visible absorbance spectra of NiO thin film

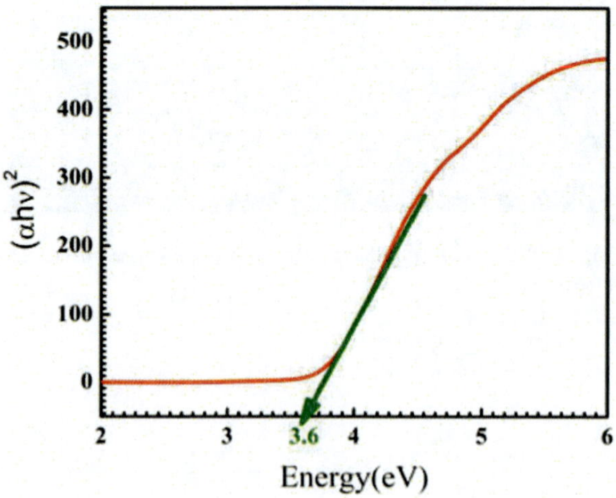


**Fig. 4** Tauc's plot for the calculation of direct band gap of NiO thin film

$$\alpha h\nu = A(h\nu - E_g)^n$$

where $A$ is energy-independent constant, $E_g$ is the band gap for a particular transition in the film, $\nu$ is the frequency of transition, and n defines the respective nature of the band transitions. $n = 1/4, 1/2, 2, 3/2, 3$ are defined for crystalline semiconductors, values corresponding to the allowed direct, allowed indirect, forbidden direct, and forbidden indirect transitions, respectively. Since NiO is a direct band gap system, the relationship between $(\alpha h\nu)^2$ and $h\nu$ is expected to show the linear behavior as shown below.

# 5 Electrical Measurements

## 5.1 I–V Characteristics

The NiO thin film (100–200 nm) was examined for its electrical conductivity using standard four-probe method at ambient temperature with (1 T) or without magnetic field. Conduction in nickel oxide film depends on the microstructure, composition, and deposition environment. I–V graphs showed a linear variation giving Ohmic behavior in both of the cases [12]. The resistance increased after application of magnetic field. NiO is ferromagnetic in nature (Figs. 5 and 6).

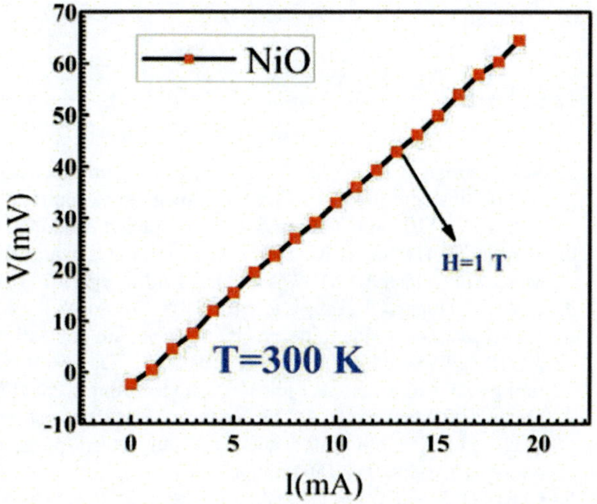


**Fig. 5** I–V curve obtained by four-probe method without magnetic field

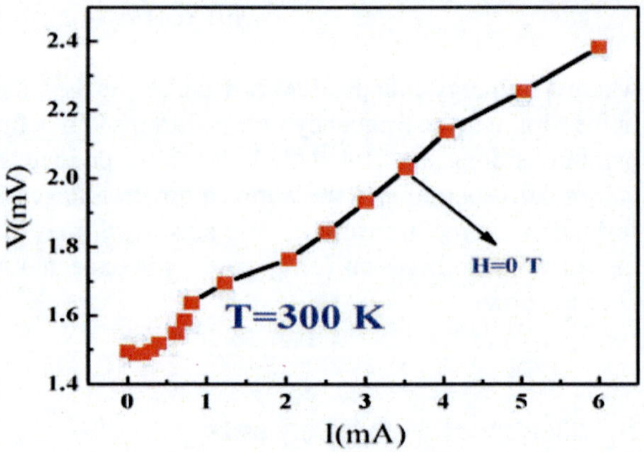


**Fig. 6** I–V curve obtained by four-probe method with 1 T magnetic field

## 6 Conclusion

It has been concluded that we successfully synthesized sputtered nickel oxide thin film on glass substrate and SEM revealed homogeneity of surface. Its optical and electrical properties were also examined. Absorbance peak was found to be at 346 nm, and band gap was calculated using Tauc's plot which comes out nearly 3.6 eV which is in agreement with the previous studies [13]. I–V characteristics of thin film with or without magnetic field showed linear behavior and resistance calculated comes in the range of 1–10 ohms. These films can be used in solar cell applications.

## References

1. Subramanian B et al (2008) Optoelectronic and electrochemical properties of nickel oxide (NiO) films deposited by DC reactive magnetron sputtering. Phys B Condens Matter 403(21–22):4104–4110. https://doi.org/10.1016/j.physb.2008.08.014
2. Murphy TP, Hutchins MG (1995) Oxidation states in nickel oxide electrochromism. Sol Energy Mater Sol Cells 39(2–4):377–389. https://doi.org/10.1016/0927-0248(96)80003-1
3. Diao CC, Huang CY, Yang CF, Wu CC (2020) Morphological, optical, and electrical properties of p-type nickel oxide thin films by non-vacuum deposition. Nanomaterials 10(4). https://doi.org/10.3390/nano10040636. Williams ACC, Davies HTO, Chadury Y (2000) Simple pain rating scales hide complex idiosyncratic meanings. Pain 85(3):457–463
4. Salunkhe P, Muhammed Ali AV, Kekuda D (2020) Investigation on tailoring physical properties of Nickel Oxide thin films grown by dc magnetron sputtering. Mater Res Express 7(1). https://doi.org/10.1088/2053-1591/ab69c5
5. Chen SC, Kuo TY, Lin HC, Chen RZ, Sun H (2020) Optoelectronic properties of p-type NiO films deposited by direct current magnetron sputtering versus high power impulse magnetron sputtering. Appl Surf Sci 508:145106. https://doi.org/10.1016/j.apsusc.2019.145106
6. Usha KS, Sivakumar R, Sanjeeviraja C, Sathe V, Ganesan V, Wang TY (2016) Improved electrochromic performance of a radio frequency magnetron sputtered NiO thin film with high optical switching speed. RSC Adv 6(83):79668–79680. https://doi.org/10.1039/c5ra27099e

7. Chouhan R, Gupta M, Sen P (2020) Nonlinear refraction in NiO thin films. In: 3rd International conference on condensed materials and applied physics 2220:090016. https://doi.org/10.1063/5.0001171
8. Predanocy M, Hotový I, Čaplovičová M (2017) Structural, optical and electrical properties of sputtered NiO thin films for gas detection. Appl Surf Sci 395:208–213. https://doi.org/10.1016/j.apsusc.2016.05.028
9. Hammad AH, Abdel-wahab MS, Vattamkandathil S, Ansari AR (2019) Influence the oxygen flow rate on the film thickness, structural, optical and photoluminescence behavior of DC sputtered NiOx thin films. Phys B Condens Matter 568(May):6–12. https://doi.org/10.1016/j.physb.2019.05.012
10. Reddy AM, Reddy AS, Lee KS, Reddy PS (2011) Growth and characterization of NiO thin films prepared by DC reactive magnetron sputtering. Solid State Sci 13(2):314–320. https://doi.org/10.1016/j.solidstatesciences.2010.11.019
11. Khalaf MK, Mutlak RH, Khudiar AI, Hial QG (2017) Influence of discharge voltage on the sensitivity of the resultant sputtered NiO thin films toward hydrogen gas. Phys. B Condens. Matter 514:78–84. https://doi.org/10.1016/j.physb.2017.03.037
12. Sushmitha V, Maragatham V, Deepak Raj P, Sridharan M (2018) Structural, electrical, optical and magnetic properties of NiO/ZnO thin films. IOP Conf Ser Mater Sci Eng 310(1). https://doi.org/10.1088/1757-899X/310/1/012022
13. Hotovỳ I, Huran J, Spiess L, Liday J, Sitter H, Hašcík Š (2002) The influence of process parameters and annealing temperature on the physical properties of sputtered NiO thin films. Vacuum 69(1–3):237–242. https://doi.org/10.1016/S0042-207X(02)00338-X

# A Review on Sources of Uncertainty in Thermal Conductivity Measurement for Thermal Transport Metrology

**Ashish Kumar, D. K. Misra, Sahiba Bano, Bal Govind, and Komal Bhatt**

**Abstract** Thermoelectric (TE) is the potential solid state-based technology that can effectively convert waste heat into electricity. The accurate determination of thermoelectric parameters is necessary to measure the true, accurate and precise determination of the thermoelectric figure of merit (ZT) to measure the thermoelectric performance of materials. In this article, we have highlighted the sources of uncertainty occurring in the measurement of thermal conductivity that includes the combined errors in measurement of density, thermal diffusivity and heat capacity.

**Keywords** Thermal diffusivity · Laser flash · Heat capacity · Density

## 1 Introduction

Thermoelectricity has gained a lot of attraction in the last few decades due to its effective utilization of heat wasted from various sources for converting into useful electricity [1]. Thermoelectric performance of a material can be gauged by the thermoelectric figure of merit, $ZT = \frac{s^2 \sigma T}{k}$, where $s$, $\sigma$, $k$, $T$ are Seebeck coefficient, electrical conductivity, thermal conductivity and temperature respectively [1–3]. For high thermoelectric performance, high ZT is required. In the last decade, many publications have shown an increased ZT value above 1 [1]. However, as a thermoelectric figure of merit ZT cannot be measured directly and evaluated by using the three parameters, i.e. Seebeck coefficient, electrical conductivity and thermal conductivity [1–3], and because of these, uncertainties up to 15% in overall ZT is noticed in most of the materials [4]. Therefore, it is difficult to find out the actual value of ZT in general. Thus, the development of a device depends upon accurate and precise measurement through measurement tools and methods. To establish a technology, calibration of equipment using standard reference is essential, and the calibrated equipment provides authentic and reliable data for the actual realization

---

A. Kumar (✉) · D. K. Misra · S. Bano · B. Govind · K. Bhatt
CSIR-National Physical Laboratory, Dr. K. S. Krishnan Marg, New Delhi 110012, India
e-mail: ashishjangra47@gmail.com

Academy of Scientific and Innovative Research (AcSIR), Ghaziabad 201002, India

 
S. Yadav et al. (eds.), *Recent Advances in Metrology*, Lecture Notes in Electrical Engineering 906, https://doi.org/10.1007/978-981-19-2468-2_16

of technology [4]. With the advent growth of computational and instrumentation tools, accuracy, repeatability and reproducibility are the fundamental requirements for precise and accurate estimation of TE parameters of materials. In order to improve the accuracy of a particular experimental setup, first step is to find out the systematic errors in measurement and need of reference standards by analysing sources of errors in measurement [3]. However, due to limited reference standards, especially at high temperature, actual accuracy in measurement is not well established, particularly in India. Therefore, statistical uncertainty in individual parameter may lead to combined large uncertainty in ZT for thermoelectric performance evaluation [4]. Thus, it is mandatory to establish metrology for TE technology for the industrialization of TE applications.

Few researchers have found scattered values of ZT, when measurements were conducted on the same materials in different laboratories [5]. This is primarily because of the lack of measurement standards. In addition to this, before conducting the thermoelectric measurement, their density and homogeneity need to be checked [3]. The measured ZT deviates when the experimental density of samples is less than 90% of the theoretical value. Therefore, the good quality of samples is also required for the accurate determination of ZT [3].

Currently, researchers in the field of thermoelectric are focussing on increasing the ZT value via the reduction in thermal conductivity. However, direct measurement of thermal conductivity is not possible and is evaluated by using the values of thermal diffusivity ($\alpha$), heat capacity ($C_\mathrm{p}$) and density ($d$), which is given as [6, 7]

$$K = \alpha C_\mathrm{p} d \tag{1}$$

Therefore, accurate measurement of these sub-parameters is also required for finding the true ZT value. Here, in this article, we are discussing the sources of uncertainties for thermal diffusivity, heat capacity and density measurement, which occurs during the estimation of thermal conductivity.

## 2 Estimation of Thermal Conductivity

Thermal conductivity is estimated by multiplying values of thermal diffusivity, heat capacity and density.

### 2.1 *Laser Flash Apparatus Setup for Measuring Thermal Diffusivity*

The laser flash apparatus has emerged as a popular method for measuring thermal diffusivity of materials to calculate the thermoelectric figure of merit. The main advantage of the flash method is its high reliability, accurate results, simplicity in measurement and its application in a wide range of conditions [8]. The flash method is used to measure the thermal conductivity of both thin films and bulk samples [9]. In this method, laser flash irradiates heat on one side of the sample and temperature rise is observed by infrared (IR) detector on the other side of the sample as shown in Fig. 1a [9–11]. Figure 1b shows an example of temperature rise vs time profile which is used to evaluate the thermal diffusivity [12]. The flash method follows general assumptions which are as follows [8]

(i) The materials are homogeneous, and their density does not show variation with temperature.
(ii) The materials are thermally insulated, and there is no loss of heat from the surface of the slab.
(iii) The laser heat pulse is evenly dispersed over the surface of the slab and is absorbed by a thin layer of material.
(iv) The laser heat pulse is instantaneous, and its duration is negligible in comparison with the slab's thermal response.

Thermal diffusivity can be calculated from the expression in terms of sample thickness ($L$) and half time ($t_{0.5}$), i.e. the time in which the temperature of the rear side reaches its half of the maximum value, as [8]

$$\alpha = 0.1388\frac{L^2}{t_{0.5}} \tag{2}$$

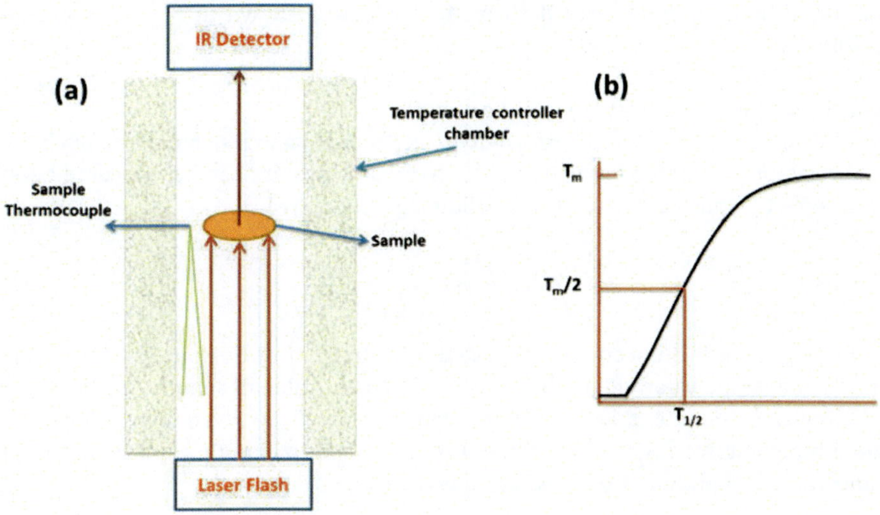


**Fig. 1** **a** Schematic of laser flash diffusivity principle **b** an example of temperature rise curve obtained at the rear face when the laser is irradiated on the front side of the sample

## 2.2 Types of Uncertainty in Thermal Diffusivity Measurement by Laser Flash Apparatus

Every measurement is associated with the errors in the measurement process which lead to the deviation in the measurement value of an estimated property from its true value. These errors in the measurement can come from several components. These components can be primarily grouped into two categories, i.e. Type A and Type B, according to the GUM method [8].

**Type A Component of Uncertainty**
Type A uncertainties are calculated from the statistical analysis of a series of measurements, i.e. by evaluating the standard deviation of series of measurements by the measurement method [8].

**Type B component of uncertainty**
A Type B uncertainty is estimated based on the scientific judgement made by considering relevant information such as a record of previous measurements, the experience of handling particular materials and instruments and uncertainty provided in the calibration [8].

## 2.3 Uncertainty Components

There are various uncertainty components associated with the measurement of thermal diffusivity by a laser flash method which is shown in Fig. 2. These are as follows

**The uncertainty associated with the sample**
According to Eq. (2), thermal diffusivity is dependent on the square of the thickness of the sample. Therefore, uncertainty in thermal diffusivity can be measured based on sample thickness as given by the following relation [8]

$$\frac{\Delta \alpha}{\alpha} = 2\frac{\Delta L}{L}.$$

For reducing this uncertainty, repeated measurement of thickness must be done and their average value should be taken as the actual value of sample thickness.

Moreover, sample thickness $L$ at temperature $T$ is given by thickness at room temperature ($L_{\mathrm{O}}$) and the correction term due to sample expansion at high temperature. Therefore, the thickness is given by

$$L = L_{\mathrm{O}} + \Delta_{\mathrm{L}}.$$

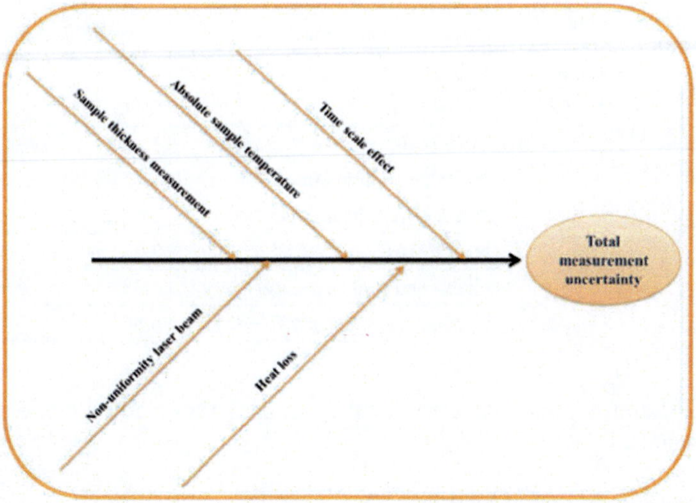


**Fig. 2** Schematic diagram showing uncertainty parameters from different possible sources for thermal diffusivity measurement

Therefore, the uncertainty in thickness arises due to uncertainty in measurement thickness and uncertainty due to expansion of sample at higher temperature [13].

$$u_C^2(L) = u^2(L + u^2(\Delta L) + 2u(+u^2(\Delta L) + 2u)$$

**The uncertainty associated with temperature**
During thermal diffusivity measurement of samples, the absolute temperature of samples is not taken into account. This makes a discrepancy in the measurement, and as a result, uncertainty occurs because of not knowing the actual temperature of samples. Uncertainty in temperature comes from uncertainty due to repeatability of measurements, calibration, resolution, drift of thermometer and emissivity, respectively [13], which is described as

$$u^2(T) = u_r^2 + u^2(C) + u^2(C_R) + u^2(C_d) + u^2(C_\epsilon)$$

**The uncertainty associated with the time-scale effect**
While recording temperature rise vs time profile, it is important to carefully check the starting point when the time starts from the origin [8, 13].

**Uncertainty due to non-uniform heating**
Laser beam uniformity tells about the uniformity in sample heating, and it depends upon the energy level of the laser. However, practically, there is always a deviation from monochromatic laser beam though it is quite minor and below the limit of

**Table 1** Uncertainty budget for thermal diffusivity measurement of Pyroceram 9606 at 25 °C estimated by Migliorini et al. [13]

| Uncertainty components | Standard uncertainty (%) |
|---|---|
| Repeatability | 0.95 |
| Sample thickness | 0.11 |
| Time scale | 1.66 |
| Finite pulse time effect | 1.45 |
| Non-uniform heat effect | 3.0 |
| Heat losses | 2.0 |
| Combined standard uncertainty | 4.33 |
| Expanded uncertainty ($k = 2$) | 8.66 |

detection. Therefore, one has to follow the standard ISO method while taking the measurement.

**Uncertainty due to loss of heat**
Loss of heat cannot be avoided during the laser flash experiment. Although heat losses have less impact on the accuracy of results at low temperatures, however, they can have an impact on the high-temperature experiment. Heat losses contribution can be given in the form of Biot number [13] which is given by $B_i = \frac{U.L_C}{k} < 0,1$, where $U$ = heat transfer coefficient, $L_C$ = characteristic length and $k$ = thermal conductivity. Thus, this coefficient of Biot number must be considered when the measurement is performed or if it is included in the software. It must be double-checked before starting the measurement. The measurement uncertainty for thermal diffusivity measurement of Pyroceram 9606 at 25 °C estimated by Migliorini et al. [13] from different sources is presented in Table 1.

## 2.4 *Error in Heat Capacity (Cp)*

For determining thermal conductivity, heat capacity is also needed to evaluate carefully. Although heat capacity can be estimated by flash systems, however, a large discrepancy in the result is often observed by the same [3, 4, 14, 15]. Therefore, differential scanning calorimetry (DSC) is often used for heat capacity measurement due to its accurate results where thermocouple standardization is of utmost importance.

## 2.5 *Errors in Density Measurement*

Density is also an important parameter required to estimate the thermal conductivity value. Density is measured based on Archimedes' principle by measuring the weight in air and weight in liquid [3]. Therefore, the weighing using the balance must be calibrated with the primary standard. However, density measured may lead to being

**Table 2** Uncertainty in the thermal conductivity parameters for the given temperature ranges for different materials

| S. No. | Parameter | Material | Uncertainty (%) | Temperature range (K) | References |
|---|---|---|---|---|---|
| 1 | Thermal diffusivity | Gadolinium | 3–4 | 287–1277 | [19] |
| 2 | Thermal diffusivity | Sandstone | 3 | 302.9–774.3 | [7] |
| 3 | Specific heat capacity | Sandstone | 1 | 308–768 | [7] |
| 4 | Thermal diffusivity | Isotropic graphite | 3–6 | 300–1200 | [20] |

overestimated if the liquid is being absorbed in the pores [3]. Therefore, the weight of materials should be measured before and after measuring weight in liquid, and the balance as suggested must be calibrated with the mass standard.

### 2.6 *Recent Research in the Thermal Conductivity Metrology*

Some recent researches in the field of thermal conductivity metrology are summarized in Table 2. Akoshima et al. [16] developed certified reference materials using isotropic graphite named NMIJ RM 1401a by measuring thermal diffusivity from laser flash apparatus, specific heat capacity from DSC and estimating density from sizes and masses. Also, they have investigated the thermal diffusivity for pyrolytic graphite from room temperature to 1000 K [17]. Recently, Breuer et al. [18] improved the thermal diffusivity measurements by including a detector that gives inherent delayed response in the laser flash method. Samoshkin et al. [19] presented the thermal diffusivity data of gadolinium in the temperature range of 287–1277 K with an error of 3–4%. Abdulagatov et al. [7] also investigated the thermal diffusivity and specific heat capacity of sandstone at high temperature, and uncertainties are found to be 3% and 1%, respectively. For better clarity, we have summarized the uncertainty evaluated by different groups as shown in Table 2.

## 3 Conclusion

This article has stressed the attention on the importance of thermal conductivity metrology to establish thermoelectric technology. The details of the laser flash apparatus principle have been discussed for the measurement of thermal diffusivity and hence the thermal conductivity. The number of sources of uncertainty occurring in the thermal conductivity measurement has also been highlighted in this review article for accurate and precise measurement for a better understanding of general readership.

**Acknowledgements This work is dedicated to our late great mentor, Dr. D. K. Misra, who has passed away recently.** One of the authors AK highly acknowledges DST-Inspire SRF Fellowship (Grant no.-IF180005) for financial support.

## References

1. Snyder GJ, Toberer ES (2008) Complex thermoelectric materials. Nature Mater 7:105–114
2. Sootsman JR, Chung DY, Kanatzidis MG (2009) New and old concepts in thermoelectric materials. Angew Chemie—Int Ed
3. Borup KA, De Boor J, Wang H et al (2015) Measuring thermoelectric transport properties of materials. Energy Environ Sci 8:423–435
4. Wei T-R, Guan M, Yu J et al (2018) How to measure thermoelectric properties reliably. Joule 2:2183–2188
5. Wang H, Porter WD, Böttner H et al (2013) Transport properties of bulk thermoelectrics: an international round-robin study, part II: thermal diffusivity, specific heat, and thermal conductivity. J Electron Mater 42:1073–1084
6. Shinzato K, Baba T (2001) A laser flash apparatus for thermal diffusivity and specific heat capacity measurements. J Therm Anal Calorim
7. Abdulagatov IM, Abdulagatova ZZ, Kallaev SN et al (2015) Thermal-diffusivity and heat-capacity measurements of sandstone at high temperatures using laser flash and DSC methods. Int J Thermophys. https://doi.org/10.1007/s10765-014-1829-4
8. Vozár L, Hohenauer W (2005) Uncertainty of thermal diffusivity measurements using the laser flash method. Int J Thermophys. https://doi.org/10.1007/s10765-005-8604-5
9. Zhao D, Qian X, Gu X et al (2016) Measurement techniques for thermal conductivity and interfacial thermal conductance of bulk and thin film materials. J Electron Packag Trans ASME. doi 10(1115/1):4034605
10. Neumann W, Wallisch K (1983) Determination of the thermal conductivity of graphite and high-temperature alloys by the laser-flash method. Measurement. https://doi.org/10.1016/0263-2241(83)90008-8
11. Parker WJ, Jenkins RJ, Butler CP, Abbott GL (1961) Flash method of determining thermal diffusivity, heat capacity, and thermal conductivity. J Appl Phys. https://doi.org/10.1063/1.1728417
12. Altun O, Boke YE, Kalemtas A (2008) Problems for determining the thermal conductivity of TBCs by laser-flash method. J Achiev Mater Manuf Eng
13. Migliorini FL, Silva EHC, Grossi PA, et al (2009) Calculated uncertainty of the thermal diffusivity measurement based on flash laser method. In: 19th IMEKO World congress 2009
14. May AF, Toberer ES, Saramat A, Snyder GJ (2009) Characterization and analysis of thermoelectric transport in n-type Ba 8 Ga 16–x Ge 30+x. Phys Rev B 80:125205
15. Toberer ES, Christensen M, Iversen BB, Snyder GJ (2008) High temperature thermoelectric efficiency in Ba 8 Ga 16 Ge 30. Phys Rev B 77:75203
16. Akoshima M, Abe H, Baba T (2015) Development of reference materials for thermal-diffusivity measurements by the flash method. Int J Thermophys 36:3272–3287
17. Akoshima M, Abe H, Baba T (2015) Thermal diffusivity of carbon materials as candidate reference materials. Int J Thermophys 36:2507–2517
18. Breuer S, Schilling FR (2019) Improving thermal diffusivity measurements by including detector inherent delayed response in laser flash method. Int J Thermophys 40:1–17

19. Samoshkin DA, Agazhanov AS, Savchenko IV, Stankus SV (2017) Thermal diffusivity of gadolinium in the temperature range of 287–1277 K. High Temp 55:221–225
20. Akoshima M, Baba T (2006) Study on a thermal-diffusivity standard for laser flash method measurements. Int J Thermophys 27:1189–1203

# Error Analysis in Measurement of Electrical Conductivity

**Sahiba Bano, Ashish Kumar, Bal Govind, Komal, and D. K. Misra**

**Abstract** Technology development primarily needs accurate and precise measurement of any parameter with stated uncertainty. Thermoelectric technology depends upon the performance of material which is gauged by figure of merit (ZT). The ZT comprises three interrelating parameters, namely Seebeck coefficient ($\alpha$), electrical conductivity ($\sigma$) and thermal conductivity ($\kappa$). Out of these three parameters, this review article focuses on only electrical conductivity and errors associated with its measurement. Factors causing uncertainty in the electrical conductivity measurement and solutions to minimize them are also discussed for general readership.

**Keywords** Electrical conductivity · Four-probe technique · And sources of errors

## 1 Introduction

The performance of TE material is governed by dimensionless parameter, figure of merit, $ZT = \frac{\alpha^2\sigma}{\kappa}T$, where $\alpha$, $\sigma$ and $\kappa$ are Seebeck coefficient, electrical conductivity and thermal conductivity respectively at absolute temperature $T$ [1]. Researchers have been focusing in improving the efficiency of material through different approaches [2–7]. Significant improvement in ZT will definitely make significant contribution to commercialization of thermoelectric technology; however, the measurement accuracy is also a critical parameter to be addressed properly. For instance, the large electrical contact in hall measurement and electrical conductivity may lead to inaccurate measurement of transport properties [8, 9]. The different geometries such as two-probe and four-probe techniques are being used for Seebeck coefficient that results discrepancy in the data obtained from these techniques [10, 11]. In context

S. Bano (✉) · A. Kumar · B. Govind · Komal · D. K. Misra
CSIR-National Physical Laboratory, Dr. K.S. Krishnan Marg, New Delhi 110012, India
e-mail: nplsahiba16@gmail.com

D. K. Misra
e-mail: misradk@nplindia.org

Academy of Scientific & Innovative Research (AcSIR), CSIR-NPL Campus, New Delhi 110012, India

 
S. Yadav et al. (eds.), *Recent Advances in Metrology*, Lecture Notes in Electrical Engineering 906, https://doi.org/10.1007/978-981-19-2468-2_17

of thermoelectric technology, accurate and precise measurement of all parameters needs to be established as they are relating to the electronic and lattice structure modifications. Here in this article, we have discussed the common measurement technique for electrical conductivity and pointed out the relevant errors associated in such measurements that can be beneficial to researcher to identify them at its early stage of measurement. A detail discussion on the factors that causes uncertainty in the measurement of electrical conductivity and also their solutions for minimizing of errors is the prime focus of this article.

## 2 Electrical Conductivity

Electrical conductivity ($\sigma$) measures the electrical current density (**J**) by Ohm's law **J = E**$\boldsymbol{\sigma}$ when electric field (**E**) is applied to the conductor. The schematic diagram for measuring electrical conductivity is presented in Fig. 1. When current (**I**) is applied to the uniform conductor of width x and thickness y, electric field is produced around it. The material's resistance (**R**) can then be determined by the potential difference (**ΔV**) generated between the two point of conductor and applied current (**I**) using relation **R** = **ΔV/I**. Resistance **R** depends on the geometric size and shape of the material such as $R = \frac{1}{\sigma} \times \frac{L}{x \times y}$, where $L$ is the distance between the probes. Hence, electrical conductivity can be determined using relation [12]:

$$\sigma = \frac{1}{R} \times \frac{L}{x \times y} \tag{1}$$

$\sigma$ is independent on electric field while it depends on temperature.

The electrical conductivity can be evaluated using Eq. (1) for homogenous sample. In most of the equipment, the electrical conductivity is determined simultaneously with thermopower which needs thermal gradient. Following these conditions, the voltage difference is given by [12]:

$$\Delta \boldsymbol{V} = \boldsymbol{R} \times \boldsymbol{I} + \boldsymbol{\alpha_{ab}} \Delta \boldsymbol{T} = \frac{1}{\boldsymbol{\sigma}} \times \frac{1}{\boldsymbol{x} \times \boldsymbol{y}} \boldsymbol{I} + \boldsymbol{\alpha_{ab}} \Delta \boldsymbol{T} \tag{2}$$

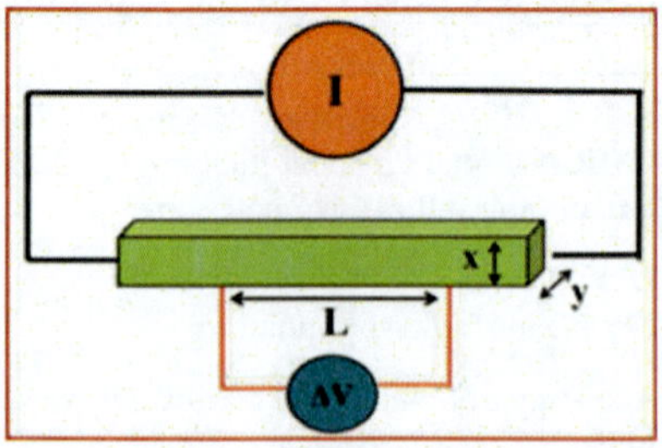


**Fig. 1** Schematic diagram of general method for the measurement of electrical conductivity

where $\boldsymbol{\alpha_{ab}}$ = relative thermoelectric power of two conductors $a$ and $b$ and $\Delta T$ = temperature difference. Basically, two methods are being used to remove the effect of thermal gradient on electrical conductivity measurement which are as follows:

(a) **DC measurement method**: In this method, two voltage measurements, one with current flow (**ΔV**) and another either without current flow or current flow in opposite direction ($\Delta \boldsymbol{V'} = -\boldsymbol{R} \times \boldsymbol{I} + \boldsymbol{\alpha_{ab}} \Delta \boldsymbol{T}$), are required to perform the measurement on the sample. On subtracting these measured value, real voltage drop across the resistor can be obtained as

$$\Delta \boldsymbol{V_r} = \Delta \boldsymbol{V} - \Delta \boldsymbol{V'} = (\boldsymbol{R} \times \boldsymbol{I} + \boldsymbol{\alpha_{ab}} \Delta \boldsymbol{T}) - (-\boldsymbol{R} \times \boldsymbol{I} + \boldsymbol{\alpha_{ab}} \Delta \boldsymbol{T}) \quad (3)$$

$$= 2\boldsymbol{R} \times \boldsymbol{I} \quad (4)$$

(b) **AC measurement method**: In this method, frequency of order from 10 to 1000 Hz with AC current is applied. Here the thermopower contribution is not possessing frequency-dependent part, and thus signal is directly proportional to R which can be determined at AC current frequency.

# 3 Measurement Techniques Used for Different Shape of the Materials

## *3.1 For Thin and Uniform Wire*

The schematic diagram for measurement of electrical conductivity of long thin and uniform wire of diameter $D$ is depicted in Fig. 2. The electrical conductivity can be determined by formula [12]:

$$\sigma = \frac{I}{\Delta V_{ab}} \times \frac{4L}{\pi D^2} \quad (5)$$

The medium, where the sample is placed, should have high thermal conductivity with non-conductive nature so that it can absorb heat generated in the sample which

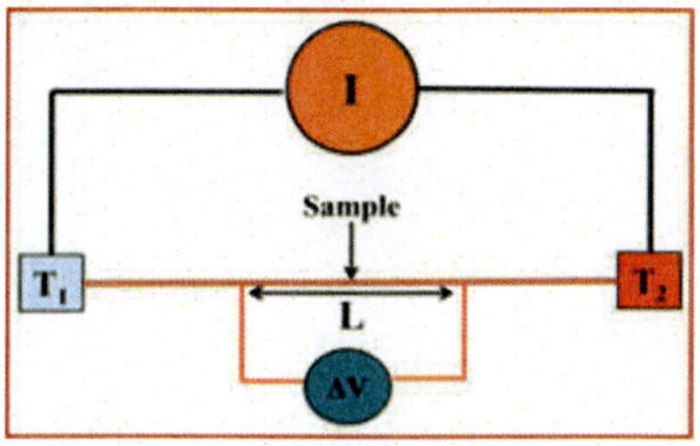


**Fig. 2** Electric circuit diagram for electrical conductivity of thin or uniform wire

resulted minimized thermal gradient. The low-current density is used for minimizing of Joule heating, and one can maintain the large $\Delta V_{ab}$ by taking the large distance $L$ between the probes. However, there are certain issues with this measurement that includes primarily (i) the measurement of long wire sample which is difficult to measure in the limited volume chamber and secondly (ii) the sample preparation in the form of thin homogenous wire which is also a tedious job and is quite difficult or sometimes impossible to prepare it.

## 3.2 Measurement of Regular Geometric Shape Sample

This method is generally used for short samples with simple geometry which allows precise and accurate measurement of potential difference and current density. The current contact is formed in such a way that the uniform current can be distributed all over the sample. In addition to this, temperature should not be varied and has to be set so that it can be determined precisely and accurately. The $\sigma$ of the sample of area A which has same value in both directions can be calculated by [12]

$$\sigma = \frac{1}{\Delta V} \times \frac{L}{A} \tag{6}$$

## 3.3 Measurement of Regular Geometric Shape Sample in Bar Shaped as Further Modification (Four-Probe Technique)

In this technique, four probes (two for voltage and two for current) are placed on the flat surface of the bar-shaped sample as shown in Fig. 3. The current is applied from one end of the sample to other end, and the voltage is measured at two points across

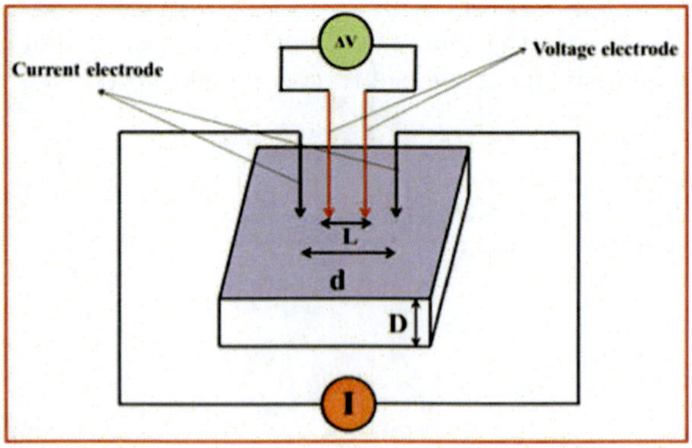


**Fig. 3** Diagram of four-probe technique for electrical conductivity measurement

the sample. When one considers the approximation, the electrodes must be placed symmetrically and distance of electrodes from the edge of sample should be much greater than the distance between the electrodes. The electrical conductivity can thus be evaluated by an expression:

$$\sigma = \frac{2I}{\pi \times \Delta V} \times \left[ \frac{1}{d - L} - \frac{1}{d + L} \right] \tag{7}$$

where $d$ = distance between current electrodes and $L$ = distance between voltages probes.

To remove the uncertainty in $L$, the voltage contacts should be narrow. This technique is considered to be the most sensitive to the geometry.

***Errors associated with four-probe technique***

The errors occurred in this technique is primarily due to inaccurate determination of distance between the probes ($L$) and sample thickness ($s$). The errors in determining $L \sim \Delta L/L$ are $\geq 1\%$, and this value is same for the case of $s$ ($\Delta s/s$). Therefore, accuracy limit for four-probe technique can be of at least 2%. The measurement of length must be measured with a scale which is calibrated with the primary standard.

## *3.4 Measurement of Irregular Geometric Shaped Sample (Vander Pauw Method)*

This method determines the electrical conductivity of sample of irregular shape [13]. The four point contacts $P$, $Q$, $R$ and $S$ are made along its circumference as shown in Fig. 4. In this method, the geometry factors causing error are less as compared to four-probe technique, because uniformity in thickness can be achieved by polishing the flat samples.

***Error associated with this technique***

In this measurement technique, errors are primarily due to non-ideal contacts which is associated with finite size and offset from the edge of the sample.

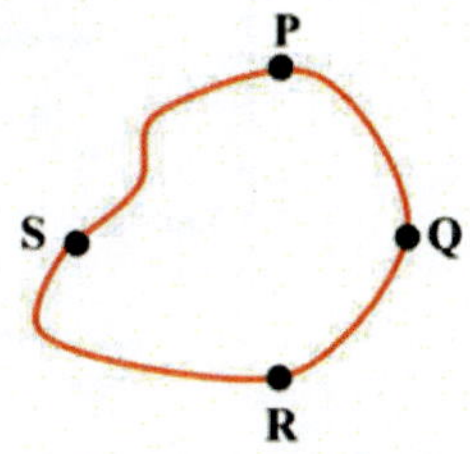


**Fig. 4** Arbitrary shape of sample for electrical conductivity measurement by Vander Pauw method

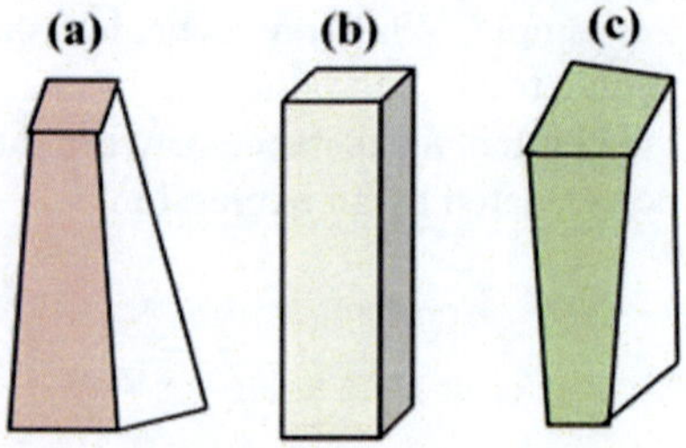


**Fig. 5** Samples of different geometry (centre is ideal one and the outer two samples shown exaggerated draft angles) [14]

## 4 Precautions Needed for Sample Preparation

Precautions for sample preparation are of utmost important and for which a representative geometry of sample is presented in Fig. 5. The sample of different geometry as depicted in Fig. 5 [14] consists of outer two samples displaying draft angles which lead to a variation of cross-sectional area along the sample length. The sample should have proper parallel faces and uniform cross-sectional area like a shape of middle sample of Fig. 5. The variation in width should be less than 0.1 mm for sample of length 1 cm. Therefore, sample preparation should be performed with utmost care to avoid the errors.

## 5 Factors Causing Uncertainty in Electrical Conductivity Measurement

The errors in electrical conductivity primarily arise due to three factors which are illustrated in Fig. 6:

(a) **Errors associated with electrical signal measurement such as voltage and current magnitude**

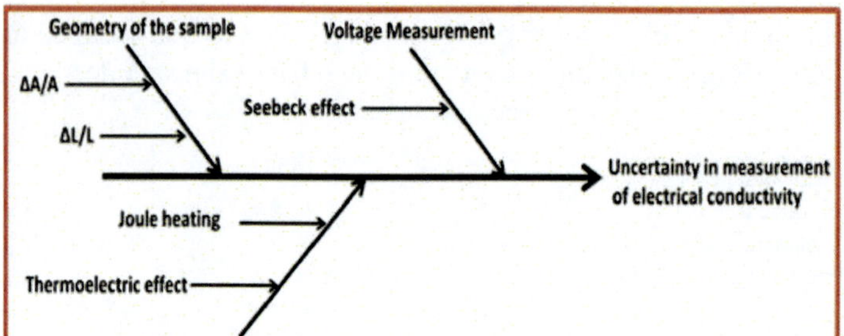


**Fig. 6** Fishbone diagram showing factors responsible for uncertainty in electrical conductivity measurement

The variation in the voltage offset measurement is generated due to Seebeck effect which causes the temperature gradient. This effect can be generated in wire, samples and also at the junction.

***Error minimization***: This error can be reduced by taking voltage as a function of applied current despite under constant flow of current.

(b) **Errors associated with shape or geometry of the specimen**

The electrical conductivity is highly sensitive to the geometry of the sample; therefore, the errors associated with the shape are basic errors in the measurement. This error arises due to imperfection of sample ($\Delta A$) or inaccurate measurement of distance ($\Delta L$), and therefore, the total error in geometric factor is equal to

$$\frac{\Delta A}{A} + \frac{\Delta L}{L} \tag{8}$$

***Error minimization***: This error can be minimized by taking care of manufacturing of sample. The dimensions should be measured precisely. Moreover to reduce uncertainty in $L$, one can take narrow voltage contact.

(c) **Errors associated with temperature measurement**

The change of temperature can be noticed due to two effects such as (i) Joule heating and (ii) thermoelectric effects [15]; which cause error in temperature measurement. The Joule heating contributes to $I^2R$, and hence, one can minimize it by taking measurement at small current ($I$). The difficult task here is the minimization of errors caused by thermoelectric effects. Heat is transferred from one end to another end through current which results in temperature gradient. This effect will cause generation of additional voltage known as Seebeck voltage [16] and hence results to overestimated value of resistivity [17, 18].

***Error minimization***: This can be reduced by using proper design of specimen holder which can make good thermal contact between specimen and environment. This special design can provide equal heat distribution on all over the sample. In such a case, environment of gas of high thermal conductivity is required to pump the heat distribution uniformly. Secondly, measurement of voltage should be performed immediately before and after switching on the current.

The estimated uncertainty value arising from different sources in Seebeck coefficient measurement is appended in Table 1 [14].

## 6 Conclusion

Different techniques for measurement of electrical conductivity have been discussed in detail for quantifying the accurate and precise value. Various factors causing errors

**Table 1** Seebeck coefficient measurement

| S. No. | Source | Typical value |
|---|---|---|
| 1 | Thermocouple tip radius | 0.25 mm |
| 2 | Thermocouple separation length | ±0.1 mm |
| 3 | Sample uniformity | ±0.1 mm/cm |
| 4 | Caliper resolution | ±0.01 mm |
| 5 | Statistical variation | Experimentally calculated |
| 6 | Wire discrepancy | Experimentally calculated |
| 7 | DAQ voltage uncertainty | 50 ppm + 1.2 μV |
| 8 | DAQ current uncertainty | 0.2% + 0.3 mA |

in different techniques are also highlighted to carefully perform the measurement prior to assign the property value of the materials. It is hoped that this review will support researchers and young scientists who have been working in the field of thermoelectric technology.

## References

1. Snyder GJ, Toberer ES (2008) Complex thermoelectric materials. Nat Mater 7:105–114. https://doi.org/10.1038/nmat2090
2. Duan S, Cui Y, Chen X, Yi W, Liu Y, Liu X (2019) Ultrahigh thermoelectric performance realized in black phosphorus system by favorable band engineering through group VA doping. Adv Funct Mater 29:1904346
3. Gao P, Lu X, Berkun I, Schmidt RD, Case ED, Hogan TP (2014) Reduced lattice thermal conductivity in Bi-doped Mg2Si0.4Sn0.6. Appl Phys Lett 105:2–7. https://doi.org/10.1063/1.4901178
4. Fu C, Bai S, Liu Y, Tang Y, Chen L, Zhao X et al (2015) Realizing high figure of merit in heavy-band p-type half-Heusler thermoelectric materials. Nat Commun 6:8144
5. Zhang Q, He J, Zhu TJ, Zhang SN, Zhao XB, Tritt TM (2008) High figures of merit and natural nanostructures in Mg 2 Si 0.4 Sn 0.6 based thermoelectric materials. Appl Phys Lett 93:102109
6. Guan M, Zhao K, Qiu P, Ren D, Shi X, Chen L (2019) Enhanced thermoelectric performance of quaternary $Cu_{2-2x}Ag_{2x}Se_{1-x}S_x$ Liquid-like Chalcogenides. ACS Appl Mater Interfaces
7. Du Z, He J, Chen X, Yan M, Zhu J, Liu Y (2019) Point defect engineering in thermoelectric study of InSb. Intermetallics 112:106528
8. Harman TC, Spears DL, Walsh MP (1999) PbTe/Te superlattice structures with enhanced thermoelectric figures of merit. J Electron Mater 28:L1-5
9. Harman TC, Taylor PJ, Spears DL, Walsh MP (2000) Thermoelectric quantum-dot superlattices with high ZT. J Electron Mater 29:L1-2
10. Martin J (2013) Protocols for the high temperature measurement of the Seebeck coefficient in thermoelectric materials. Meas Sci Technol 24. https://doi.org/10.1088/0957-0233/24/8/085601
11. Martin J, Wong-Ng W, Green ML (2015) Seebeck coefficient metrology: do contemporary protocols measure up? J Electron Mater 44:1998–2006. https://doi.org/10.1007/s11664-015-3640-9

12. Burkov AT, Fedotov AI, Novikov S V (2016) Methods and apparatus for measuring thermopower and electrical conductivity of thermoelectric materials at high temperatures. Thermoelectr Power Gener Look Trends Technol 353–389
13. Philips'Gloeilampenfabrieken O (1958) A method of measuring specific resistivity and Hall effect of discs of arbitrary shape. Philips Res Rep 13:1–9
14. Mackey J, Dynys F, Sehirlioglu A (2014) Uncertainty analysis for common Seebeck and electrical resistivity measurement systems. Rev Sci Instrum 85. https://doi.org/10.1063/1.4893652
15. Rowe DM. Thermoelectrics handbook: macro to nano. CRC press; 2018.
16. Bhandari CM, Rowe DM (1995) CRC handbook of thermoelectrics. CRC Press, Boca Raton, FL, p 49
17. Kirby CGM, Laubitz MJ (1973) The error due to the Peltier effect in direct-current measurements of resistance. Metrologia 9:103
18. Cheremisin MV (2001) Peltier-effect-induced correction to ohmic resistance. J Exp Theor Phys 92:357–360

# Errors in Measurement of Magnetic Field and Magnetic Moment with its Associated Uncertainty

**Bal Govind, D. K. Misra, S. P. Khanna, Ashish Kumar, and Sahiba Bano**

**Abstract** Evaluation of precise and accurate value of magnetic moments of materials is difficult without knowing the sources of errors associated with the measurements. Uncertainty in measurements usually occurs due to the methods adopted for measurement of a particular parameter. Since, magnetic moments are sensitive to the strength of applied magnetic field. Therefore, accurate value of magnetic field at any observation point for the actual realization of magnetic technology is essential. In this article, we have stressed on importance of magnetic measurement and associated uncertainty with it in a magnetic field as well as magnetic moment of the materials. In addition to this, the actual realization of the unit of magnetic flux density, calibration of equipment and magnetic characteristics of specified magnetic materials are also elaborated.

**Keywords** Realization of unit Tesla · Assessment and dissamitaion · Assigned value of magnetic moment · Gyromagnetic ratio

## 1 Introduction

Metrology in science, known as scientific metrology, is important for the development of new measurement standards, their maintenance and dissemination [1]. Improvements in the preciseness and accurateness of the measurements dictate the quality of any product. For example, accurate measurements of weight for making tablets of medicine may lead to prevent various side effects in the living body due to excess medicine added due to the major value of uncertainty. In magnetic measurement system, half-metallic ferromagnetic property governs due to one type of electron (spin up or down) present in the system which enables to provide integer value of spin magnetic moment. Slight uncertainty in magnetization measurements leads

B. Govind (✉) · D. K. Misra · S. P. Khanna · A. Kumar · S. Bano
CSIR-National Physical Laboratory, Dr. K.S. Krishnan Marg, New Delhi 110012, India
e-mail: bgovind32@gmail.com

B. Govind · S. Bano
AcSIR, Academy of Scientific & Innovative Research, Ghaziabad, Uttar Pradesh 201002, India

 
S. Yadav et al. (eds.), *Recent Advances in Metrology*, Lecture Notes in Electrical Engineering 906, https://doi.org/10.1007/978-981-19-2468-2_18

to produce integer/non-integer value of magnetic moment which losses the half-metallic property. Therefore, to make the spintronic devices for working standard, accurate and precise value of magnetization is necessary.

Magnetic induction (B) must be accurately specified for measuring the magnetic properties of the materials [2]. Precise measurement of magnetic field strength at any specific point leads to control the uncertainty in the results obtained for particular magnetic materials. Therefore, we have focused on measurement of ferromagnetic samples with the uncertainty evaluation. Material's selection for the calibration of equipment is crucial due to the specific magnetization losses which provides the prediction of loss of flux densities and wide range of magnetization frequencies as discussed in details.

In order to realize the magnetic technology, the measured value must be assigned with uncertainty value [3]. Lack of any precise measurements hinders the development of new technology. Finally, keeping all these facts in view, we have stressed attention on the importance of assigning property value of magnetic properties with its associated uncertainties. Further, in this article, various sources of errors in measurements particularly magnetic field and magnetic moment are highlighted for better understanding on the measurement issues. In addition to this, quantitative assessment of the position and spatial orientation of the magnetic moment vector has been specifically described.

## 2 Realization of Unit of Magnetic Flux Density

Unit of magnetic flux density (B), Tesla, was realized by determining of gyromagnetic ratio ($\boldsymbol{\gamma}$). Nuclear magnetic resonance (NMR) technique was used to determine the gyromagnetic ratio of pure water with precise value of frequency with an uncertainty better than $10^{-10}$. This was done at PTB Germany and a unit Tesla with this uncertainty is finally maintained as the primary standard. Magnetic flux density (B) is linearly dependent on precession frequency ($\omega$) of magnetic moment around magnetic flux density

$$B = \omega I \boldsymbol{\gamma} \tag{1}$$

The gyromagnetic coefficients ($\gamma$) are well known for several nuclei and is known as fundamental constant and evaluated for proton in pure water with small uncertainty of $1.3 \times 10^{-8}$. Precession frequency (ω) of the nuclei is measured accurately in the DC magnetic field. From this precession frequency, one measures the magnetic field strength precisely. When oscillator used for particular frequency, oscillates with same frequency as precession frequency of the given nuclei, and then absorption of energy takes place in general. The magnetometer based on the Hall-effect sensors is calibrated using NMR magnetometer. Field coils are usually calibrated using NMR magnetometer in the magnetic field range from 1 to 100 mT with the relative uncertainty of $10^{-4}$.

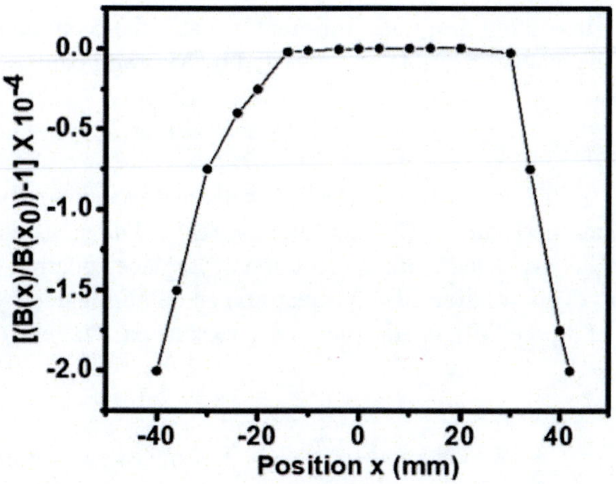


**Fig. 1** Variation of flux density with the position along the axis of pickup coils

Field coil is used to generate the magnetic stray field at the central point of the coil by applying the current with the standard voltage source [4, 5]. It is worth mentioning here that the homogeneity in magnetic field must be maintained around the center. Calibration of field coils as standard direct current (DC) and alternating current (AC) fields is primarily performed to produce magnetic field. Homogeneity of the field has been found around the center and the same is plotted by taking data from source [6] as shown in Fig. 1.

In such experiment, position is measured by an optical means and magnetic flux density using a fluxgate magnetometer. The temperature coefficient of the coil constant ($K_C$) is determined by following relation

$$\frac{K_c(I_c)}{K_c(I_c = 0)} - 1 = -(84.0 \pm 2).10^{-6}.\left(\frac{I_c}{A}\right) \tag{2}$$

where $I_c$ and $A$ are the current and the cross section of the wire, respectively. Magnetic field is assumed to be resulting from the collection of large number of magnetic dipoles/multipoles which leads to produce field lines with strong enough intensity. Although, the theory of magnetic monopole is being developed but has not been established in reality. Therefore, uncertainty evaluation in the magnetic dipole moment can directly provide the uncertainty in the magnetic field at an specific point, i.e., point of measurements in space. This will give the information about structure and spatial configuration of the field lines. There are several reports [7] which describes the distribution of magnetic field in space, importance of development of measurement system to increase the accuracy of the instruments [8], an alternative method and disadvantage of the model of measuring system. Implementation of measuring system is significantly simplified by using inductive sensors as primary measuring transducers with the point methods system [9].

**Measuring magnetic moment's method** The static magnetic field induced due to the magnetic dipole as suggested by Maxwell equation is given as [9]

$$\nabla \times H = 0 \text{ or } H = -\nabla U \tag{3}$$

Mathematical function ($U$) is defined in such a way that non-swirling field ($H$) is produced due to U. This function ($U$) is known as magnetic potential and value of $U$ is evaluated from the solution of Laplace equation ($\nabla^2 U = 0$).

This solution of this equation can be eliminated by considering the contribution of dipole [10], quadruple..., etc. expansion. Therefore, the solution can be obtained as:

$$U = \frac{1}{4\pi}\sum_{i=1}^{\infty}\frac{1}{r^{i+1}}\sum_{j=0}^{i}(g_{ij}\cos j\varphi + h_{ij}\sin j\varphi).P_i^j \cos\theta \tag{4}$$

where $r$, $\theta, \varphi$ is the spherical coordinates at the point of observation, $g_{ij}$, $h_{ij}$ is the coefficients of zonal, axial and sectoral harmonics of the multipole magnetic moments, $i$ is the order of spatial harmonics of the Gaussian series, j is the order of harmonic of elementary multipole, $P_i^j \cos\theta$ is the Legendre's polynomials. Partial differentiation of U with respect to the spherical coordinates gives the component of electric field in the direction of their respective coordinates. For example

$$H_\theta = -\frac{1}{r}\frac{\partial U}{\partial\theta} = \frac{1}{4\pi}\sum_{i=1}^{\infty}\frac{1}{r^{i+2}}\sum_{j=0}^{i}(g_{ij}\cos j\varphi + h_{ij}\sin j\varphi).\frac{\partial P_i^j\cos\theta}{\partial\theta} \tag{5}$$

Signal ($V_n$) is measured from the induction sensor by assuming the equatorial plane ($\theta = 0$) with $n = (2n\text{-}1)45°$ and $m = (2\,m\text{-}1)45°$, $n = \ldots 4$, $m = 5\ldots 8$. Therefore, the resulting signal produced by the magnetic fled intensity is given by

$$\begin{aligned} Vz = {} & \frac{8}{m_f r^3} g_{10} - -\frac{4}{m_f}\sum_{i=5}^{\infty}\frac{1-(-1)^i}{r^{i+2}}\frac{(r_2/r_1)^{3-i}-1}{(r_2^2-r_1^2)-1} \\ & .\sum_{j=1}^{i}(g_{ij}\cos j45° + \sin j45°).\frac{\partial P_i^j\cos\theta(\theta = 90°)}{\partial\theta} \end{aligned} \tag{6}$$

From this Eq. (6), axial magnetic moment can be simplified as

$$Vz = \frac{8}{m_f r^3} g_{10} \tag{7}$$

Therefore,

$$M = g_{10} = \frac{V_z m_f r^3}{8} \tag{8}$$

This is the model equation for measuring the axial component of magnetic moment.

For the estimation of the total uncertainty of the axial magnetic moment, we can expand in the view of Taylor's series

$$\begin{aligned} Uc(Mz) = \Big[ & a^2\big(\overline{V}_{ind}\big)u^2\big(\overline{V}_{ind}\big) + a^2\big(\hat{r}_{ind}\big)u^2\big(\hat{r}_{ind}\big) \\ & + a^2\big(\hat{m}_f\big)u^2\big(\hat{m}_f\big) + \frac{1}{2}\left[\left(\frac{\partial_2 M_z}{\partial V^2}\right)^2 + \left(\frac{\partial_2 M_z}{\partial r^2}\right)^2 + \left(\frac{\partial_2 M_z}{\partial m_f^2}\right)^2\right]\Big]^{1/2} \end{aligned} \tag{9}$$

where $a_l = \frac{\partial f}{\partial y_l}$, $l = v, r, m_f$, $\overline{V}_{ind} = \frac{1}{d}\sum_{n=1}^{d} \hat{V}_{ind,d}$, $\overline{r}_{ind} = \frac{1}{d}\sum_{n=1}^{d} \hat{r}_{ind,d}$, $\overline{m}_f = \frac{1}{d}\sum_{n=1}^{d} \hat{m}_{f,d}$, $d$- no. of observations, $u$- related uncertainty.

Now, axial magnetic moment is given by

$$M = \frac{1}{p}\sum_{p=1}^{i} f\big(V_{1,p}, r_{1,p}\big)$$

Coverage ratio $p = t_{0.95}.(D_{\text{eff}})$, $D_{\text{eff}}$-effective degree of freedom.

Finally, the extended uncertainty is given by

$$U = p.U_c.\widehat{M} \tag{11}$$

## 3 Measurements of Magnetic Moment

Vibrating sample magnetometer has become commonly accepted technique for the measurement of saturation magnetization, remanence magnetization and coercivity mostly for ferromagnetic/ferrimagnetic materials. Voltage generated in the pickup coil due the magnetized sample in the presence of external magnetic field is given by [11] the relation

$$V_C = K V_S M_S \tag{12}$$

This shows that the proportional relationship between the voltages ($V_C$) induced in the coil and magnetization (Ms) of the sample with proportionality factor $K$, $V_S$ is the volume of the sample. Uncertainty in the magnetization of the sample can be determined by the evaluation of uncertainty in the volume and voltage of the coil. Therefore, the combined uncertainty is given by:

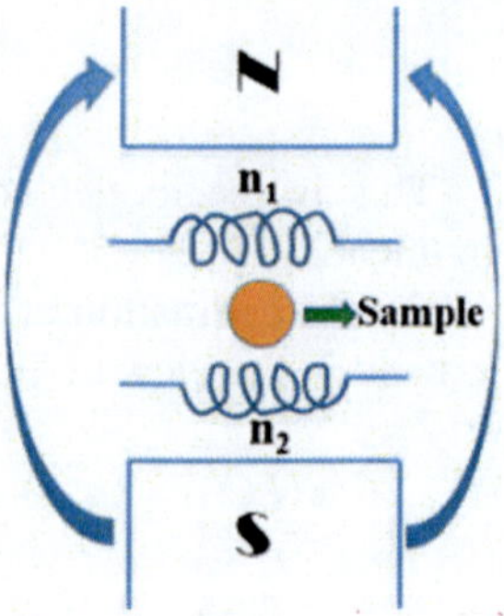


**Fig. 2** Schematic diagram of setup of magnetic measurements showing that sample between the magnetic dipole, pickup coils for measurement of the sample's magnetization

$$U_{Ms} = \sqrt{U_{Vs}^2 + U_{Vc}^2} \tag{13}$$

where $U_{Ms}$ is the uncertainty in the magnetization due to the volume uncertainty in the sample and voltage uncertainty in the pickup coil. Measurement of mass is more suitable rather than volume of the sample. Unknown conventional mass would be then concluded in equation which is as follows

$$M_y = M_s + \delta m_d + \delta m + \delta m_c + \delta a \tag{14}$$

where $M_s$—conventional mass of the standard, $\delta m_d$—Drift value of standards since its last calibration, $\delta m$—Observed difference between unknown and standard mass, $\delta m_c$—Correction for magnetic effect and eccentricity, $\delta a$—Air buoyancy correction (Fig. 2).

In order to identify the specific magnetization losses, compensation term is used. The compensation loss term is given by the relation

$$C_l = \frac{\left(\frac{n_1}{n_2}\right)(V_1 V_2 R_s cos\varphi/m)(4l/l_m)}{\beta\left(\frac{f}{f_p}\right)^2 (ff/1.1107)^2 + (1-\beta)\left(\frac{f}{f_p}\right)} \cdot \left(\frac{f n_2 m \hat{J}_p\left(1 + \tau\mu_0\hat{H}/\hat{J}A_t/A\right)}{V_2 l\rho}\right)^{\delta} - (T - T_q)\delta_\tau \tag{15}$$

where $n_1$ and $n_2$ is the number of turns in the primary and secondary coils, $V_1$ and $V_2$ are voltages in the primary and secondary coils, shunt resistance $R_s$, specimen mass m, length $l$, cross section $A$, density $\rho$, effective area of secondary winding $A_t$, mean magnetic path length lm, frequency $f$, temperature $T$, temperature coefficient $\delta_\tau$, magnetic polarization $\hat{J}_p$, magnetic field strength $\hat{H}$, $\varphi$ is the phase between primary and secondary windings, $\tau$ is the error of air flux compensation,$\beta$ is a proportionality constant introduced due to the loss separation. Here, specific losses strongly influenced with the polarization, $\delta$ is used for the correction in the deviation of resulting specific losses. The value of $\delta$ is given as

$$\delta = \frac{\partial P_s}{\partial \hat{J}} \frac{\hat{J}}{P_s} \tag{16}$$

For each measured polarization, both are identified because of the fact that performed loss calculation would lead to rising the model error.

Dispersion from the mean of observed quantities is characterized by the type, while mathematical model considering influence factors into measuring system results to the evaluation of type B uncertainty.

Combined standard uncertainty can be calculated according to type B evaluation from Eq. 15, which is as follows:

$$Uc2(Cl) = \sum_{i=1}^{N} \left( \frac{\partial f}{\partial y_i} \right)^2 U2(yi) \tag{17}$$

Linear approximation will be valid with assumption that derivation is small around the desirable measuring point for the evaluation of combined standard uncertainty. It is assumed to be the normal distribution of the repeated measurements. During the materials characterization, different parameters influence the results and uncertainty propagates through all the steps. Details of sources of parameters are taken from the Ishikawa Diagram [12] as shown Fig. 3.

From the above diagram, one can clearly see that various sources of errors such as repeated measurements, materials type, and method adopted, personnel who are

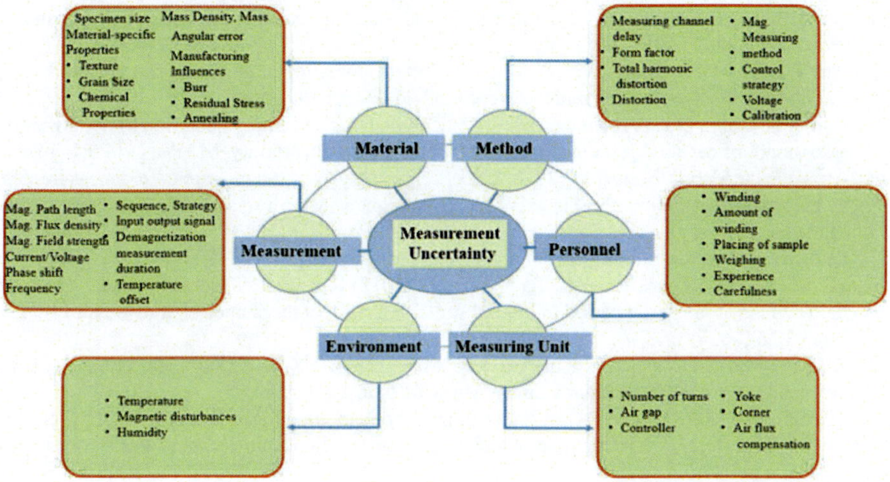


**Fig. 3** Diagram of influencing parameters during the materials characterization (Ishikawa Diagram [12])

performing the experiments, measuring unit and environment contributes to the errors in the measured value. Therefore, while reporting any property value, measurements may include carefully uncertainty based associated results from the different sources for reliable and accurate data presentation. Accurate and reliable data will support as a backbone for developing the new technology.

## 4 Conclusion

We have summarized about the necessity of realization of unit of magnetic field, range of homogenous field in the magnetic measurement system and uncertainty determination in the evaluation of magnetic moment and magnetic field at a point from the point method for the actual development of spintronic devices. Various factors causing errors in magnetic field and magnetic moment during measurement are addressed and highlighted its importance prior to assign these values. In addition to this, importance of metrology for these parameters are also discussed to establish the spintronic technology and its assessment and dissemination.

**Acknowledgements** This work is dedicated to our late great mentor, Dr. D. K. Misra who has passed away recently and has been corresponding author in this work. One of the authors Bal Govind highly acknowledges CSIR-SRF fellowship (Budget head-P-90807) for financial support.

## References

1. Aswal DK (2020) Quality infrastructure of India and its importance for inclusive national growth. Mapan 35:139–150
2. Park PG, Kim W-S, Khorev VN, Shifrin VY (2009) Generation of uniform magnetic field using a single-layer solenoid with multi-current method. Mapan 24:9–14
3. Kumar A, Misra DK. A review on the statistical methods and implementation to homogeneity assessment of certified reference materials in relation to uncertainty. MAPAN 2020:1–14.
4. Arpaia P, De Vito L, Kazazi M (2016) Uncertainty analysis of the magnetic field measurement by the translating coil method in axisymmetric magnets. Metrologia 53:1317
5. Matsubara R, Takahashi Y, Fujiwara K, Ishihara Y, Azuma D (2018) Distribution of magnetic field strength inside exciting coil of single sheet tester. AIP Adv 8:47209
6. Ma B, Huang Z, Guan Z, Zu X, Jia X, Xiao Q (2018) Research of the axial strong magnetic field applied at the initial period of inertial stretching stage of the shaped charge jet. Int J Impact Eng 113:54–60
7. Zakharov IP, Vodotyka SV, Klimova KA, Shevchenko NS (2013) Some examples of the evaluation of measurement uncertainty. Meas Tech 56:591–598
8. Degtiarov O, Alrawashdeh R (2017) Development of a method for measurements of the parameters of the external magnetic field of technical means. Технологический Аудит и Резервы Производства 6
9. Degtiarov OV, Alrawashdeh RS (2019) Development of the theoretical basis of magnetic measurement uncertainty evaluation. In: 2019 IEEE International conference on advanced optoelectronics and lasers, IEEE; 2019, pp 671–674

10. Lupikov VS (2008) Magnetic moment as a function of the parameters of the source of the magnetic field for various types of electrical equipment. Scientific Works of National Technical University "Kharkiv Polytechnic Institute", No. 25, pp 67–80
11. Case WE, Harrington RD (1966) Calibration of vibrating-sample magnetometers. Res NBS (Engr Znstr) C 70:255–262
12. Elfgen S, Franck D, Hameyer K (2018) Characterisation of soft magnetic materials by measurement: Evaluation of uncertainties up to 1.8 T and 9 kHz. AIP Adv 8:47208

# Generation of Tunable Perfect Vortex Beam by Varying Focus of Axicon-Lens Assembly

**Swati Gangwar, Rajeev Dwivedi, Shibu Saha, Ranjana Mehrotra, V. K. Jaiswal, and Parag Sharma**

**Abstract** Vortex beam possess spiral phase distribution with singularity at the center. Perfect vortex (PV) beam, which is a special kind of vortex field, has garnered attention due to its distinctive spatial properties and finds application in laser machining, optical communication, and trapping. In practice, a PV beam can be obtained from a vortex beam by passing the later through an axicon-lens assembly. However, tuning the characteristics of the PV beam is important for their application. In this article, a theoretical study supported by experimental data is presented in which the impact of tuning the focus of the axicon-lens assembly for controlling the radius of the PV beam has been analyzed. The analysis warrants the controlled tunability of radius of the PV beam ring.

**Keywords** Orbital angular momentum · Diffraction · Vortex beam · Singularity

## 1 Introduction

A vortex beam has orbital angular momentum (OAM) [1], which arises due to the twisted wavefront, defined by a helical phase $\exp(-il\phi)$ [2, 3]. The number of twisting helices corresponds to the topological charge ($l$), i.e., each photon possesses OAM equal to $l\hbar$, while $\phi$ denotes the azimuthal angle [4]. Theoretically, $l$ can attain infinite values, enabling a single photon to carry unlimited information while performing optical communication [5]. Typically, a traditional vortex beam, like Laguerre–Gaussian (LG) or Bessel–Gaussian (BG) beams, has a radial dependence

S. Gangwar · R. Dwivedi · S. Saha
Academy of Scientific and Innovative Research (AcSIR), Kamla Nehru Nagar, Ghaziabad, UP 201002, India

S. Gangwar · S. Saha · R. Mehrotra · V. K. Jaiswal · P. Sharma (✉)
CSIR-National Physical Laboratory, Dr. K. S. Krishnan Marg, New Delhi 110012, India
e-mail: sharmap2@nplindia.org

*Present Address:*
R. Dwivedi
Laboratoire Hubert Curien, Université Jean Monnet, Saint-Étienne, France

 
S. Yadav et al. (eds.), *Recent Advances in Metrology*, Lecture Notes in Electrical Engineering 906, https://doi.org/10.1007/978-981-19-2468-2_19

upon the value of the topological charge [6, 7], i.e., the ring diameter strongly depends on the topological charge. This introduces difficulties in coupling multiple OAM beams into a single fiber for multiple data transmission and also restricts the particle trapping ability of higher-order vortex beams [5, 8].

The difficulty of multiple OAM beam coupling was resolved with the introduction of the perfect optical vortices as its radial profile remains independent of topological charge [9]. The experimental generation of perfect vortex (PV) beam was first reported in 2013 by Ostrovsky et al. [9]. The unique properties of PV beams warrant its application in medical science [10], particle trapping [11], and optical communication [4, 12, 13]. As the PV beam is used in these diverse fields, tunability of its ring radius is very important as it increases the span of the application.

One of the popular methods for generation of PV beams is the axicon-lens assembly [14]. The most convenient and cost-effective way of achieving a controlled variation in the PV beam radius would be the variation of the focal length of the converging lens. In the present manuscript, we report the study carried out to understand the impact of variations of the focal length, of converging lens in the axicon-lens assembly, on the generation of perfect vortex beam.

## 2 Theoretical Study of PV Beams

When a Gaussian beam is passed through a vortex phase plate Laguerre—Gaussian (LG) beam is generated which when passed through the axicon-lens assembly produces the PV beam. PV beam has been mathematically simulated from a Gaussian beam through application of the functions of the optical components as mentioned above.

The amplitude of the LG beam on the incident plane of axicon, (i.e., at $z = 0$; $z$ being the propagation distance) can be defined as

$$E_0^{|l|}(\rho, \theta, 0) = \sqrt{\frac{2}{\pi |l\ |!}}(1/\omega_0)\left(\frac{\sqrt{2}\rho}{\omega_0}\right)^{|l|} \exp\left(\frac{-\rho^2}{\omega_0^2}\right) \exp(-il\theta) \tag{1}$$

where $\omega_0$ is the beam waist of the Gaussian beam. $(\rho, \theta)$ are the cylindrical coordinates. The intensity and phase plots of the $E_0^{|l|}(\rho, \theta, 0)$ for different values of $l$ are shown in Fig. 1. As expected, the radius of the LG beam increases with topological charge. Moreover, the number of $2\pi$ shifts in the phase plots also corresponds to the topological charge of the beam.

The passing of the LG beam through an axicon is mathematically simulated by solving the Fresnel diffraction integral to attain the electric field distribution at a propagation distance z. Therefore, in the focal depth region of the axicon, we obtained the amplitude distribution of the BG beam as

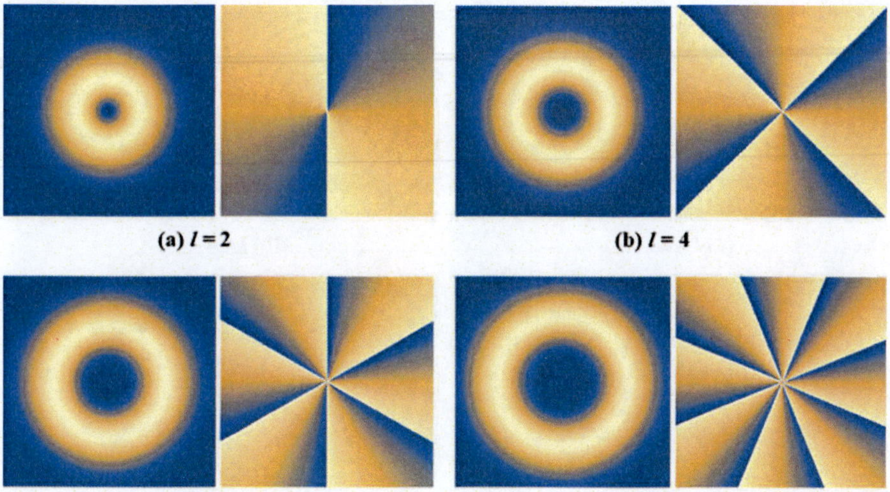

(a) $l = 2$ (b) $l = 4$ (c) $l = 6$ (d) $l = 8$

**Fig. 1** Intensity and phase profiles of the LG beam at the incident plane of the axicon for **a** $l = 2$, **b** $l = 4$, **c** $l = 6$, and **d** $l = 8$

$$E(\rho', \theta') = C_1 J_l\left(\frac{k_r r_p \rho'}{z}\right) \exp(-il\theta') \exp\left(\frac{-ik\rho'^2}{2z}\right) \tag{2}$$

where $C_1$ is a constant, $J_l$ refers to the first-order Bessel function of order $l$, and k is the propagation constant. Figure 2 shows the intensity and phase plots of the BG beam, having different values of topological charge, at a propagation distance of $z =$ 150 mm from the axicon. From Fig. 2, it can be seen that the radius of the BG beam also increases with the order of the beam. This changing radius of the LG and BG beams with topological charge limits their practical applications.

The perfect vortex beam is obtained at the focal plane of the converging lens through the Fourier transform of the BG beam. The Fourier transformation of the Eq. 2 yields the solution defined as

$$E(r, \phi) = C_2 \frac{1}{f'} \exp(-il\phi) \exp\left(-\frac{r^2 + r_1^2}{\omega_0^2}\right) I_l\left(\frac{2rr_1}{\omega_0^2}\right) \tag{3}$$

The approximation $\omega_0 \ll r_1$ leads to $I_l\left(\frac{2rr_1}{\omega_0^2}\right) \approx \exp\left(\frac{2rr_1}{\omega_0^2}\right)$. Hence, the Eq. 3 becomes [15]

$$E(r, \phi) = C_2 \frac{1}{f'} \exp(-il\phi) \exp\left(-\frac{(r - r_1)^2}{\omega_0^2}\right) \tag{4}$$

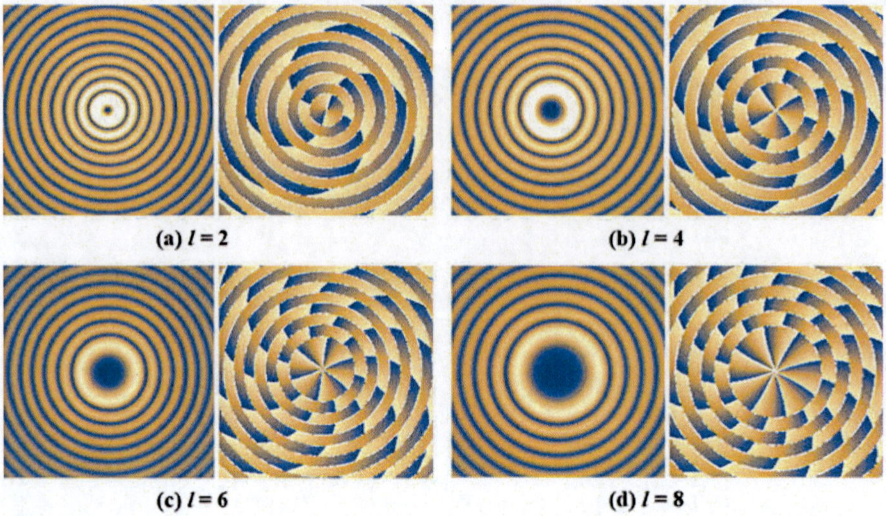


**Fig. 2** Intensity and phase profiles of the BG beam for **a** $l = 2$, **b** $l = 4$, **c** $l = 6$, and **d** $l = 8$ at $z =$ 150 mm

where $r_1$ is the radius of PV beam, $\omega_0$ is the ring width of the PV beam, and $C_2$ is constant. The radius of the PV beam is thus defined as

$$r_1 = \frac{k_r r_p f'}{k_1 z} \tag{5}$$

where $k_r$ is the radial wave vector, $k_1$ is the wave vector of the incident beam, z is the distance between axicon and lens, $f'$ is the focal length of the lens, and $r_p$ is the radial distance, which is defined by

$$r_p = -\frac{(n-1)\beta}{\left(\frac{1}{q(0)} + \frac{1}{z}\right)}$$

Here, $n$ is the refractive index of an axicon, $\beta$ is the angle of axicon, and $q(0)$ is the complex beam parameter. It can be seen from Eq. 5 that the radius of the PV beam, $r_1$, is independent of the topological charge and the radius varies linearly with the focal length $f'$.

The intensity of the PV beam at the focal plane of the axicon-lens assembly is now defined as

$$I = |E(r, \phi)|^2 = |C_2|^2 \frac{1}{f'^2} \exp\left(-\frac{2(r - r_1)^2}{\omega_0^2}\right) \tag{6}$$

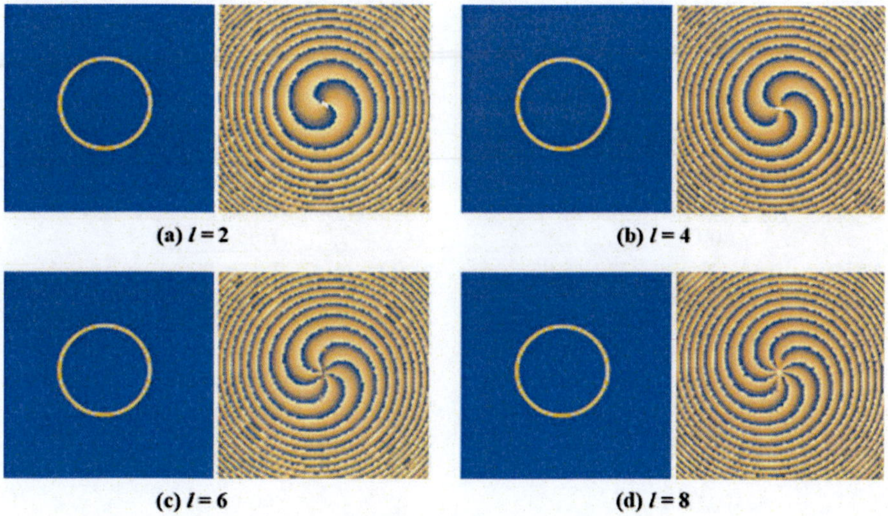


**Fig. 3** Intensity and phase profiles of the PV beam for **a** $l = 2$, **b** $l = 4$, **c** $l = 6$, and **d** $l = 8$

The intensity and phase plots of PV beam for different values of $l$, as calculated from Eq. 6, are shown in Fig. 3. It can be observed that the radial profile of the PV beam remains invariant of the topological charge, while the phase profile indicates the topological charge through the increasing number of spirals.

## 3 Results

PV beams were theoretically simulated for different values of topological charges and focal length of the converging lens as explained in the previous section. The intensity patterns and phase patterns of PV beam obtained for different focal lengths ($f' = 15$, 20, 25, and 30 cm) at a fixed topological charge of $l = 2$ are shown in Fig. 4(i), while that for topological charge of $l = 6$ are shown in Fig. 4(ii). It can be inferred that the radius of the PV beam is strongly dependent on the focal length of the converging lens but remains independent of the topological charge.

The radial intensity pattern for the PV beam rings (shown in Fig. 5) is plotted for a closer analysis. The two peaks in figure represents the two intensities obtained as one traverses across the PV rings along a diameter. Hence, the distance between the two peaks is indicative of the diameter of the PV ring. Each graph in Fig. 5 contains respective patterns for topological charge of $l = 2$, 4, 6, and 8. It can be observed that the graphs show no effect of any variation in the topological.

Figure 5 shows the theoretically and experimentally obtained radial intensity patterns are in close correspondence with each other. Further, it may be observed from Fig. 5 that the separation between the peaks increases with the focal length of

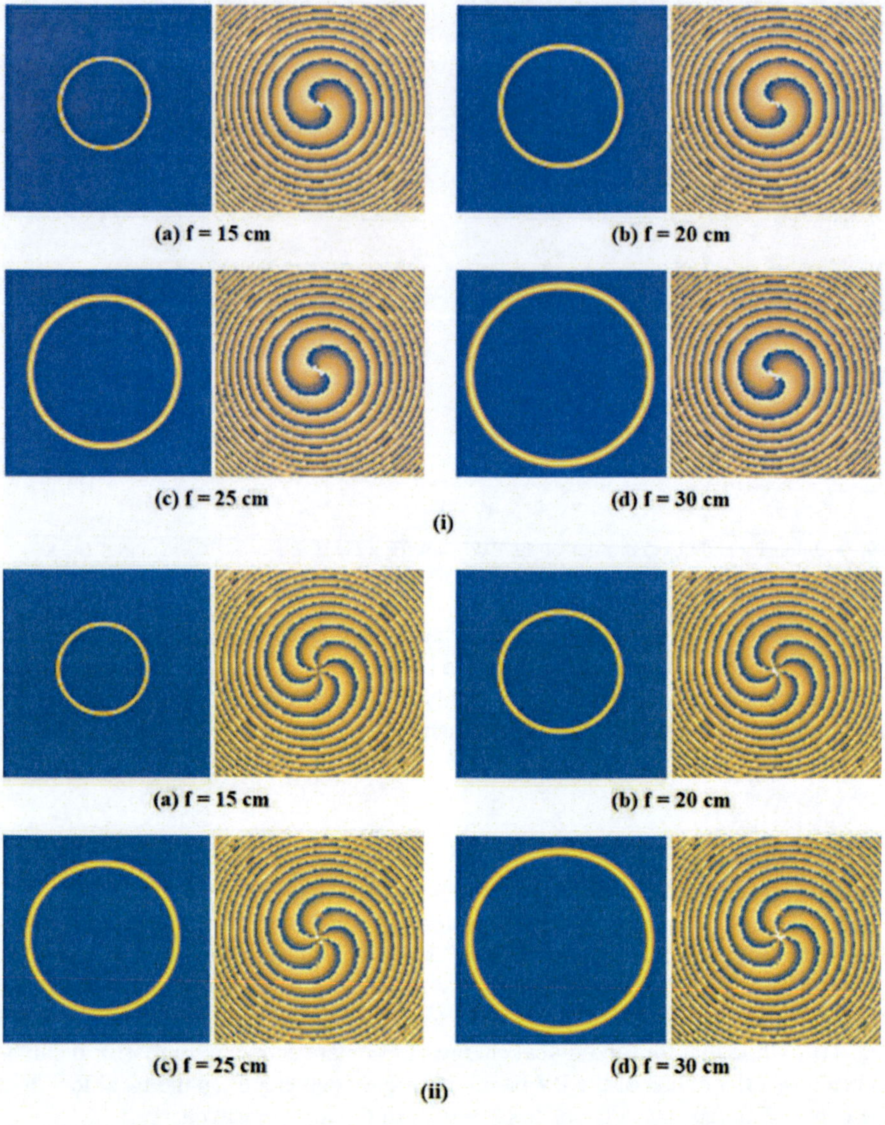


**Fig. 4** Intensity and phase profiles of PV beam at (i) $l = 2$ and (ii) $l = 6$ for different focal length of converging lens

the converging lens in the axicon-lens assembly. It may be noted that the scales of all the *x*-axis in Fig. 5 are to the same scale. Thus, the radius of the PV beam can be tuned by changing the focal length irrespective of the topological charge.

The radius of the PV beam (theoretically simulated and experimentally obtained) is plotted against focal length of the converging lens, for different values of topological charge is shown in Fig. 6. The theoretical and line experimental data closely

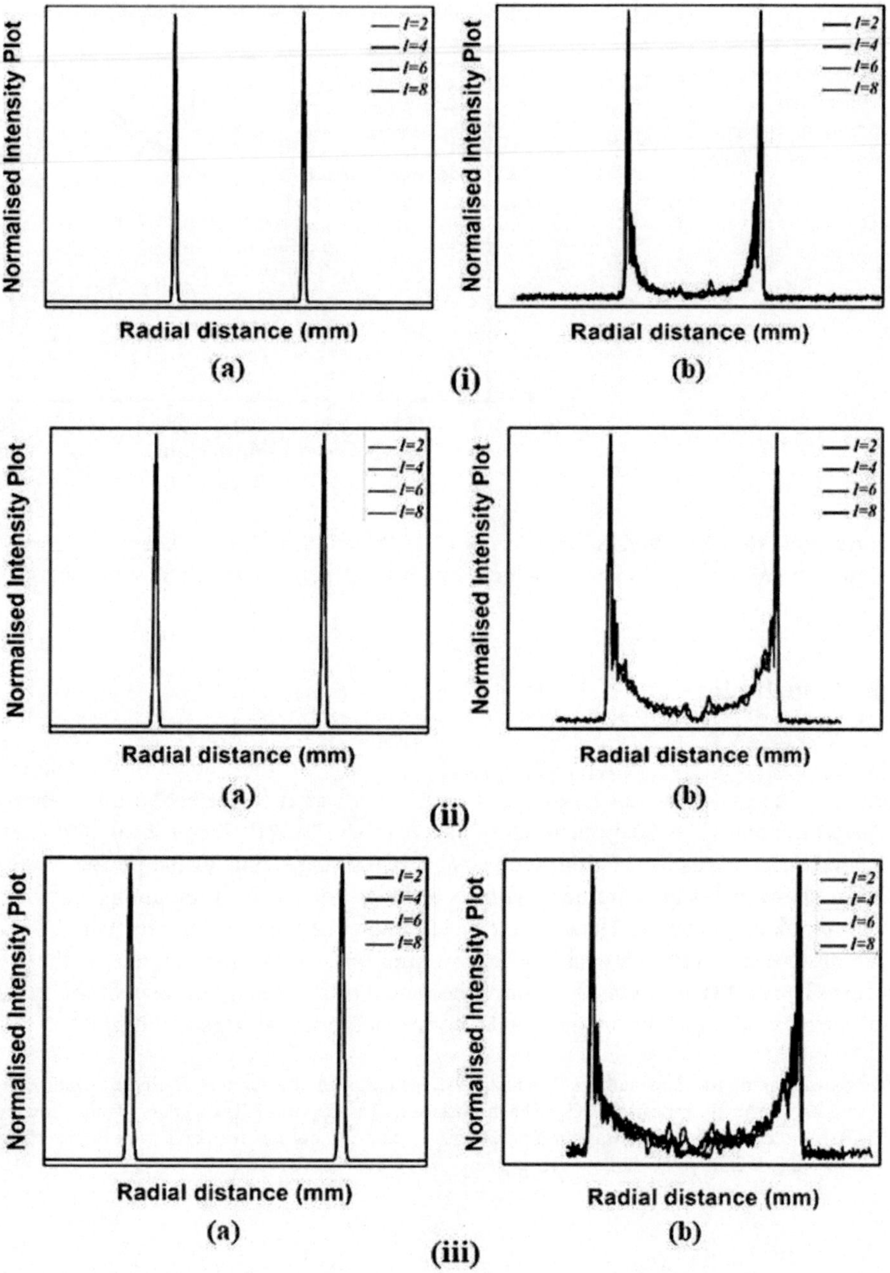


**Fig. 5** Radial intensity pattern (**a** theoretical and **b** experimental) of PV for different focal lengths of the converging lens, (i) $f = 150$ mm, (ii) $f = 200$ mm, and (iii) $f = 250$ mm

**Fig. 6** Variation of PV beam radius (experimental and theoretical) with respect to the focal length of the converging lens in axicon-lens assembly

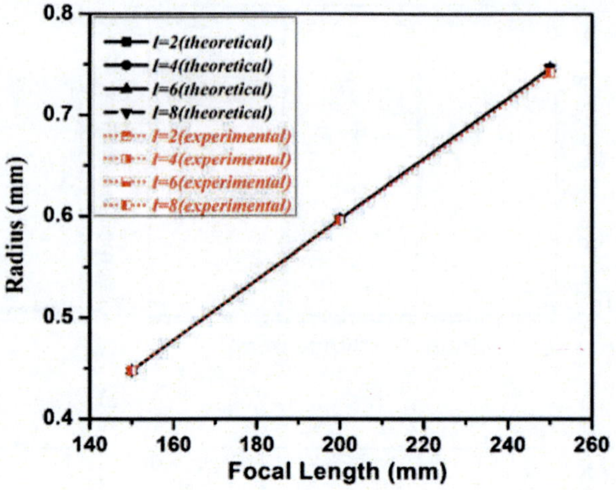


correspond to each other. A linear variation in the radius of the PV beam is observed with change in the focal length, which remains independent of topological charge.

## 4 Conclusion

In the present study, we have theoretically investigated the impact of variation in the focal length of the converging lens on the generation of perfect vortex beam by axicon method. The theoretical study has also been substantiated through experimental data. It was interesting to observe that the ring radius of the perfect vortex beam shows a linear increment with focal length of the lens remaining invariant with topological charge. Hence, we can conclude that changing the focusing of the axicon-lens assembly plays a key role in precisely controlling the size of the PV beam. The ability to precisely control the radius of PV beam will benefit the fields of optical communication, non-contact cornea surgeries, and laser cutting.

**Acknowledgements** The authors thank all the members of the Optical Radiation Metrology Group for technical support and scientific discussions. The support of Head, DU#1.0 and Director, CSIR-National Physical Laboratory, New Delhi, for carrying out this research work is thankfully acknowledged. One of the authors, SG is thankful to CSIR for SRF.

## References

1. Allen L, Beijersbergen MW, Spreeuw RJC, Woerdman JP (1992) Orbital angular momentum of light and the transformation of Laguerre-Gaussian laser modes. Phys Rev A 45(11):8185
2. Coullet P, Gil L, Rocca F (1989) Optical vortices. Opt Commun 73(5):403–408

3. Yao AM, Padgett MJ (2011) Orbital angular momentum: origins, behavior and applications. Adv Opt Photonics 3(2):161–204
4. Wang J, Yang JY, Fazal IM, Ahmed N, Yan Y, Huang H, Willner AE (2012) Terabit free-space data transmission employing orbital angular momentum multiplexing. Nat Photonics 6(7):488–496
5. Ramachandran S, Gregg P, Kristensen P, Golowich SE (2015) On the scalability of ring fiber designs for OAM multiplexing. Opt Express 23(3):3721–3730
6. Simpson NB, Allen L, Padgett MJ (1996) Optical tweezers and optical spanners with Laguerre-Gaussian modes. J Mod Opt 43(12):2485–2491
7. Curtis JE, Grier DG (2003) Structure of optical vortices. Phys Rev 90(13):133901
8. Yan H, Zhang E, Zhao B, Duan K (2012) Free-space propagation of guided optical vortices excited in an annular core fiber. Opt Express 20(16):17904–17915
9. Ostrovsky AS, Rickenstorff-Parrao C, Arrizón V (2013) Generation of the "perfect" optical vortex using a liquid-crystal spatial light modulator. Opt Lett 38(4):534–536
10. Mair A, Vaziri A, Weihs G, Zeilinger A (2001) Entanglement of the orbital angular momentum states of photons. Nature 412(6844):313–316
11. Chen M, Mazilu M, Arita Y, Wright EM, Dholakia K (2014) Optical trapping with a perfect vortex beam. In Opt Trapping Opt Micromanipulation XI, 9164:91640K
12. Djordjevic IB, Arabaci M (2010) LDPC-coded orbital angular momentum (OAM) modulation for free-space optical communication. Opt Express 18(24):24722–24728
13. Bozinovic N, Yue Y, Ren Y, Tur M, Kristensen P, Huang H, Ramachandran S (2013) Terabit-scale orbital angular momentum mode division multiplexing in fibers. Science 340(6140):1545–1548
14. Liu Y, Ke Y, Zhou J, Liu Y, Luo H, Wen S, Fan D (2017) Generation of perfect vortex and vector beams based on Pancharatnam-Berry phase elements. Sci Rep 7(1):1–8
15. Durnin JJJA (1987) Exact solutions for nondiffracting beams. I. The scalar theory. J Opt Soc Am A 4(4):651–654

# A Brief Discussion on GMR and TMR Effect and Importance of Metrology for Accurate Measurements of Its Parameters

**Komal Bhatt, D. K. Misra, Bal Govind, Sahiba Bano, and Ashish Kumar**

**Abstract** Heusler-based materials have been shown to be promising for spintronics application and gradually attracting a tremendous amount of research interest due to exciting properties. This article deals with the details about the giant magnetoresistance and tunnel magnetoresistance used for especially in magnetic switching, spin valves and spin switches based on GMR. In addition to this, we have stressed the importance of accurate and precise measurement for the actual realization of magnetic-based technology.

**Keywords** Heusler · Spintronics · Giant magnetoresistance (GMR) · Tunnel magnetoresistance (TMR)

## 1 Introduction

Heusler family of compounds have been shown promising materials for various applications such as solar cell, thermoelectric applications, and semiconductors (SCs) with tunable band gaps to multifunctional topological insulators and spintronics. In addition to this, Heusler compounds exhibit interesting magnetic behaviour and multifunctional magnetic properties, such as magneto-optical, magneto-caloric, and magneto-structural characteristics for various applications. In the family of magneto-electrical Heusler compounds, the half-metallic ferromagnetic behaves as a metallic in one spin channel and exhibits an insulating band gap in another spin channel. These materials have been found to exhibit fully spin-polarized conduction electrons near the Fermi level. Tunable band gap and high operating temperature provide a key role in the multifunctional properties of these materials and make them viable materials for spintronic applications.

Half-Heusler NiMnSb has been explored by De Groot et al., and they found that NiMnSb exhibits half-metallic nature having Curie temperatures of 730 K [1].

K. Bhatt (✉) · D. K. Misra · B. Govind · S. Bano · A. Kumar
CSIR- National Physical Laboratory, Dr. K. S. Krishnan Marg, New Delhi 110012, India
e-mail: komal44154d@gmail.com

Academy of Scientific and Innovative Research (AcSIR), Ghaziabad 201002, India

 
S. Yadav et al. (eds.), *Recent Advances in Metrology*, Lecture Notes in Electrical Engineering 906, https://doi.org/10.1007/978-981-19-2468-2_20

Several other HH such as NiMnGe, NiMnSi have also been explored theoretically and experimentally and exhibit to be half-metallic ferromagnetic nature generally above the room temperature [2, 3]. Due to the presence of four vacant tetrahedral sites in HH alloys, inadequate stability has been observed in the formation of a stable structure of HH rather than FH alloys.

Therefore, in this article, we have focused mainly on the characteristics of the FH alloys. In addition to this, Half-metallic $Co_2$-based Heusler alloys have been found to be increasing interest for developing magnetic tunnel device due to their high Curie temperatures as well as magnetic moment. $Co_2MnSi$ and $Co_2FeSi$ have been found to have the highest magnetic moment (~6.0 $\mu_B$) as well as high Curie temperature (1100 K) [4]. Half-metallic nature has been assisted by the significant value of magnetoresistance obtained in their thin films. Several Ni- and Fe-based FH alloys have also been investigated and found to be promising materials for device fabrication. In this endeavour, several materials such as $Ni_2MnGe$, $Ni_2MnSn$, $Fe_2VAl$, and $Fe_2TaGa$ FH have been investigated in recent year.

## 2 Crystal Structure and Atomic Ordering of Full-Heusler

Heusler alloys belong to a group of ternary intermetallic with the stoichiometric composition XYZ for half-Heusler or $X_2YZ$ for full-Heusler which crystallizes in an ordered face centred cubic structure. There are three interpenetrating sublattices in HH and four interpenetrating sublattices in FH (space group no. 216, $F\overline{4}3m$, $C1_b$ and space group no. 225, $Fm\overline{3}m$ $L2_1$, respectively). In HH matrix, transition metal atoms ($X$, $Y$) occupy the 4a (0, 0, 0), 4b (1/2, 1/2, 1/2) site, while sp-atom ($Z$) occupies the 4c (1/4, 1/4, 1/4) site. However, one tetragonal 4d (3/4, 3/4, 3/4) site is vacant in HH, whereas this vacant site is filled by $X$ or $Y$ in FH matrix. Valence electron count is the key parameter for tuning the magnetic properties as suggested by the Slater–Pauling rule which is given below

$$M_{\text{total}} = Z_{\text{total}} - 18 \quad \text{for HH} \tag{1}$$

$$M_{\text{total}} = Z_{\text{total}} - 24, \;\; M_{\text{total}} = Z_{\text{total}} - 28 \text{ for FH} \tag{2}$$

where $M_{\text{total}}$ and $Z_{\text{total}}$ are the total magnetic moment and total valance electron count. One can estimate the magnetic moment value from the above equation by changing the valence electron count. HH materials exhibit only one magnetic sublattice since the atoms on the octahedral sites can carry a finite magnetic moment, whereas FH compounds exhibit two different magnetic sublattices due to two $X$ atoms at the tetrahedral sites which allows interacting both $X$ atoms with each other. Due to which $X_2YZ$ Heusler compounds can show all types of magnetic phenomena, such as ferromagnetism, ferrimagnetism, antiferromagnetic, and half-metallic ferromagnetism (Fig. 1).

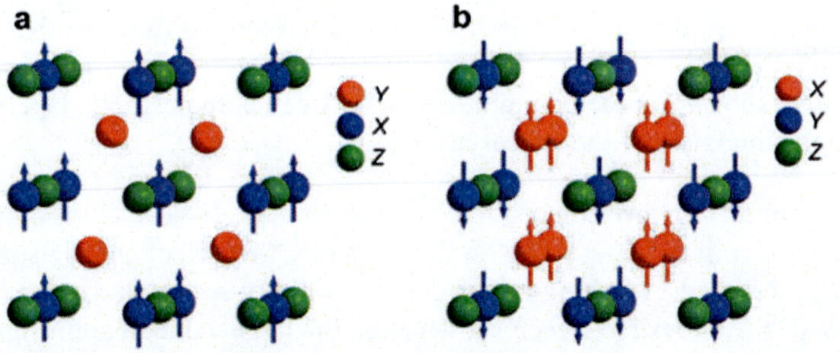


**Fig. 1** **a** Half-Heusler alloy (*XYZ*) having one magnetic sublattice. **b** Full-Heusler ($X_2YZ$) alloy having two magnetic sublattices which can couple ferromagnetically or antiferromagnetically [5]

# 3 Application of Heusler Alloys for Spintronic Devices

In the past decade, spintronics have gained interest to improve the reliability, switching speed, the performance of computers by shrinking the sizes of electronic components on a silicon chip. Non-volatility of data storage, the increased speed of data processing, the high-storage density, and the low-energy consumption are the main advantages of this new technology. To exploit the full potential of spintronics, the development of new magnetic materials, magnetic SCs, and HMF are essentially required. Due to their exceptional electronic structure, HMF meets all the requirements of spintronics.

## 3.1 *Giant Magnetoresistance (GMR)*

Giant magnetoresistance (GMR) property has been used as an application in the field of information technology. GMR effect is based primarily on electron scattering that governs by the applied magnetic field and leads to enhance the resistance of the carriers in the materials. GMR depends upon the relative orientation of the magnetization direction of the two magnetic layers and is separated by the nonmagnetic thin spacer layer [6]. To realize the significant value of GMR, one of the magnetic layers is "pinned" by an antiferromagnetic material, and while another layer is free to rotate. The layers which are pinned off are, therefore, insensitive to moderate magnetic fields. The free layers facilitate the magnetization to be rotated by the application of a small magnetic field. Usually, the resistance of the device is low if the magnetization of both ferromagnetic layers is aligned in parallel direction, whereas if the ferromagnetic layers are aligned antiparallel, resistance is high. The interlayer exchange coupling exists between magnetic layers and can be adjusted by varying the thickness of the nonmagnetic spacer layer [7] The current can be allowed to flow either perpendicular (CPP) or parallel (CIP) to the interfaces. CPP configuration gives the lower resistivity value and leads to improve the data rates (Figs. 2 and 3).

This curve shows that the communication between both magnetic layers strongly related to the thickness of the spacer layer.

**Spin valves and Spin switches**

Spin valve and spin switches work on the basis of GMR effect and is usually applicable for high-speed switching-based devices. Highly spin-polarized full-Heusler $Co_2FeGe_{0.5}Ga_{0.5}$ alloys have been shown to be the large value of spin signal ($\Delta R$ ~ 12.8 mΩ) using copper (Cu) wire [9]. High spin polarization leads to induce large $\Delta R$-value by expensing the spin absorption. This result shows that the high spin polarization is necessary for a high value of resistance. However, the stoichiometric composition needs not to be included for high spin polarization as well as for high magnetoresistance ratio. It is worth mentioning that $Co_2FeGeAl$ also exhibits high GMR ratio of ~40% at room temperature and ~120% at 10 K temperature [10].

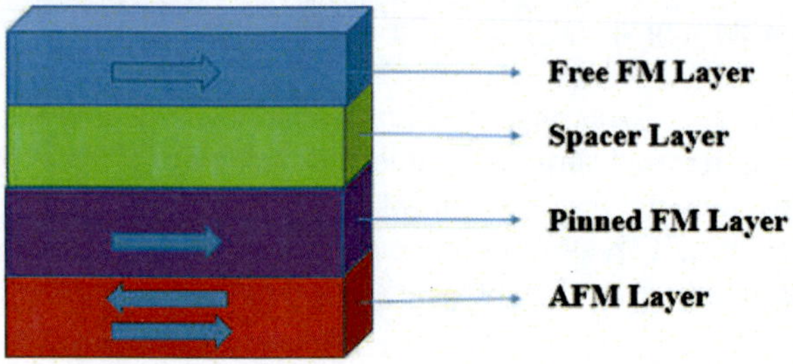


**Fig. 2** Schematic diagram of multilayer having one free layer separated by a spacer layer and one pinned layer pinned by AFM layers

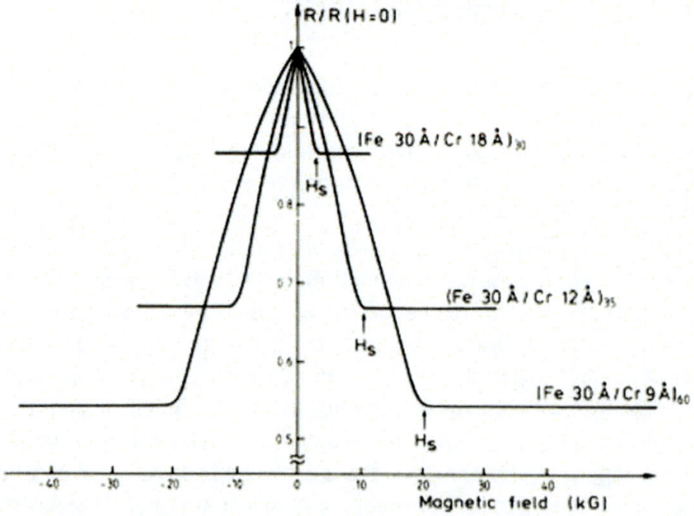


**Fig. 3** Variation of magnetoresistance with the applied magnetic field for different thickness of Cr spacer layer Fe/Cr superlattice at 4.2 K [8]

The ultimate goal of spintronics can be reached in two different ways, one is to engineer the insulation barrier, and the other is to develop new electrode materials with 100% spin polarization. Half-metallic ferromagnetic metals, such as Heusler compounds especially, $Co_2$-based Heusler materials were studied due to their high potential as new electrode materials in spintronic devices, such as magnetic tunnel junctions (MTJs), giant magnetoresistance (GMR) devices and for spin injection from ferromagnetic electrodes into the semiconductor.

## 4 Tunnelling Magnetoresistance

It is worth mentioning here that the replacement of the metallic spacer by an insulating material leads to a rise in magnetoresistance by a factor of ten compared to GMR spin valves. These new devices are known as magnetic tunnel junctions (MTJs) or as tunnelling magnetoresistance (TMR) devices [11] because this effect is based on tunnelling of electrons through the insulating barrier (Fig. 4).

This equation shows that the polarization depends upon the relative number of up and down spins, and TMR value depends upon the polarization of both channels.

First large magnetoresistance in magnetic tunnel junctions was observed at room temperature by J. S. Moodera [12], and the TMR ratio of the junction is related to the spin polarization P of the electrodes as suggested in Jullière model [13], can be given as

$$\frac{\Delta R}{R_{TMR}} = \frac{2P_1P_2}{1+P_1P_2}, \mathrm{P} = \frac{N_{\uparrow} - N_{\downarrow}}{N_{\uparrow} + N_{\downarrow}} \tag{3}$$

$N\uparrow$ and $N\downarrow$ are the densities of the majority and the minority electrons at the Fermi energy level. The largest TMR ratio of 1800% was measured by Fert's group in a tunnel junction with a manganite electrode. This corresponds to an electrode spin polarization of at least 95%, but unfortunately only at 4 K [14]. The first prediction of half-metallicity in MnNiSb provides research interest, aiming at the utilization of Heusler compounds in MTJs.

In this work, it was suggested that sufficient crystallinity of the thin films plays a major role in MTJs apart from the surface roughness and the interface morphology between the Heusler electrode and the barrier. Thus, it is advisable to take care of these issues before developing spintronic devices.

Based upon the measurement involved for making these magnetic devices, accurate, and precise measurement of magnetic moment, and other related parameters are

**Fig. 4** Schematic diagram showing the parallel and antiparallel alignment of the magnetic layer using in TMR devices

very important. For example, all half-metallic ferromagnetic materials have an integral magnetic moment. However, a small deviation from this integral value causes to reduce in the spin polarization. Therefore, accurate measurement is important for the half-metallic property. Moreover, in memory devices, the orientation of magnetization is used as a bit in thin films for data storage. Every bit should be well separated from each other for saving the data.

## 5 Conclusion

We have highlighted the different properties observed in thin-film suitable for spintronic devices. GMR and its importance from the application point of view have been discussed in detail. TMR has also been described for the development of spintronic devices. Finally, we have highlighted the importance of metrology for accurate and precise measurement of magnetic properties for the improvement in the technology.

## References

1. De Wijs GA, De Groot RA (2001) Towards 100% spin-polarized charge-injection: the half-metallic NiMnSb/CdS interface. Phys Rev B 64(2):20402
2. An Dinh V, Sato K, Katayama-Yoshida H (2008) New high-Tc half-Heusler ferromagnets NiMnZ (Z = Si, P, Ge, As). J Phys Soc Jpn 77(1):14705
3. Zhang R-J, Eckern U, Schwingenschlögl U (2014) Fate of half-metallicity near interfaces: the case of NiMnSb/MgO and NiMnSi/MgO. ACS Appl Mater Interfaces 6(16):14516–14521
4. Qin G, Ren W, Singh DJ (2020) Interplay of the local moment and itinerant magnetism in cobalt-based Heusler ferromagnets: Co2 TiSi, Co2 MnSi and Co2 FeSi. Phys Rev B 101(1):14427
5. Graf T, Winterlik J, Müchler L, Fecher GH, Felser C, Parkin SSP (2013) Magnetic heusler compounds. In: Handbook of magnetic materials, vol 21, Elsevier, pp 1–75
6. Grünberg P, Schreiber R, Pang Y, Brodsky MB, Sowers H (1986) Layered magnetic structures: evidence for antiferromagnetic coupling of Fe layers across Cr interlayers. Phys Rev Lett 57(19):2442
7. Barthelemy A et al (1990) Magnetic and transport properties of Fe/Cr superlattices. J Appl Phys 67(9):5908–5913
8. Baibich MN et al (1988) Giant magnetoresistance of (001) Fe/(001) Cr magnetic superlattices. Phys Rev Lett 61(21):2472
9. Takahashi YK, Kasai S, Hirayama S, Mitani S, Hono K (2012) All-metallic lateral spin valves using Co2Fe (Ge0. 5Ga0. 5) Heusler alloy with a large spin signal. Appl Phys Lett 100(5):52405
10. Nakatani TM, Hase N, Goripati HS, Takahashi YK, Furubayashi T, Hono K (2012) Co-based Heusler alloys for CPP-GMR spin-valves with large magnetoresistive outputs. IEEE Trans Magn 48(5):1751–1757
11. Moodera JS, Nassar J, Mathon G (1999) Spin-tunneling in ferromagnetic junctions. Annu Rev Mater Sci 29(1):381–432
12. Moodera JS, Kinder LR, Wong TM, Meservey R (1995) Large magnetoresistance at room temperature in ferromagnetic thin film tunnel junctions. Phys Rev Lett 74(16):3273
13. Julliere M (1975) Tunneling between ferromagnetic films. Phys Lett A 54(3):225–226
14. Bowen M et al (2003) Nearly total spin polarization in La 2/3 Sr 1/3 MnO3 from tunnelling experiments. Appl Phys Lett 82(2):233–235

# Indian Standard Time Synchronization Via NTP Server Over Different Networks

**Divya Singh Yadav, Ashish Agarwal, R. C. Jha, and Abhilash Dwivedi**

**Abstract** In this paper, Indian Standard Time (IST) Dissemination Technique and synchronization method over different network architecture and their respective delay measurements have been presented. Presently, one of the robust terrestrial time dissemination method over the network is via Network Time Protocol (NTP). CSIR—NPL as the time keeper of India, is disseminating IST in public domain through NTP servers which can be accessed by the users with domain names "time.nplindia.in" and "time.nplindia.org". These NTP servers have been synchronized with primary atomic time scale through 1pps external input and attain stratum1 status in network hierarchy for IST dissemination services. Authentic delay measurement is required for synchronization of another NTP server over the network for reliable time dissemination service to the users. The work presented in this paper is toward the authentic IST dissemination measurements via the NTP servers. In this experiment, a NTP server has been synchronized with the stratum-1 server and time delay has been measured for a period of 36,000 s over different network architecture viz. direct link network, hybrid link network and wide area network. Since CSIR—NPL is working toward the nationwide legalization of IST, this experimental analysis will strengthen the confidence in building authentic and reliable IST dissemination infrastructure via NTP servers in the nation.

**Keywords** Indian standard time · Time dissemination techniques · Network time protocol · NTP server · Stratum1

## 1 Introduction

Network Time Protocol (NTP) servers are maintained by CSIR—NPL for distributing official Indian Standard Time (IST). These servers are ideally synchronized to a

D. S. Yadav · A. Agarwal (✉) · R. C. Jha · A. Dwivedi
CSIR—National Physical Laboratory, New Delhi, India
e-mail: ashish@nplindia.org

D. S. Yadav
e-mail: divya.yadav@nplindia.org

 
S. Yadav et al. (eds.), *Recent Advances in Metrology*, Lecture Notes in Electrical Engineering 906, https://doi.org/10.1007/978-981-19-2468-2_21

1pps signal from the official time scale. The dependency of time distribution via Global Navigation Satellite System (GNSS) and other references can be eliminated when NTP servers are synchronized with primary atomic time scale through 1pps external input. These servers are available for general public via public Internet with domain name “time.nplindia.org” and “time.nplindia.in”. Some sectors, including stock market and financial exchanges, are often required to utilize authentic time [1]. The NTP [2, 3] protocols have been synchronized computer clocks in packet switched variable latency network. In this paper, the experiments of synchronized NTP servers up to startum2 level had been done for measuring time delay with respect to UTC (NPLI) over different network [4, 5]. The time delay dependency due to network elements and network hierarchy of NTP servers had been investigated. The associated uncertainty due to symmetric and asymmetric packet transmission path of each network configuration such as free running NTP servers, startum1 NTP server and startum2 NTP server with three different network configurations, i.e., synchronized with respect to stratum1 via crossover cable, hybrid network and wide area network (WAN) [6, 7]. These three-network configuration has created the different path for packet transmission of NTP request with respective time offset and associated uncertainty. This experimental study had provided enough evidence of network dependency of NTP server and its network hierarchy which further helps us to design a robust mechanism of calculating the associated uncertainty in time distribution over terrestrial network. This work would help us to further investigate the method for authentic time synchronizations via NTP server to end users.

## 2 Measurement System

The experimental measurements were performed by using commercially available NTP servers installed at timing laboratory in CSIR—NPL. Two NTP servers were used in this experiment for establishing 2 level hierarchical NTP network. 1pps output from official primary time scale was used to synchronized and measuring the time offset of NTP servers with respect to UTC (NPLI). Time offset had been recorded by the time/frequency counter (Model T3200U). Network elements such as cables, crossover cables, optical fibers, light interface unit (LIU), network switches and routers were used for various measurement configuration.

## 3 Measurement Configurations

The measurement system was configured to record offset of NTP server with the network hierarchical status of stratum 1, 2 and 15 with respect to UTC (NPLI). Time offset measurement of NTP server with respect to UTC (NPLI) has been recorded for 36000 seconds with the help of Time Interval Counter (TIC). The following sections describe the network architecture for the measurements.

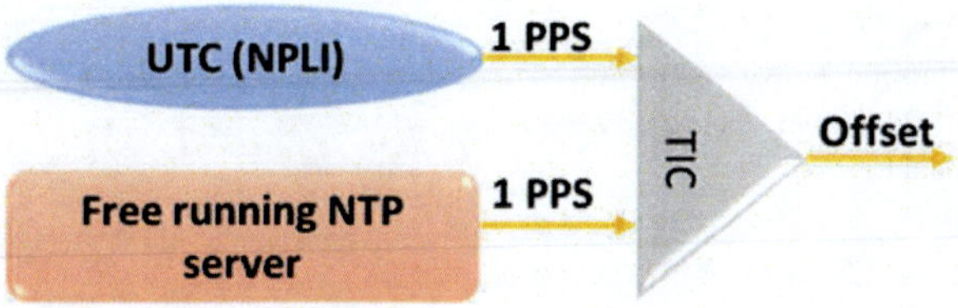


**Fig. 1** Measurement configuration for free running NTP Server

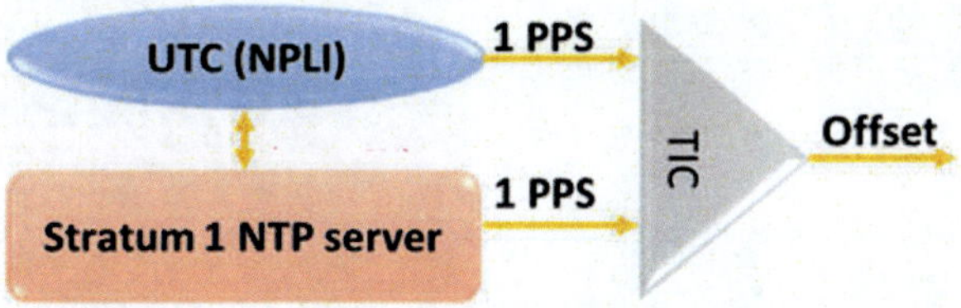


**Fig. 2** Measurement configuration for startum1 NTP server

### 3.1 Free Running NTP Server

In this configuration, NTP server is not synchronized by any external reference. The behavior of internal clock of NTP server has been investigated with respect of UTC (NPLI) output through TIC as shown in Fig. 1.

### 3.2 Startum1 NTP Server

In this configuration, NTP server is synchronized with respect to UTC (NPLI) and attains the stratum1 status in network hierarchy. The offset of strtaum1 NTP server had been measured with respect to UTC (NPLI) through TIC as shown in Fig. 2.

### 3.3 Direct Connection Via Crossover Cable

In this network configuration, NTP server is synchronized with respect to stratum1 NTP server using category 6 (cat 6) crossover cable and attains stratum2 status in network hierarchy. This representation is the simplest possible LAN configuration. Stratum1 NTP server had a unique Internet Protocol (IP) address which assured guaranteed delivery of the packets. The offset measurement of stratum2 NTP server had been measured with respect to UTC (NPLI) through TIC as shown in Fig. 3.

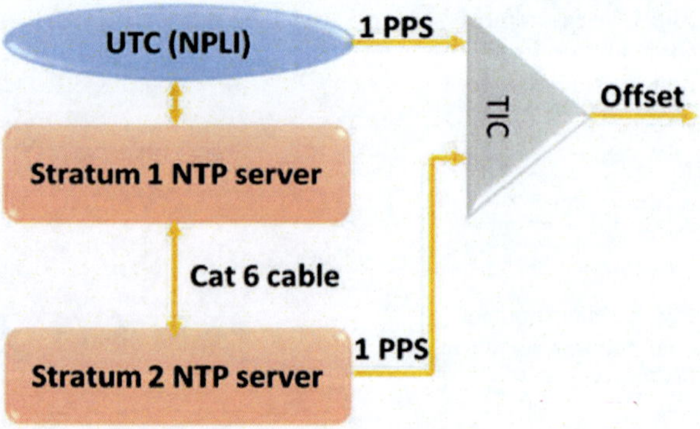


**Fig. 3** Measurement configuration for startum2 NTP server via crossover cable

## *3.4 Hybrid Connection Via Crossover Cable and Optical Fibers*

In this network configuration, NTP server is synchronized with respect to stratum1 NTP server using cat6 crossover cable and optical fiber network. This NTP server is also attains startum2 status in network hierarchy with two network switch arrangement. This is considered to be a data link layer path for packet transmission. The offset measurement of startum2 NTP server had been measured with respect to UTC (NPLI) through TIC as shown in Fig. 4.

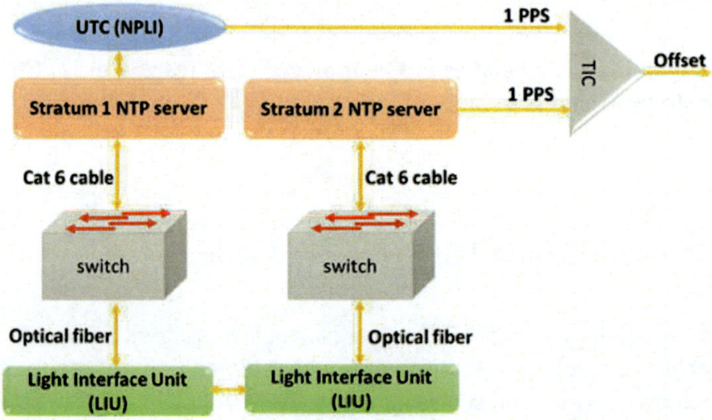


**Fig. 4** Measurement configuration for startum2 NTP server via crossover cable and optical fiber

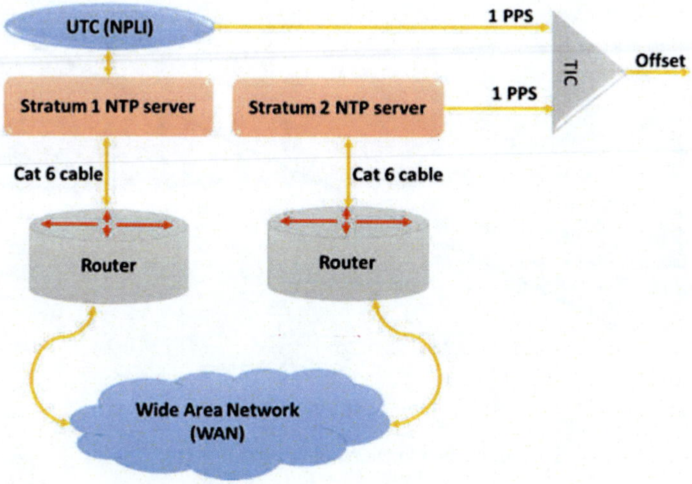


**Fig. 5** Measurement configuration for startum2 NTP server via WAN

### 3.5 *Wide Area Network (WAN) Connection Via Various Network Elements*

In this network configuration, NTP server is synchronized with respect to startum1 NTP server via WAN. In this configuration, public Internet is being used to get synchronized and NTP server attains startum2 status in network hierarchy with multiple switches and routers arrangement. The path for packet transmission is not fixed for this network configuration. The offset measurement of startum2 NTP server had been measured with respect to UTC (NPLI) through TIC as shown in Fig. 5.

## 4 Measurement Results

The following section shows the time offset of each configuration and the change in time offset with respect to UTC (NPLI) as the network architecture changes. As noted in the Sect. 3, the time offset has been recorded at each second for 10 hours.

### 4.1 *Free Running NTP Server Measurement Result*

Figure 6 shows the measurement result of free running NTP server. The external 1 PPS had been removed for more than 24 hours, this NTP server is behaving like a free running clock. The time offset of this free running NTP server is measured with respect to UTC (NPLI), it is showing a drift of 2.6 μs/s.

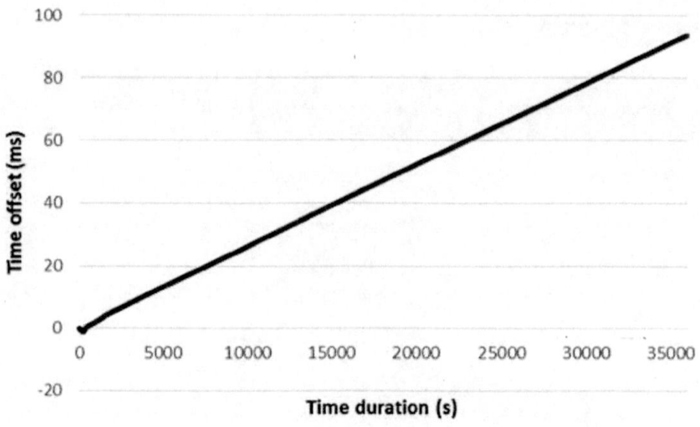


**Fig. 6** Free running NTP server measurement result

## 4.2 *Stratum1 NTP Server Measurement Result*

Figure 7, shows the measurement result of startum1 NTP server. The NTP server is called the stratum1 when the internal clock of NTP server is synchronized with respect to reliable stratum0 source. Here in this case, the startum0 source is the primary atomic time scale UTC (NPLI). The time offset had been measured with respect to UTC (NPLI) which is not greater than ± 0.8 μs which is the maximum measured uncertainty for stratum1 NTP server.

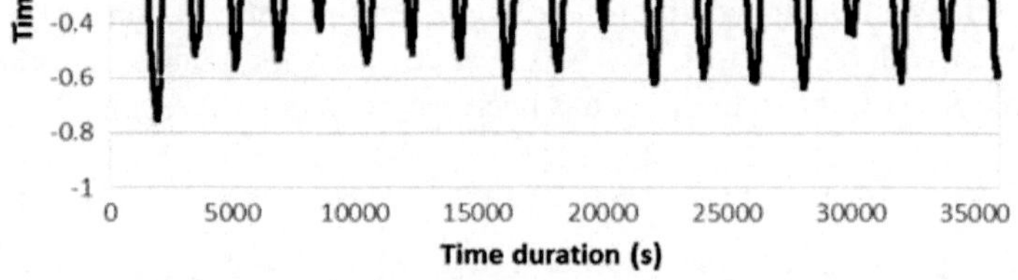


**Fig. 7** Stratum1 NTP server measurement result

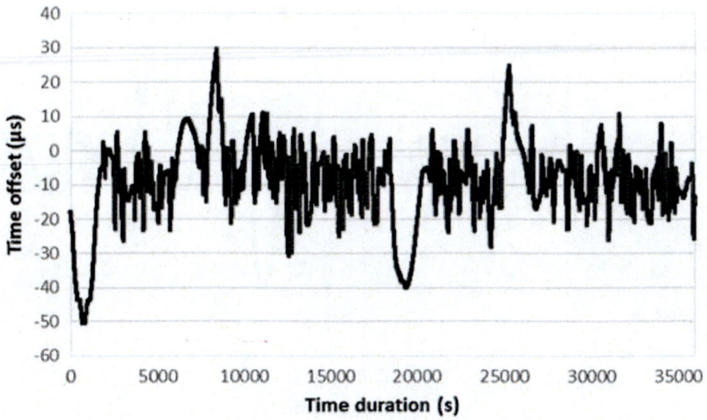


**Fig. 8** Direct connection via crossover cable measurement result

## 4.3 Direct Connection Via Crossover Cable Measurement Result

Figure 8, shows the measurement result of the time offset of stratum2 NTP server with respect to UTC (NPLI). NTP server is synchronized with respect to startum1 NTP server via crossover cable and attained the starum2 status in network hierarchy. The crossover cable eliminates the network asymmetry for packet transmission and achieve better synchronization for ability at stratum2 level. The time offset is measured with respect UTC (NPLI). The maximum uncertainty is achieved is ± 51 μs at startum2 network hierarchy of NTP servers.

## 4.4 Hybrid Connection Via Crossover Cable and Optical Fibers Measurement Result

Figure 9, shows the measurement result of hybrid connection via crossover cable and optical fibers. The NTP server is synchronized with respect to stratum1 NTP server and attains the status of starum2 in network hierarchy. The introduction of layer 2 switch and physical layer network element such as crossover cables, Light Interface Units (LIU) and optical fibers had been done. When the time offset had been measured with respect to UTC (NPLI), it is found that these network elements provided the maximum uncertainty of ± 70μs while the asymmetry of network for packet transmission had still not introduced.

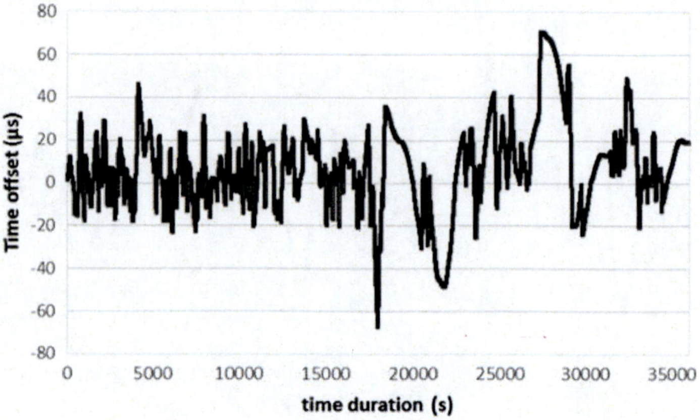


**Fig. 9** Hybrid connection via crossover cable and optical fibers measurement result

### *4.5 Wide Area Network (WAN) Connection Via Various Network Elements Measurement Result*

Figure 10, shows the measurement result of Wide Area Network (WAN) connection. When a NTP server is synchronized with respect to stratum1 NTP server over public internet and attained stratum2 status in network hierarchy. There were various network elements must be present in public network which introduce path asymmetry for packet transmission. The time offset of stratum2 NTP server with respect to UTC (NPLI) had been measure with maximum uncertainty of ± 1ms.

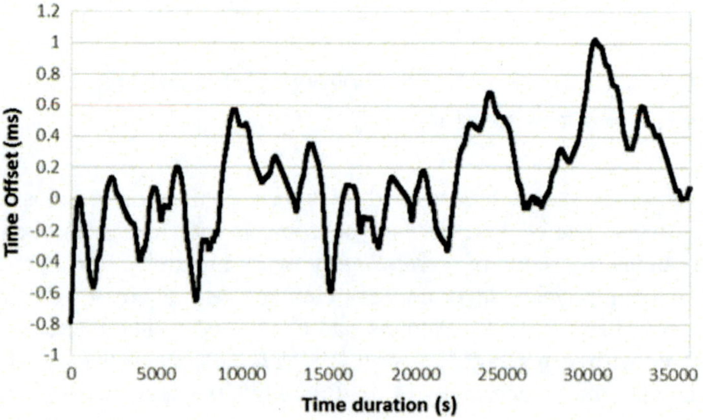


**Fig. 10** Wide area network (WAN) connection via various network elements measurement result

**Table 1** Comparison of uncertainty with respect to various network configuration

| NTP server hierarchy | Network configuration | Reference | Network symmetry | Uncertainty |
|---|---|---|---|---|
| Stratum 15 | Free running | Internal clock | – | 2.6 μs/s |
| Stratum 1 | Direct | Stratum 0 | – | ± 0.8 μs |
| Stratum 2 | Cat 6 | Stratum 1 | Yes | ± 51 μs |
| Stratum 2 | Hybrid | Stratum 1 | Yes | ± 70 μs |
| Stratum 2 | WAN | Stratum 1 | No | ± 1 ms |

## 5 Conclusion

Indian Standard time synchronization via NTP servers had been investigated up to stratum 2 network hierarchy. Time offset with respect to UTC (NPLI) had been recorded. Time offset is contributed the maximum uncertainty and as the network elements had been introduced in various network configuration, the uncertainty increases. When the startum2 NTP server is synchronized to startum1 NTP server via Local Area Network (LAN), i.e., Cat 6 cable, switches and optical fiber, it creates a dedicated symmetric network path for packet transmission. On the other hand, when the startum2 NTP server is synchronized to startum1 NTP server via public Internet, it creates an asymmetric network path for packet transmission and uncertainty also increases in the range of milliseconds as shown in Table 1.

Thus it is concluded that NTP network hierarchy introduces uncertainty due to various network configurations and as the number of stratum increases the uncertainty increases.

## References

1. Lombardi M, Novick A, Neville-Neil G, Cooke B (2016) Accurate, traceable, and verifiable time synchronization for world financial markets. J Res Nat Inst Stand Technol 121:436. https://doi.org/10.6028/jres.121.023
2. Mills D (2006) Computer network time synchronization: the network time protocol. Taylor and Francis, Boca Raton, FL
3. Mills D, Martin J, Burbank J, Kasch W (2010, June) Network time protocol Version 4: protocol and algorithms specification, Internet engineering task force RFC-5905
4. Lombardi M, Levine J, Lopez J, Jimenez F, Bernard J, Gertsvolf M et al (2014, December) International comparisons of network time protocol servers. In: Proceedings of the precise time and time interval meeting (ION PTTI), Boston, Massachusetts, USA, pp 57–66
5. Novick AN, Lombardi MA (2015) Practical limitations of NTP time transfer. In: 2015 joint conference of the IEEE international frequency control symposium & the European frequency and time forum, pp 570–574. https://doi.org/10.1109/FCS.2015.7138909

6. Novick A, Lombardi M, Franzen K, Clark J (2018) Improving packet synchronization in an NTP server, ION/precise time and time interval conference, Reston, VA [online]. https://tsapps.nist.gov/publication/get_pdf.cfm?pub_id=925159
7. Novick A, Lombardi M (2015) Practical limitations of NTP time transfer. In: Proceedings of the 2015 joint conference of the IEEE international frequency control symposium & European frequency and time Forum, Denver, CO [online]. https://doi.org/10.1109/FCS.2015.7138909

# Recent Developments in Surface Acoustic Wave (SAW) Sensors and Their Applications in Different Areas

**Ashish Lakra, Deepak Kumar Lakra, Rakesh Kumar Chhikara, and Pradeep Lakra**

**Abstract** Nowadays, chemical sensors are used in detecting the presence of certain gases and even liquid analytes. Due to the recent researches in the area of chemical sensors, their sensitivity, reliability and environmental stability are improving. This is due to the improvements in the two components (receptors and transducers) of the sensors. This paper is a critical review of recent advancements in chemical sensors, particularly the surface acoustic wave (SAW) sensors. SAW sensors are attracting researchers due to their short detection time, high sensitivity and high reliability. Metal oxide semiconductors are important option for preparing a sensitive film of SAW sensors. The film prepared from metal oxide semiconductors also has high thermal and chemical stability; one example is ZnO which can be used for making the film of SAW sensors for detecting different types of gases. The use of pulsed laser deposition (PLD) for depositing sensitive porous films has increased the ability of SAW sensors to detect different gases at very low concentrations. Monolayer films of $TiO_2$ and bilayer films of $Pd/TiO_2$ deposited on quartz substrates of SAW sensors are the examples of this. Considerable improvement in the sensing properties is observed when porous gold layers are used in place conventional dense gold layer. Nowadays, high-temperature SAW sensor (based on ZnO films) is developed which can work at high temperature (as high as 850 °C) with good stability. These high temperature gas sensors would be useful in engines and combustion system. There is a broad spectrum of SAW sensors which have a variety of uses ranging from detection of small gas molecules to large bio-molecules or sometimes even the whole cell structures.

A. Lakra
Department of Chemistry, PDM University, Bahadurgarh 124507, India

D. K. Lakra
SBV, New Delhi, Delhi, India

R. K. Chhikara (✉)
Department of Applied Sciences, MSIT (IP University), New Delhi, Delhi 110058, India
e-mail: rakeshkumar@msit.in

P. Lakra
Directorate of Education, TGT Natural Science, RPVV, Raj Niwas Marg, New Delhi, Delhi, India

 
S. Yadav et al. (eds.), *Recent Advances in Metrology*, Lecture Notes in Electrical Engineering 906, https://doi.org/10.1007/978-981-19-2468-2_22

**Keywords** Chemical sensors · SAW sensors · Metal oxide semiconductors · PLD · Porous gold layers · Thermal stability · Chemical stability

# 1 Introduction

Generally, a particular sensor consists of a chemical or biological receptor which is designed to recognize the desired analyte and the transducer converts the information in a signal, producing a quantitative and/or qualitative output. Chemical and biological sensors can be used for both screening and quantitative applications depending on the need. Due to the recent researches in the area of chemical and biological sensors, their reliability, sensitivity and stability in adverse conditions are rapidly improving.

A SAW sensor is based on the principle of piezoelectric effect. ZnO, AIN, GaAs, $LiNbO_3$, etc., are the common materials used for this purpose. The resonator is the main component of a SAW sensor. It is generally made up of IDTs deposited on the surface of the substrate, piezoelectric substrate and reflection gratings on both sides. A material that has zero temperature coefficient and having high values of $K^2$ are ideal for SAW sensors designed for chemical detection.

Sensors have a number of uses in monitoring temperature, pressure, chemicals and bio-molecules in a particular environment. These sensors can be employed for detection of pollution, detection of volatile organic compounds, chemical warfare agents and many inorganic gases.

# 2 Recent Advancements in SAW Sensors

A SAW sensor is based on the principal of piezoelectric effect. Some of the most common materials that are used for this purpose are quartz, ZnO, AlN, GaAs, $LiNbO_3$, $LiTaO_3$. IDT is used as the electrodes in order to increase the surface area, and their primary task is convert the mechanical wave to the electrical signal and vice versa [1]. IDT is also associated with the frequency of the sensor for which is designed [2]. The AlN material can stand against high level of mechanical deformations [3]. QCM is also used as a chemical or biosensor as it is very small, low temperature coefficient, low fabrication cost, easy availability and high level of sensitivity [4].

The main component of a SAW sensor is the resonator or the SAW delay line which is normally made up of IDTs deposited on the surface of the substrate, piezoelectric substrate and reflection gratings on both sides. During the working of a SAW sensor, the response arise from the propagation perturbations of a wave by external disturbances can be detected which is normally shown by the shifting in phases amplitudes or frequencies. The physical property or quantity can be measured by establishing an interactive relationship between the response and the disturbance [5]. These sensors are widely used in the forms of two port resonators or delay lines.

The readout of these sensors is achieved by electronic systems operating either in an open or in a closed loop configuration. The mode of operation is chosen based on the requirement such as costs, immunity against environmental influences dynamic range, bandwidth and linearity [6].

The materials that have zero temperature coefficient and posses high values of $K^2$ are ideal for the SAW sensors for chemical detection as smaller values attenuates the signals and the conversion rate is slow [7]. Rayleigh is one of the three modes that is used to SAW sensors. According to the applications, the frequency of the SAW sensor has to be of higher range [2]. Nowadays, Love-wave SAW sensors have been developed using either delay-line structure or resonant which are highly sensitive towards magnetic field sensing. These sensors are used for bio-chemical measurements [8]. In order to accurately investigate the disturbance of complex distributed mass loading on SAW sensors propagation characteristics, 2D coupling of modes and finite element method was used to stimulate the response of the SAW sensors. Four SAW resonators have been modelled with the loads of different distribution patterns by using the PDE mode of FEM software. The results obtained showed that the 2D coupling of modes theory combined with the FEM was able to stimulate the transverse mode of the device and the disturbance of the mass loading on the transverse mode effectively, making the stimulate more accurate [9].

## 3 Applications of SAW Sensors

Surface acoustic wave (SAW) sensors have a wide range of applications starting from medicine and life safety to temperature and pressure control and detection systems as well as detection of gases [3], and even they are also used for the real-time monitoring of torque in a rotating shaft [10]. Surface acoustic wave (SAW) sensors can be used to detect liquids, gases and the changes in the environment like humidity, presence of ultraviolet radiations, pressure, etc. These sensors have biological as well as chemical applications also. These sensors can be used for the detection of pollution and emission control, combustion exhausts and also for the detection of volatile organic compounds, chemical warfare agents and various inorganic gases [7]. These sensors can be used to detect ammonia ($NH_3$), hydrogen ($H_2$) and some of the polar as well as non-polar volatile organic compounds (polar: acetone, ethanol, chloroform) (non-polar: n-hexane, toluene and isoprene). For the detection of these gases and the organic compounds, $WO_3$, $SnO_2$, $Cu_2O/CuO$, $Co_3O_4$, ZnO, $In_2O_3$, $TiO_2$, etc. are used as the reference metal oxide semiconductor used in the sensors. For example, Miu et al. developed a sensor for the detection of hydrogen at room temperature with platinum (Pt) and a multilayer film of $WO_3$. A surface acoustic wave sensor was developed by Wang et al. in which copper ion is doped with tungsten oxide/polyaniline and is used for the detection of NO [1].

Surface acoustic wave sensors which are based on Rayleigh waves are also developed. These sensors can be used for the detection of a large number of gases such as nitrogen ($N_2$), hydrogen ($H_2$), ozone ($O_3$), oxygen ($O_2$), methane ($CH_4$), carbon monoxide (CO) and carbon dioxide ($CO_2$) and many organic vapours such as acetone, methanol, styrene and dimethyl methyl phosphonate. These sensors can also be used for the detection of various explosive chemicals by coating the sensor with a suitable polymer [7]. There are two main problems in the high temperature sensing. One is that at high temperature using a wired method it is not easy to power the sensor electrically, and the second is that the devices are apt to become invalid when exposed to overheating [11]. The ability of SAW sensors to work in wireless mode is of great importance. Due to this property, these sensors can have an ability to work in extreme environmental conditions such as high pressure, high or low temperature and toxic environment. This property makes the sensor more useful in the detection of chemical produced in mines, automobiles, combustion engines and oil industries. For instance, a wireless sensor for the detection of carbon dioxide ($CO_2$) was developed which is coated with Teflon by Wen et al. in 2007. The humidity and temperature dependence of that sensor were also determined. Similarly, Lim et al. in 2011 designed a sensor for detecting carbon dioxide ($CO_2$) and nitrogen dioxide ($NO_2$) which can even measure the temperature simultaneously [7]. Another industrial application of these sensors is that some sensors which are based on reflective delay lines were designed that allow the parallel read out of four independent temperature sensors in the 2.45 GHz ISM-band [12]. A SAW-based implantable sensor, equipped with a well-defined dimension and with lithium niobate membrane was made to characterize the sensors vs pressure and temperature. The achieved pressure accuracy is (+, − 0.56 mbar) [13]. One of the most important parameters of the SAW sensors is the sensitivity. The physical and electrical properties of the SAW delay line are simulated by using finite element method simulating using COMSOL Multi-physics software [14]. The sensitivity of a SAW sensor towards the absorption capacity of the gas molecules can be increased using quantum dots (QDs) and polymers [15].

## 4 Conclusion

Due to the recent advancements in SAW sensors, their sensitivity is improved. So, they can be used to detect the analyte even in very low concentrations. Due to the ability of SAW sensors to work in wireless mode, they can work in extreme environmental conditions like high pressure, low or high temperature and toxic environment. For example, many sensors based on zinc oxide (ZnO) can work above 200 °C and can detect an analyte at very low concentrations (as low as 2.5 ppb). Love-wave SAW sensors have been developed, which are highly sensitive towards magnetic field sensing. These sensors are used for bio chemical measurements. Surface acoustic wave sensors which are based on Rayleigh waves are also developed. These sensors can be used for the detection of a large number of gases.

**Table 1** Some selected applications of SAW sensors

| Application | Frequency (MHz) | Structure |
|---|---|---|
| Multi-channel microfluidic module for multi-analyte detection | 155 | Quartz/PMMA |
| Fluid and particle manipulation | 8–16 | $LiNbO_3$/Ti/Au |
| Rotational microfluidic motor | 20 | $LiNbO_3$/Cr/Au, |
| Acoustic micropump | 142 | $LiNbO_3$ |
| Pathogenic microorganism detection | 163.2 and 159.7 | Quartz/Al |
| Nanoliter droplet acoustic streaming | 50, 100, 200, 400, 833, and 1250 | $LiNbO_3$/Ti/Au |
| Digital microfluidic heating | 50, 100, 200, 400, 600, and 900 | $LiNbO_3$/Ti/Au |
| Versatile SAW toolbox for microfluidic applications | 11.9 | $LiNbO_3$/Ti/Au |
| Droplet transportation | 22.44 | Si/ZnO/Al |

**Table 2** Sensors based on zinc oxide (ZnO) [1]

| S. no | Analyte | Sensitive material | Minimum concentration | Operating condition |
|---|---|---|---|---|
| 1 | Acetone | ZnO microspheres | 100 ppm | 280 °C |
| 2 | Ethanol, nitrogen dioxide, benzene | ZnO nanoflowers | 5 ppm, 250 ppb, 2.5 ppm | 250 °C, 200 °C, 250 °C |
| 3 | Hydrogen sulphide | Zno-$Al_2O_3$ composite | 20 ppm | Room temperature |
| 4 | Ammonia | Polyvinyl pyrrolidine-ZnO nanofibers | 20 ppm | Room temperature |
| 5 | Hydrogen | ZnO microspheres | 100 ppm | 225 °C |
| 6 | Nitrogen dioxide | ZnO/Pd | 2.5 ppb | Room temperature |
| 7 | Ethanol | Pd/ZnO | 500 ppm, | 260 °C, |
| 8 | Ethanol | C/ZnO | 100 ppm | 300 °C, |
| 9 | Acetone | Au-doped ZnO Pd-doped ZnO | 5 ppb | 150 °C |
| 10 | Nitrogen dioxide | ZnO nanowire-integrated film | 50 ppb | Room temperature, UV activation |

## References


1. Constantinoiu I, Viespe C (2020) ZnO metal oxide semiconductor in surface acoustic wave sensors: a review. Sensors 20(18):5118
2. Fine GF et al (2010) Metal oxide semi-conductor gas sensors in environmental monitoring. Sensors 10(6):5469–5502
3. Mikhailenko DA, Markelov OA (2020) Comparison of AlN vs. $SIO_2/LiNbO_3$ membranes as sensitive elements for the SAW-based acceleration measurement: overcoming the anisotropy effects. Sensors 20(2):464

4. Mujahid A, Afzal A, Dickert FL (2019) An overview of high frequency acoustic sensors—QCMs, SAWs and FBARs—chemical and biochemical applications. Sensors 19(20):4395
5. Jiang C, Chen Y, Cho C (2019) A three-dimensional finite element analysis model for SH-SAW torque sensors. Sensors 19(19):4290
6. Durdaut P et al (2019) Equivalence of open-loop and closed-loop operation of SAW resonators and delay lines. Sensors 19(1):185
7. Devkota J, Ohodnicki PR, Greve DW (2017) SAW sensors for chemical vapors and gases. Sensors 17(4):801
8. Schmalz J, Kittmann A, Durdart P, Spetzeler B, Faupel F, Hoft M, Quandt E, Gerken M, (2020) Multi-mode love-wave SAW magnetic-field sensors. MDPI
9. You R et al (2020) Simulation of SAW sensors with various distributed mass loadings using two-dimensional coupling-of-modes theory. Sensors 20(24):7260
10. Silva D et al (2017) Measuring torque and temperature in a rotating shaft using commercial saw sensors. Sensors 17(7):1547
11. Shu L et al (2019) The investigation of high-temperature SAW oxygen sensor based on ZnO films. Materials 12(8):1235
12. Bruckner G, Bardong J (2019) Wireless readout of multiple SAW temperature sensors. Sensors 19(14):3077
13. Nicolay P et al (2018) A ln/si-based saw pressure sensor. Sensors 18(10):3482
14. Achour B et al (2020) Simulation/experiment confrontation, an efficient approach for sensitive SAW sensors design. Sensors 20(17):4994
15. Constantinoiu I, Viespe C (2019) Hydrogen detection with SAW polymer/quantum dots sensitive films. Sensors 19(20):4481

# An Adaptive Color Image Watermarking Scheme Using Combination of PSO-LTSVR in Wavelet Domain for Copyright Protection

**Rajesh Mehta and Ravinder Agarwal**

**Abstract** With the advancement of internet technology, copyright protection and ownership assertion of multimedia content (audio, images and videos) is becoming a more challenging task. During last two decades various researchers raised the issue of watermarking for copyright protection and ownership assertion of multimedia contents, but still trying to find the better solution than the existing approaches. In order to handle the issues of imperceptibility and robustness evolved in color image watermarking, a novel color image watermarking approach in YCbCr space using discrete wavelet transform (DWT) and combination of particle swarm optimization (PSO) and Lagrangian twin support vector regression (LTSVR) is proposed. The low frequency wavelet coefficients of selected regions based on the human visual characteristics (HVS) is selected to embed the watermark. Considering the watermarking problem as an optimization and regression problem, PSO is used to find the optimal value of watermark strength to embed the watermark and LTSVR is used to extract the watermark against image processing attacks. With the combination of PSO-LTSVR in wavelet domain, the proposed approach reduces the trade-off between imperceptibility and robustness. The experimental results tested on different textured color images clearly reveal the imperceptibility of the watermarked images and robustness of the proposed scheme against JPEG compression attacks with different quality factors (QF) as quantified by peak-signal-to-noise ratio (PSNR) and bit correct ratio (BCR).

**Keywords** DWT · PSO · LTSVR · PSNR · BCR

---

R. Mehta (✉) · R. Agarwal
Thapar Institute of Engineering and Technology, Patiala, Punjab, India
e-mail: rajesh.mehta@thapar.edu

R. Agarwal
e-mail: ravinder_eeed@thapar.edu



S. Yadav et al. (eds.), *Recent Advances in Metrology*, Lecture Notes in Electrical Engineering 906, https://doi.org/10.1007/978-981-19-2468-2_23

# 1 Introduction

Due to the rapid development of multimedia and internet technology in today's era, the need of copyright protection of these multimedia contents is required from preventing illegal copying and sharing of data [1, 2]. It is observed from the existing research work done by various researches that digital watermarking provides a secure and robust system against such type of issues [3]. Digital watermarking is the process of embedding the watermark which may be an image or logo in an host multimedia content (image, audio and video) [4] which consist of owner's identification or copyright of data. Based on the insertion of watermark, digital watermark is separated into two categories: spatial and frequency domain watermarking. For copyright protection application, transform (frequency) domain schemes are more preferred as comparison to spatial domain [5] since it provides more robustness against attacks. Various transform domain schemes includes DFT [1, 4], DCT [5, 6], DWT [7, 8] along with its different variants [8, 9] along with some linear algebra-based techniques such as SVD [8], QR decomposition [10] in the field of digital watermarking. Currently researchers are more focused on optimization and machine learning-based color image watermarking schemes [7, 10, 11]. Thus in the proposed work, the main focused is on the color images with high imperceptibility and more robustness against attacks. Here we have focused only on the JPEG compression attacks with different quality factor (QF) ranging from 10 to 90. The proposed scheme comprises features extraction using low frequency sub band obtained using DWT and then selection of blocks based on statistical feature followed by QR factorization method [12]. Then extracted features form the dataset to embed the watermark through the training process of LTSVR and PSO.

The rest of the paper is organized as follows: the preliminaries are explained in Sect. 2. The proposed color image watermarking scheme which include embedding and extracting procedure is described in Sect. 3. The imperceptibility and robustness against JPEG compression attacks are briefly explained in results and analysis described in Sect. 4. Finally the conclusion of the proposed scheme and experimental results along with future directions is drawn in Sect. 5.

# 2 Preliminaries

In this section, the techniques used for embedding and extracting the watermark in proposed approach are DWT, PSO and LTSVR is explained as:

## 2.1 DWT

DWT is a popular signal processing technique used by many researchers in the field of digital watermarking [12]. In the present work, Daubechies filter (*db2*) for DWT is used to decompose the Y channel of the image. It divides the image onto low frequency subband (*LL*) and details subbands (*LH, HL, HH*). Due to its multiresolution and energy compaction property, *LL* subband is used to embed the watermark. DWT decomposes the image into low and high frequency subband using Eqs. 1 and 2 defined as:

$$t_{\text{low}}[k] = \sum_{n} s[n]h[2k - n] \quad (1)$$

$$t_{\text{high}}[k] = \sum_{n} s[n]g[2k - n] \quad (2)$$

In this work, LL subband is used for feature extraction followed by QR decomposition [10] to embed the watermark. The LL subband is used due to energy compaction property and then upper triangular matrix *R* is used to form the dataset. In *R* matrix the significant coefficient as explained in embedding procedure is used for inserting the watermark bits as it provides more robustness as demonstrated through the experimental results.

## 2.2 Particle Swarm Optimization (PSO)

Particle swarm optimization is a swarm intelligence technique inspired from flocking of birds or fish schooling. The two important properties of swarm intelligence are self-organization and division of labor. Initially, PSO was proposed by J. Kennedy and R. Elberhart [8]. The steps involved in PSO algorithm are:

1. Position and velocity of the particles are randomly generated within search space.
2. Every particle is associated with its velocity determined as:

$$v_i(t+1) = wv_i(t) + c_1 r_1 \big(p_{\text{best,i}}(t) - X_i(t)\big) + c_2 r_2 (g_{\text{best}}(t+1) - X_i) \quad (3)$$

   where $w, c_1, c_2, r_1, r_2$ is the inertia, acceleration coefficients and random numbers. $v_i$ is the velocity of each particle, $p_{\text{best},i}$ represent the personal best of $i$th particle and $g_{\text{best}}$ is the global best.
3. The position of the particle is modified as:

$$X_i(t+1) = X_i + v_i \quad (4)$$

4. Evaluate the objective function defined by Eq. 5 in the proposed approach and update the population irrespective of fitness.
5. Update $P_{best}$ and $g_{best}$.

In this work PSO is used to find the scaling factor (watermark strength) $\gamma$ to embed the watermark in the selected regions of luminance component. The multiple scaling factor decided by PSO is used to reduce the trade-off between imperceptibility and robustness. The fitness function which is to be minimized is defined as:

$$\text{ming} = \frac{M}{\sum_{i=1}^{M} \text{NC(org\_ WM,rec\_ WM)}} - \text{NC(host\_ img,signed\_ img)} \quad (5)$$

Here $M$ ($M = 9$) is the number of JPEG compression attacks, org_ WM,rec_ WM,host_ img and signed_ img represents the original watermark, recovered watermark, host image and signed image, respectively.

### 2.3 *Lagrangian Twin Support Vector Regressions (LTSVR)*

LTSVR is a simple iterative algorithm which converges for any given starting point [11]. The main advantage of this method [11] is that the e-sensitive up-or-down bound regressor is determined using iterative method instead of using QPP. Considering the watermarking problem as a regression problem it is used to learn the image characteristics and due to its good generalization ability, the successful extraction of watermark is done against JPEG compression attack with quality factor ranging from 10 to 90.

## 3 Proposed Approach

The proposed approach consists of watermark embedding and extracting procedure described as:

### 3.1 *Watermark Embedding Procedure*

1. Transform the RGB image onto YCbCr space which consist of three channels Y, Cb and Cr. Y channel is selected for watermark embedding. Perform the Arnold transformation [13] on the original watermark to get the scrambled watermark which enhances the security of original watermark.
2. Apply one level DWT to Luminance channel (Y component) and obtained low frequency subband (*LL*) is further divided into blocks with dimension $4 \times 4$.

3. Fuzzy entropy [14] of every block (obtained in Step 2) is computed and arranges the blocks in descending order of fuzzy entropy.
4. Selected blocks based on the fuzzy entropy underwent through the process of QR factorization and a dataset of 2048 × 10 is formed represented by Eq. 6 using feature extracted from the upper triangular matrix R of each selected block.

$$\mathrm{DS} = \left\{ \begin{array}{c} (x_i, t_i) \in R^9 \times R : i = 1, 2, \ldots, m \\ = \left\{ \left( r_{1,1}, r_{1,2}, r_{1,3}, r_{2,2}, r_{2,3}, r_{2,4}, r_{3,3}, r_{3,4}, r_{4,4} \right), r_{1,4} \right\} \end{array} \right\} \tag{6}$$

5. The feature comprising dataset is partitioned and odd number of samples are used to train the LTSVR and even number of samples are used to embed the watermark corresponding to the output obtained by LTSVR on comparing with the original target value using the following formula:

$$\text{if wm\_ bit} = 1$$

$$r_{1,4}{}' = \max(r_{1,4}, r_{1,4}^{\mathrm{LTsvr}} + \gamma)$$

else

$$r'_{1,4} = \min\left(r_{1,4}, r_{1,4}^{\mathrm{LTsvr}} - \gamma\right) \tag{7}$$

here $\gamma$ is the optimal value of the watermark strength obtained by the PSO, scm_bit is the scrambled watermark bit obtained using Arnold transformation, $\mathrm{r}_{1,4}$ and $\mathrm{r}_{1,4}^{\mathrm{LTsvr}}$ is the target value and LTSVR output value, respectively.
6. After embedding the watermark onto luminance component, watermarked image is formed by combining the watermarked Y channel, Cb and Cr channel followed by inverse DWT. Finally, the imperceptibility of the watermark and quality of the watermarked image is evaluated by PSNR.

## 3.2 *Watermark Extraction Procedure*

Extracting the watermark from the watermarked image is the inverse process of watermark embedding which include following steps:

1. Transform the watermarked RGB image to YCbCr space to get the luminance component.
2. Perform one level DWT to the Y channel and using the key index, the selected blocks underwent through QR decomposition to obtain the upper triangular matrix R.
3. Using the R matrix, similar to Step 4 of embedding as explained in Sect. 3.1, dataset is formed and even number of samples are supplied to the trained LTSVR to get the desired output. Then the LTSVR output is compared with the target

and scrambled watermark bits are extracted according to the following rule:

$$\text{ext\_ WM}' = \begin{cases} 1 \text{ if } r_{1,4}' > r_{1,4}^{\text{LTsvr}} \\ 0 \quad \text{otherwise} \end{cases} \tag{8}$$

here $r_{1,4}'$ is the target value and $r_{1,4}^{LT\text{svr}}$ is the LTSVR output.

4. Inverse Arnold transformation [13] is applied to the extracted scrambled watermark obtained in Step 3 to obtain the extracted watermark.
5. Finally, the quality of extracted watermark is computed using BCR and NC.

## 4 Experimental Results and Analysis

The proposed method is evaluated by performing extensive experiments on different kinds of standard images and the results of images shown in Fig. 1 with dimension 512 × 512 are explained in this paper. A binary watermark image shown in Fig. 2 is used as watermark to embed in the color host image. After insertion of watermark using the embedding procedure in the host images, the corresponding watermarked images are shown in Fig. 3.

From Fig. 3 it is inferred that there is no visual quality degradation in the quality as measured by PSNR value defined as:

a b

**Fig. 1** Original host images **a** *Lena* **b** *Pepper*

**Fig. 2** Original watermark image

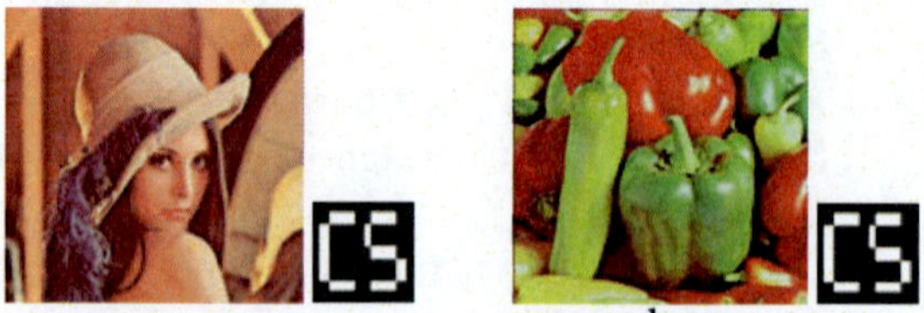

a b

PSNR: 53.1554, NC=1, BCR=1 PSNR:53.3320, NC:1, BCR:1

**Fig. 3** Watermarked Images along with extracted watermark without attack **a** *Lena* **b** *Pepper*

$$\text{PSNR}(\text{host\_ img,signed\_ img}) = 10\log_{10}\frac{255^2}{\text{MSE}}(\text{dB}) \tag{9}$$

where mean square error (MSE) is computed using:

$$\text{MSE} = \frac{\sum_{i=1}^{3}\sum_{j=1}^{M}\sum_{k=1}^{N}(\text{host\_ img}(i,j,k) - \text{signed\_ img}(i,j,k))^2}{3 \times M \times N} \tag{10}$$

The high PSNR value **53.1554** (dB) and **53.3320** (dB) corresponding to *Lena* and *Pepper* images shows that the good quality of watermarked images. Also the extracted watermark from these watermarked images as shown in Fig. 3 clearly reveals that there is no visually difference between original and extracted watermark as quantified by the NC and BCR values defined as:

$$\text{BCR}(\text{org\_ WM,rec\_ WM}) = \frac{\sum_{i=1}^{d}\sum_{j=1}^{d}(\text{org\_ WM}(i,d) \otimes \text{rec\_ WM}(j,d))}{d \times d} \tag{11}$$

$$\text{NC}\left(\text{org\_ WM,rec\_ WM} = \frac{\sum_{i=1}^{d}\sum_{j=1}^{d}\text{org\_ WM}(i,j) * \text{rec\_ WM}(i,j)}{\sum_{i=1}^{d}\sum_{j=1}^{d}\left[\text{org\_ WM}(i,n)\right]^2}\right) \tag{12}$$

Here org_WM represent the original watermark image, rec_WM indicate the extracted watermark and $d$ represent the dimension of the watermark $d = 32$. The NC and BCR value without performing signal processing operations (attacks) is described in Fig. 3. The NC and BCR values equal to one indicates the good visual quality of recovered watermark. The performance in terms of robustness of the proposed approach is verified by performing JPEG compression attack with quality factor ranging from 10 to 90 on *Lena* and *Pepper* images. The results evaluated against JPEG compression attacks are listed in Table 1.

From Table 1, it is observed that the proposed approach is robust against JPEG compression as the visual quality of extracted watermark is recognizable and it is degraded with JPEG compression at QF = 10. Since the robustness of the proposed scheme is evaluated on JPEG compression, so in this work a comparison only on the JPEG compression against different QF = 40, 60 and 80 is performed with the technique proposed by Su et al. [15]. The NC value of extracted watermark against QF = 40, 60 and 80 is 0.7338, 0.8270 and 0.9146 whereas using proposed scheme NC values against same QF on Lena image is 0.9883, 1 and 1. This clearly reveals the significant improvement using the combination of PSO and LTSVR in wavelet domain. In future, this work can be further extended to validate the proposed scheme against different kinds of signal processing attacks and geometric attacks.

**Table 1** BCR value of extracted watermark on *Lena* and *Pepper* images

| JPEG Compression attack (QF) | *Lena* | | *Pepper* | |
|---|---|---|---|---|
| | BCR | Visual quality | BCR | Visual quality |
| JPEG (QF = 90) | 1 |  | 1 |  |
| JPEG (QF = 80) | 1 |  | 1 |  |
| JPEG (QF = 70) | 1 |  | 0.9941 |  |
| JPEG (QF = 60) | 1 |  | 0.9941 |  |
| JPEG (QF = 50) | 0.9961 |  | |  |
| JPEG (QF = 40) | 0.9883 |  | 0.9883 | 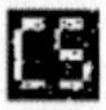 |
| JPEG (QF = 30) | 0.9590 |  | 0.9463 |  |
| JPEG (QF = 20) | 0.9275 |  | 0.8848 |  |
| JPEG (QF = 10) | 0.7783 |  | 0.7780 |  |

## 5 Conclusion

In the present work, a secure and robust color image watermarking scheme for copyright protection applications is proposed. The scheme uses the key features of wavelet transform and QR decomposition to embed the watermark in an imperceptible way. Through the combination of PSO and LTSVR, an optimized watermark strength is found and then robustness is verified by successfully extraction of watermark against JPEG compression attacks with different quality factors. Through the extensive experiments and results discussed in Sect. 4 reveals that the proposed scheme is imperceptible and robust as quantified by PSNR, NC and BCR values. Also the efficacy of the proposed scheme is validated by comparing with state of art method against JPEG compression attack. In future, the robustness of the proposed scheme can be enhanced against another signal processing attacks which includes filtering attacks, contrast enhancement and geometric operations including rotation, scaling and translation which is prime requirement of copyright protection applications.

## References

1. Zhang X, Su Q, Yuan Z, Liu D (2020, October) An efficient blind color image watermarking algorithm in spatial domain combining discrete Fourier transform. Optik 1(219):165272
2. Sharma C, Bhaskar A (2020, December 23) A review on video watermarking techniques for compressed domain with optimization algorithms. Mater Today Proc
3. Baba SE, Krikor LZ, Arif T, Shaaban Z (2010) Watermarking of digital images in frequency domain. Int J Autom Comput 7(1):17–22
4. Cedillo-Hernández M, García-Ugalde F, Nakano-Miyatake M, Pérez-Meana HM (2014) Robust hybrid color image watermarking method based on DFT domain and 2D histogram modification. SIViP 8(1):49–63
5. Singh RK, Singh AK (2019, June 15) A recent survey of DCT based digital image watermarking theories and techniques: a review. In: International conference on advanced informatics for computing research. Springer, Singapore, pp 431–440
6. Amer I, Hishmat P, Badawy W, Jullien G (2010) Comparisons and analysis of DCT-based image watermarking algorithms. In: Advanced techniques in computing sciences and software engineering. Springer, Dordrecht, pp 55–58
7. Karajeh H, Khatib T, Rajab L, Maqableh M (2019) A robust digital audio watermarking scheme based on DWT and Schur decomposition. Multimedia Tools Appl 78(13):18395–18418
8. Nandi S, Santhi V (2016) DWT–SVD-based watermarking scheme using optimization technique. In: Artificial intelligence and evolutionary computations in engineering systems 2016. Springer, New Delhi, pp 69–77
9. Chen HY, Zhu YS (2012) A robust watermarking algorithm based on QR factorization and DCT using quantization index modulation technique. J Zhejiang Univ Sci C 13(8):573–584
10. Song W, Hou JJ, Li ZH, Huang L (2011) Chaotic system and QR factorization based robust digital image watermarking algorithm. J Cent South Univ Technol 18(1):116–124
11. Balasundaram S, Tanveer M (2013) On Lagrangian twin support vector regression. Neural Comput Appl 22(1):257–267
12. Shensa MJ (1992) The discrete wavelet transform: wedding the a trous and Mallat algorithms. IEEE Trans Signal Process 40(10):2464–2482
13. Wu L, Zhang J, Deng W, He D (2009, December 26) Arnold transformation algorithm and anti-Arnold transformation algorithm. In: 2009 first international conference on information science and engineering. IEEE, pp 1164–1167
14. Lee HM, Chen CM, Chen JM, Jou YL (2001) An efficient fuzzy classifier with feature selection based on fuzzy entropy. IEEE Trans Syst Man Cybern Part B (Cybern) 31(3):426–432
15. Su Q, Wang G, Lv G, Zhang X, Deng G, Chen B (2017) A novel blind color image watermarking based on Contourlet transform and Hessenberg decomposition. Multimedia Tools Appl 76(6):8781–8801

# Study of Electromagnetic Induced Transparency and Its Dependence on Probe Decay for Cascade and Lambda Models

**Swaraj Shekhar Nande, Monika, Harish Singh Rawat, and Satya Kesh Dubey**

**Abstract** In this work interaction of the electromagnetic wave with a three-level cascade and lambda atomic system has been discussed analytically. The steady-state solution of optical Bloch equations for each atomic model is obtained to get the absorption and dispersion profile of the probe field. Furthermore, the phenomenon of Electromagnetic Induced Transparency (EIT) has been observed when the coupling field is applied along with the probe field. We also discuss the dependency of the EIT window on probe decay rate and establish the condition at which probe absorption in the presence of coupling field matches the one, without coupling field. Atom-based metrology techniques utilize EIT for microwave electric and magnetic field sensing. Results reported in this work may be proved beneficial in fixing the parametric regime for optimum results in EIT-based microwave metrology.

**Keywords** Electromagnetic Induced Transparency (EIT) · Three-level Cascade and Lambda system · Susceptibility

## 1 Introduction

In this modern era, atom-based metrology techniques are growing at a greater pace in every section of science and technology to make all physical quantities directly traceable to SI units or universal constants. The physics exploiting the atoms and their properties have found applications in various advanced fields such as quantum information processing [1], microwave E-field and H-field sensing [2–4], quantum communication [5], atomic clock [6] and many more. The reason behind this is that atom-based standards have the advantage of self-calibration and can be made

S. S. Nande
Visva-Bharati University, Shantiniketan, West Bengal 731235, India
e-mail: swarajshekhar324@gmail.com

Monika · H. S. Rawat · S. K. Dubey (✉)
Academy of Scientific and Innovative Research (AcSIR), Ghaziabad 201002, India

CSIR-National Physical Laboratory, Dr. K. S. Krishnan Marg, New Delhi 110012, India
e-mail: dubeysk@nplindia.org



S. Yadav et al. (eds.), *Recent Advances in Metrology*, Lecture Notes in Electrical Engineering 906, https://doi.org/10.1007/978-981-19-2468-2_24

directly SI-traceable due to the invariance of atomic parameters. The length, time and frequency measurement atomic standards have already being used for several years [7, 8]. However, work is still going on in many laboratories for the development of SI-traceable electric and magnetic field sensors.

It has been reported in many experimental studies that E-field and H-field can be sensed by making use of the quantum phenomena of Electromagnetically Induced Transparency (EIT) [9, 10], a technique in which all the measurements are done in terms of frequency instead of amplitude measurement making it SI-traceable. The Autler-Townes splitting [11, 12] of EIT peak in the presence of microwaves can be used for E-field sensing, as the splitting of EIT peak is directly proportional to the applied E-field strength [9, 13]. Furthermore, the magnetic field can be sensed by comparing the EIT resonance peak in the absence and presence of applied magnetic field strength and measuring the shift in the resonance peak [10] which is proportional to the applied field strength. Thus, the whole concept of E-field and H-field sense is driven by the phenomenon of Electromagnetic induced Transparency [14–18] for precise sensing. Therefore, in this work, we present the basic theoretical model to obtain the phenomenon of Electromagnetic Induced Transparency (EIT) in a three-level atomic system. The method used in these models can also be applied to higher-level atomic systems to get promising results. Also, the dependence of the EIT window on probe decay rate has been reported that would help set the limit on decay rate to get optimal results.

The article is organized as follows. In section 2, out of three widely accepted cascade, lambda and vee models of the three-level atomic system, the first two are described and their corresponding density matrix equations are presented [19]. The steady-state solution of the above density matrix equations is obtained to get the density matrix element for probe absorption. Further, the expression for susceptibility is obtained to get the absorption and dispersion profiles of the probe field. Also, gives a detailed explanation of how the transparency appears in the probe absorption with the introduction of the coupling field, i.e., the phenomenon of Electromagnetic Induced Transparency (EIT) for both the cascade and lambda models. In section 3, we have shown the dependency of the EIT window on the probe decay rate. Finally, we sum up the main results of the work.

## 2 Models

The master equation involving intermediate state decay terms for the three-level atomic system is given by [20, 21]

$$\frac{\mathrm{d}\rho}{\mathrm{d}t} = -\frac{i}{\hbar}[H, \rho] - \frac{\Gamma(\sigma_+\sigma_-\rho - 2\sigma_-\rho\sigma_+ + \sigma_+\sigma_-)}{2} \tag{1}$$

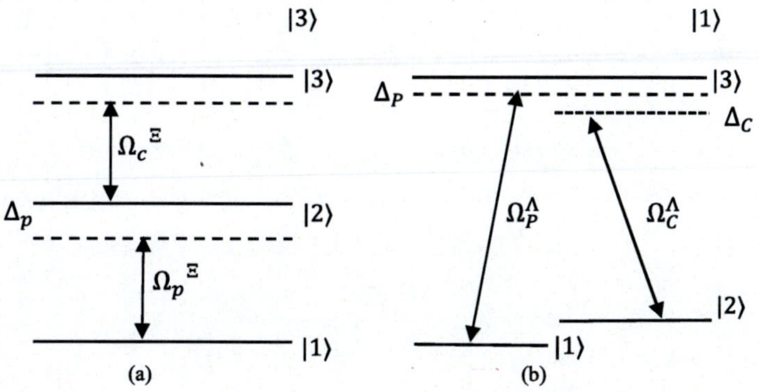


**Fig. 1** Energy level diagrams of the **a** three-level cascade model, and **b** three-level lambda model

where H is the Hamiltonian of the atomic system, ρ is the density matrix of the corresponding atomic system, $\hbar$ is the Plank's constant, Γ is the atomic decay rate and $\sigma_+ = |j><i|$ and $\sigma_- = |i><j|$ are the atomic projection operators ($i, j =$ 1, 2, 3 and $j > i$). The cascade and lambda models of the three-level atomic system are shown in Fig. 1. with their levels following the same energy hierarchy condition, i.e., energy state $|3\rangle$ is the highest energy state, and energy state $|1\rangle$ is the lowest energy state.

## 2.1 Cascade Model

The time-dependent Hamiltonian [22] for the three-level cascade atomic system can be expressed as

$$H^{\Xi} = \begin{bmatrix} \hbar\omega_1 & -\frac{\hbar\Omega_p^{\Xi}}{2}e^{i\omega_p t} & 0 \\ -\frac{\hbar\Omega_p^{\Xi}}{2}e^{-i\omega_p t} & \hbar\omega_2 & -\frac{\hbar\Omega_c^{\Xi}}{2}e^{i\omega_c t} \\ 0 & -\frac{\hbar\Omega_c^{\Xi}}{2}e^{-i\omega_c t} & \hbar\omega_3 \end{bmatrix}. \quad (2).$$

where $\Omega_p^{\Xi}$ and $\Omega_c^{\Xi}$ are the Rabi frequencies associated with probe and coupling field, respectively; and $\hbar\omega_1$, $\hbar\omega_2$, and $\hbar\omega_3$. represent the stationary state energy levels of the cascade system. Substituting $H^{\Xi}$ into Eq. (1), and applying slowly varying amplitude approximation, i.e., $\rho_{11} = \tilde{\rho}_{11}$; $\rho_{22} = \tilde{\rho}_{22}$; $\rho_{33} = \tilde{\rho}_{33}$; $\rho_{12} = \tilde{\rho}_{12}e^{-i\omega_p t}$; $\rho_{23} = \tilde{\rho}_{23}e^{-i\omega_c t}$; $\rho_{13} = \tilde{\rho}_{13}e^{-i(\omega_p+\omega_c)t}$, the optical Bloch equations for the density matrix elements to study the evolution of three-level cascade system are given by [23]

$$\dot{\tilde{\rho}}_{11} = -i\frac{\Omega_p^{\Xi}}{2}(\tilde{\rho}_{12} - \tilde{\rho}_{21}) + \Gamma_{12}\tilde{\rho}_{22} + \Gamma_{13}\tilde{\rho}_{33} \quad (3)$$

$$\dot{\tilde{\rho}}_{22} = i\frac{\Omega_p^{\Xi}}{2}(\tilde{\rho}_{12} - \tilde{\rho}_{21}) - i\frac{\Omega_c^{\Xi}}{2}(\tilde{\rho}_{23} - \tilde{\rho}_{32}) - \Gamma_{12}\tilde{\rho}_{22} + \Gamma_{23}\tilde{\rho}_{33} \tag{4}$$

$$\dot{\tilde{\rho}}_{33} = i\frac{\Omega_c^{\Xi}}{2}(\tilde{\rho}_{23} - \tilde{\rho}_{32}) - \Gamma_{13}\tilde{\rho}_{22} - \Gamma_{23}\tilde{\rho}_{33} \tag{5}$$

$$\dot{\tilde{\rho}}_{12} = -i\frac{\Omega_p^{\Xi}}{2}(\tilde{\rho}_{11} - \tilde{\rho}_{22}) - i\frac{\Omega_c^{\Xi}}{2}\tilde{\rho}_{13} + C_{12}\tilde{\rho}_{12} \tag{6}$$

$$\dot{\tilde{\rho}}_{13} = i\frac{\Omega_p^{\Xi}}{2}\tilde{\rho}_{23} - i\frac{\Omega_c^{\Xi}}{2}\tilde{\rho}_{12} + C_{13}\tilde{\rho}_{13} \tag{7}$$

$$\dot{\tilde{\rho}}_{23} = -i\frac{\Omega_c^{\Xi}}{2}(\tilde{\rho}_{22} - \tilde{\rho}_{33}) + i\frac{\Omega_p^{\Xi}}{2}\tilde{\rho}_{13} + C_{23}\tilde{\rho}_{23} \tag{8}$$

while $\dot{\tilde{\rho}}_{21}$, $\dot{\tilde{\rho}}_{31}$ and $\dot{\tilde{\rho}}_{32}$ are given by the complex conjugate of $\dot{\tilde{\rho}}_{12}$, $\dot{\tilde{\rho}}_{13}$ and $\dot{\tilde{\rho}}_{23}$, respectively. The parameters $C_{12,}$ $C_{13}$ and $C_{23}$ represent the complex detuning rates and are given by $\left(-i\Delta_P - \frac{\Gamma_{12}}{2}\right)$, $\left(-i(\Delta_P + \Delta_C) - \frac{\Gamma_{13}}{2} - \frac{\Gamma_{23}}{2}\right)$ and $\left(-i\Delta_C - \frac{\Gamma_{12}}{2} - \frac{\Gamma_{13}}{2} - \frac{\Gamma_{23}}{2}\right)$, respectively. The detunings are given by $\Delta_p = \omega_p - \omega_{12}$ and $\Delta_c = \omega_c - \omega_{23}$, where $\omega_{ij} = \omega_j - \omega_i$ with $(j > i)$ gives the on-resonance transition frequency between the states $i$ and $j$. Further the terms $\Gamma_{ij}$ represent the transition decay rates from upper state $j$ to lower state $i$.

The steady-state solution (i.e., $\dot{\tilde{\rho}} = 0$) of the above density matrix equations under weak probe limit ($\Omega_P^{\Xi} \ll \Omega_C^{\Xi}$) gives [23]

$$\begin{aligned} \tilde{\rho}_{12} &\approx \frac{2i\Omega_P^{\Xi}C_{13}}{4C_{12}C_{13} + \left(\Omega_C^{\Xi}\right)^2}\ ; \ \tilde{\rho}_{13} \approx -\frac{\Omega_P^{\Xi}\,\Omega_C^{\Xi}}{4C_{12}C_{13} + \left(\Omega_C^{\Xi}\right)^2}\ ; \\ \tilde{\rho}_{23} &\approx \frac{i\ (\Omega_P^{\Xi})^2\ \Omega_C^{\Xi}}{2C_{23}(4C_{12}C_{13} + (\Omega_C^{\Xi})^2)}\ ; \end{aligned} \tag{9}$$

Since the main focus of this paper is on the absorption profile of the probe beam, so linear optical susceptibility can be expressed as [24]

$$\chi^{\Xi} = \frac{2N\mu_{21}^2\tilde{\rho}_{12}}{\hbar\varepsilon\Omega_P^{\Xi}}; \tag{10}$$

whose real part describes the dispersive properties and the imaginary part describes the absorption properties of the probe beam. Here $N$ is the local atomic density, $\hbar$ is the Plank's constant, $\varepsilon$ is the permittivity of the atomic medium, and $\mu_{21}$ is the transition dipole moment corresponding to atomic level $|1\rangle$ and $|2\rangle$ between which probe field is being applied.

The real and imaginary part of electrical susceptibility as a function of probe detuning, $\Delta_p$ with or without the application of the coupling field is shown in Fig. 2.

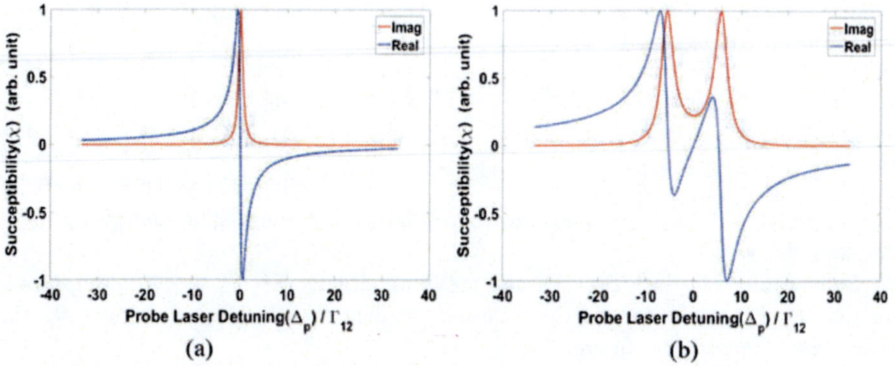


**Fig. 2** Variation of the real (blue) and imaginary (red) part of electrical susceptibility with normalized probe detuning when **a** $\Omega_C^\Xi = 0$ MHz and **b** $\Omega_C^\Xi = 5$ MHz for the Cascade atomic model

It can be observed from Fig. 2a. that without coupling frequency the imaginary part (i.e., absorption) of susceptibility shows Lorentzian shape while the real part (i.e., refractive index) shows a steep dispersive profile (because of material nonlinear refractive properties). But as the coupling field is applied, then at a particular coupling frequency (i.e.,$\Omega_p^\Xi \ll \Omega_c^\Xi$) a dip starts appearing in the absorption profile making the medium transparent to probe beam, for the particular frequency range shown in Fig. 2b. This reshaping of the absorption peak can be explained using the destructive quantum interference phenomenon between the excitation pathways as proposed by Harris et al. [18] in his work. This transparency in the probe beam is called Electromagnetic Induced Transparency [18, 19]. The other parameters are chosen to be $\Gamma_{12} = 0.3$ MHz, $\Gamma_{13} = 0.2$ MHz, $\Gamma_{23} = 3$ MHz and $\Omega_p^\Xi = 0.5$ MHz to obtain these profiles. The decay term $\Gamma_{13}$ is chosen to be small as it is dipole forbidden transition in cascade atomic system. There exist many ways to explain the physical mechanism of EIT [25]. They are equivalent under certain conditions. In the simplest case, a direct population transfer $|1\rangle \rightarrow |2\rangle$ can be estimated. However, the coupling between the states $|2\rangle$ and $|3\rangle$ by the strong beam leads to an alternative pathway to achieve the state $|2\rangle$:$|1\rangle \rightarrow |2\rangle \rightarrow |3\rangle \rightarrow |2\rangle$. It should be noticed that transition to state $|1\rangle$ is not taking place. Hence, the destructive interference of the probability amplitudes for these two routes leads to the transparency for the probe beam. Further, the dispersion profile which is positive below resonance and negative above resonance with no dispersion at resonance shows an additional peak in the presence of coupling field making refractive index zero at two additional points as observed in Fig. 2a, b. This shows the anomalous behavior of the dispersion profile of the probe beam [23].

Similarly, the time-dependent Hamiltonian [22] for the three-level lambda atomic system can be expressed as

$$H^{\Lambda} = \begin{bmatrix} \hbar\omega_1 & 0 & -\frac{\hbar\Omega_P^{\Lambda}}{2}e^{i\omega_P t} \\ 0 & \hbar\omega_2 & -\frac{\hbar\Omega_C^{\Lambda}}{2}e^{i\omega_C t} \\ -\frac{\hbar\Omega_P^{\Lambda}}{2}e^{-i\omega_P t} & -\frac{\hbar\Omega_C^{\Lambda}}{2}e^{-i\omega_C t} & \hbar\omega_3 \end{bmatrix} \quad (11)$$

where $\Omega_P^{\Lambda}$ and $\Omega_C^{\Lambda}$ are the Rabi frequencies associated with probe and coupling field, respectively; and $\hbar\omega_1$, $\hbar\omega_2$ and $\hbar\omega_3$ represent the stationary state energy levels of the lambda system.

Substituting $H^{\Lambda}$ into Eq. (1), and applying slowly varying amplitude approximation as in cascade system, the time-independent optical Bloch equations for the three-level lambda system are [24]

$$\dot{\tilde{\rho}}_{11} = -i\frac{\Omega_P^{\Lambda}}{2}(\tilde{\rho}_{13} - \tilde{\rho}_{31}) + \Gamma_{12}\tilde{\rho}_{22} + \Gamma_{13}\tilde{\rho}_{33} \quad (12)$$

$$\dot{\tilde{\rho}}_{22} = -i\frac{\Omega_C^{\Lambda}}{2}(\tilde{\rho}_{23} - \tilde{\rho}_{32}) - \Gamma_{12}\tilde{\rho}_{22} + \Gamma_{23}\tilde{\rho}_{33} \quad (13)$$

$$\dot{\tilde{\rho}}_{33} = i\frac{\Omega_P^{\Lambda}}{2}(\tilde{\rho}_{13} - \tilde{\rho}_{31}) + i\frac{\Omega_C^{\Lambda}}{2}(\tilde{\rho}_{23} - \tilde{\rho}_{32}) - \Gamma_{13}\tilde{\rho}_{33} - \Gamma_{23}\tilde{\rho}_{33} \quad (14)$$

$$\dot{\tilde{\rho}}_{12} = i\frac{\Omega_P^{\Lambda}}{2}\tilde{\rho}_{32} - i\frac{\Omega_C^{\Lambda}}{2}\tilde{\rho}_{13} + \mathcal{L}_{12}\tilde{\rho}_{12} \quad (15)$$

$$\dot{\tilde{\rho}}_{13} = -i\frac{\Omega_P^{\Lambda}}{2}(\tilde{\rho}_{11} - \tilde{\rho}_{33}) - i\frac{\Omega_C^{\Lambda}}{2}\tilde{\rho}_{12} + \mathcal{L}_{13}\tilde{\rho}_{13} \quad (16)$$

$$\dot{\tilde{\rho}}_{32} = i\frac{\Omega_C^{\Lambda}}{2}(\tilde{\rho}_{22} - \tilde{\rho}_{33}) + i\frac{\Omega_P^{\Lambda}}{2}\tilde{\rho}_{12} + \mathcal{L}_{32}\tilde{\rho}_{32} \quad (17)$$

$$\dot{\tilde{\rho}}_{21} = \dot{\tilde{\rho}}_{12}^{*},\ \dot{\tilde{\rho}}_{31} = \dot{\tilde{\rho}}_{13}^{*},\ \text{and}\ \dot{\tilde{\rho}}_{23} = \dot{\tilde{\rho}}_{32}^{*} \quad (18)$$

where the parameters $\mathcal{L}_{12}$, $\mathcal{L}_{13}$ and $\mathcal{L}_{32}$ represents the complex detuning rates and are given by $\left(-i(\Delta_P - \Delta_C) - \frac{\Gamma_{12}}{2}\right)$,$\left(-i\Delta_P - \frac{\Gamma_{13}}{2} - \frac{\Gamma_{23}}{2}\right)$, and $\left(i\Delta_C - \frac{\Gamma_{12}}{2} - \frac{\Gamma_{13}}{2} - \frac{\Gamma_{23}}{2}\right)$, respectively. The probe and coupling detuning are given by $\Delta_P = \omega_P - \omega_{13}$ and $\Delta_C = \omega_C - \omega_{23}$, respectively, where $\omega_{ij} = \omega_j - \omega_i$ with $(j > i)$ gives the on-resonance transition frequency between the states $i$ and $j$. Further the term $\Gamma_{ij}$ corresponds to transition decay rates from upper state $j$ to lower state $i$.

The approximate steady-state analytical solution (i.e., $\dot{\tilde{\rho}} = 0$) for the above-mentioned density matrix equations of Lambda model under weak probe limit ($\Omega_P^{\Lambda} \ll \Omega_C^{\Lambda}$) gives

$$\tilde{\rho}_{13} \approx \frac{2i\Omega_P^{\Lambda}\mathcal{L}_{12}}{4\mathcal{L}_{12}\mathcal{L}_{13} + (\Omega_C^{\Lambda})^2};\ \tilde{\rho}_{12} \approx -\frac{\Omega_P^{\Lambda}\Omega_C^{\Lambda}\mathcal{L}_{12}}{(4\mathcal{L}_{12}\mathcal{L}_{13} + (\Omega_C^{\Lambda})^2)\left(\mathcal{L}_{12} + \frac{(\Omega_P^{\Lambda})^2}{4\mathcal{L}_{32}}\right)}; \tag{19}$$

$$\tilde{\rho}_{32} \approx \frac{i(\Omega_P^{\Lambda})^2\Omega_C^{\Lambda}\mathcal{L}_{12}}{2\mathcal{L}_{32}(4\mathcal{L}_{12}\mathcal{L}_{13} + (\Omega_C^{\Lambda})^2)\left(\mathcal{L}_{12} + \frac{(\Omega_P^{\Lambda})^2}{4\mathcal{L}_{32}}\right)}; \tag{20}$$

Since the probe beam as shown in Fig. 1b. is associated with the density matrix element $\rho_{13}$, the dispersive and absorption profiles of the probe beam are given by the real and imaginary part of linear optical susceptibility [24]

$$\chi^{\Lambda} = \frac{2N\mu_{31}^2\tilde{\rho}_{13}}{\hbar\varepsilon\Omega_P^{\Lambda}}; \tag{21}$$

where $\mu_{31}$ corresponds to transition dipole moment between atomic states along which probe frequency, $\Omega_P^{\Lambda}$ is being applied.

Similar to the cascade model, the real and imaginary parts of electrical susceptibility are also plotted for the lambda model by scanning the probe laser over a frequency range of − 40 MHz to + 40 MHz as shown in Fig. 3. The other numerical parameters are considered to be the same as in the cascade model, i.e., $\Gamma_{12} = 0.2\,\text{MHz}$, $\Gamma_{13} = 0.3\,\text{MHz}$,$\Gamma_{23} = 3\,\text{MHz}$ and $\Omega_p^{\Lambda} = 0.5\,\text{MHz}$. The probe decay rate $\Gamma_{12}$ is taken to be small as it is dipole forbidden transition for the Lambda system. It can be observed from Fig. 3a. that without the application of coupling frequency the absorption peak shows Lorentzian profile while the real part (i.e., refractive index) shows a steep dispersive profile. But as the coupling field is applied, then after a particular coupling frequency (i.e.,$\Omega_p^{\Xi} \ll \Omega_c^{\Xi}$) a transparent window appears at the perfect resonance of the probe beam (i.e., at $\Delta_P = 0$) as shown in Fig. 3b. These results are similar to that of the cascade model. This transparency at perfect

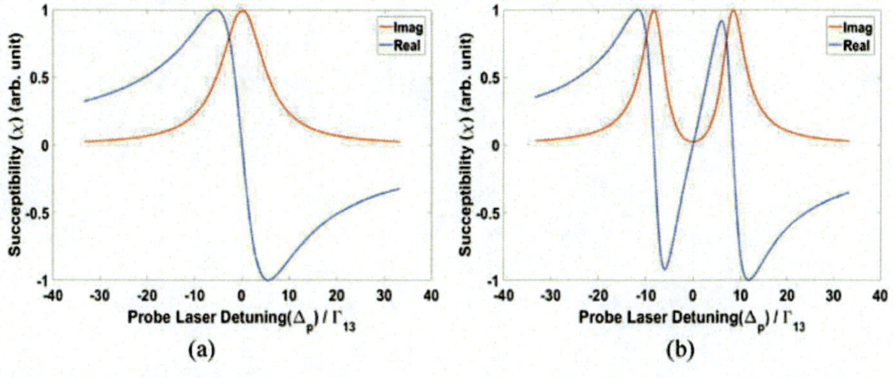


**Fig. 3** Variation of the real (blue) and imaginary (red) part of electrical susceptibility as a function of normalized probe detuning at **a** $\Omega_C^{\Lambda} = 0$ MHz and **b** $\Omega_C^{\Lambda} = 5$ MHz for the Lambda atomic system

resonance is due to destructive quantum interference between excitation pathways, called the Electromagnetic Induced Transparency [26]. However, the refractive index is showing odd function behavior having a positive value below resonance and negative value above resonance with no dispersion at perfect resonance as observed in Fig. 3b. This shows the anomalous behavior of the dispersion profile of the probe field.

## 3 Numerical Results

In this section, the dependence of the absorption profile on the probe decay rate for all the above-mentioned three-level atomic models has been discussed.

The blue and red color curves, shown in Fig. 4 represent the absorption profile of the probe beam for the cascade model in the absence and presence of a coupling field, respectively. It can be observed from Fig. 4a that when probe decay rate ($\Gamma_{21}$) equals to 0.3 MHz, a sharp transparency window appears at the center of the absorption profile at $\Omega_c^{\Xi} = 5$ MHz. It can be observed from the x-axis of Fig. 4b. that as the probe decay rate, ($\Gamma_{21}$) is increased from 0.3 MHz to 3 MHz, the transparency

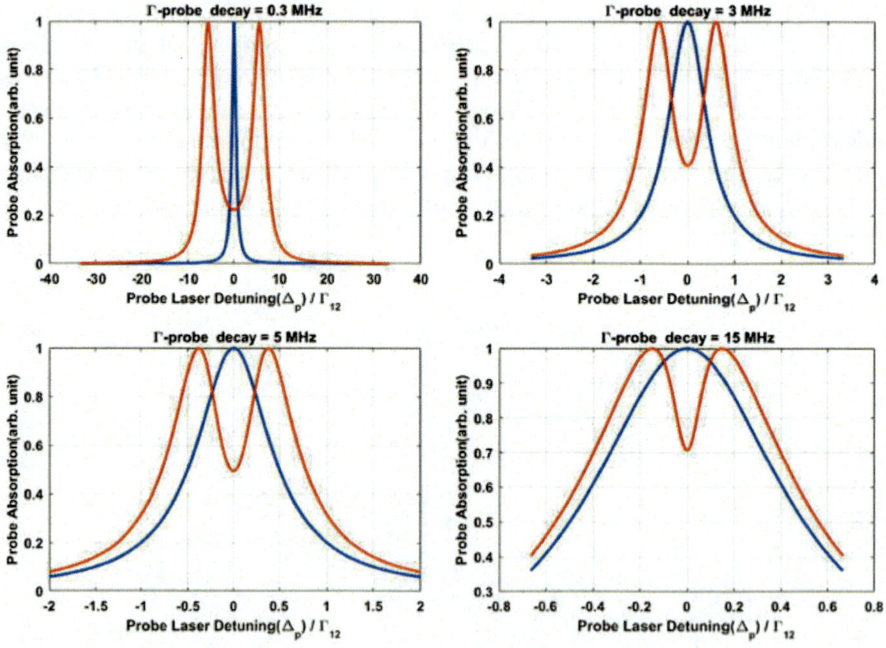


**Fig. 4** Probe absorption as a function of the probe decay rate, $\Gamma_{12}$ **a** 0.3 MHz **b** 3 MHz **c** 5 MHz and **d** 15 MHz for the Cascade atomic system. The blue and red curves representing the probe absorption at $\Omega_c^{\Xi} = 0$ MHz, and $\Omega_c^{\Xi} = 5$ MHz, respectively

window decreases along with the decrease in the absorption profile linewidth. When the probe decay rate is increased to 5 MHz, the transparency dip and linewidth further decrease, and finally, at the probe decay rate of 15 MHz, the transparency decreases to 30% from 70%, thereby increasing probe absorption. This can be interpreted to disturb the resulting quantum interference phenomenon [26] between the excitation pathways created by the coupling laser. As with the increase of probe decay rate for a fixed coupling laser frequency, excitation of electrons from the ground state to the higher state gradually decreases. This phenomenon destroys the transparency which was being created by the strong coupling laser for the weak probe laser. Therefore, one should not work in a regime where the decay rate of the atomic states is so high that EIT is not recognizable anymore. This study can be very essential for EIT-based applications utilizing cascade systems like high-resolution spectroscopy [27], SI -traceable RF and microwave Sensing [9], etc.

A little different type of variations in absorption profile can also be observed in Fig. 5. for the lambda model on increasing probe decay rate (i.e., $\Gamma_{13}$ incase of lambda model), both in the absence and presence of a coupling beam. The other numerical parameters are taken to be the same as given for Fig. 4. plots. These parameters are chosen to be sensible but non-specific to study the qualitative nature of the imaginary part of electrical susceptibility.

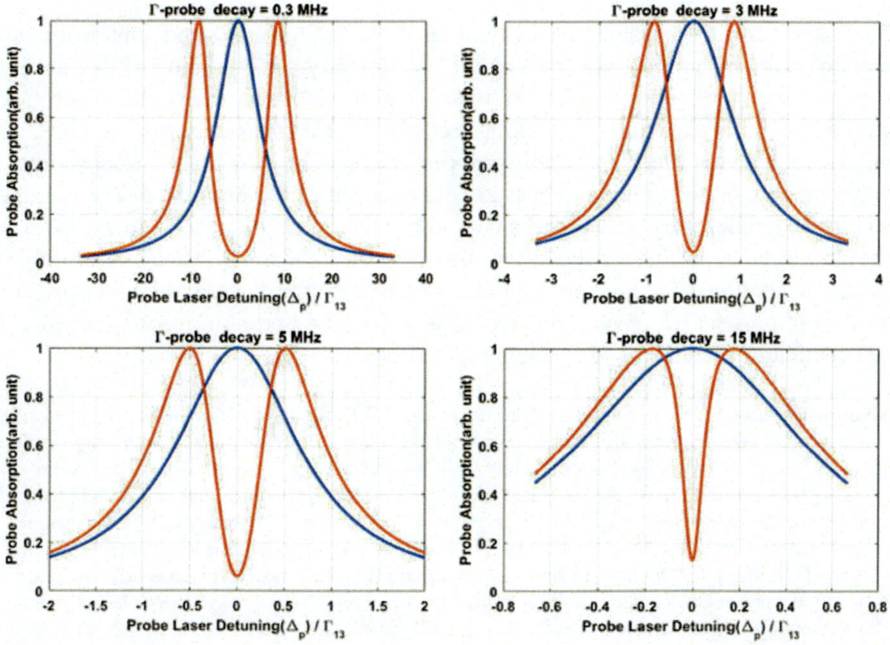


**Fig. 5** Probe absorption as a function of the probe decay rate, $\Gamma_{13}$ **a** 0.3 MHz **b** 3 MHz **c** 5 MHz and **d** 15 MHz for the Lambda atomic system. The blue and red curves representing the probe absorption at $\Omega_C^\Lambda = 0$ MHz and $\Omega_C^\Lambda = 5$ MHz, respectively

It can be observed from Fig. 5. that the linewidth of probe absorption both in the presence and absence of coupling strength decreases, similar to the case shown in Fig. 4. for the cascade system. Whereas, as the probe decay rate is increased from 0.3 MHz to 15 MHz, the transparency dip remains the same for the Lambda models as compared to the Cascade model, where a significant decrease in the amplitude of EIT dip appears. This signifies that fewer atoms participate in probe absorption in the case of the lambda model as compared to the cascade model in the presence of coupling strength. In other words, it explains the presence of a more significant dark state in the case of the Lambda as compared to the Cascade model of a three-level atomic system. Thus, an atomic vapor system containing alkali atoms for which probe decay is significantly less can be very useful for studying and understanding the physics of atomic spectroscopy and the metrology associated with it. This study can be very essential for EIT-based applications utilizing Lambda systems like lossless propagation of light pulse [28], all-optical switching [29] quantum storage devices, logic gates, and the generation of photonic qubits [30], slowing down group velocity of light [31], even stopping and storage of light pulse in the EIT medium [32].

## 4 Conclusion

This paper presents a systematic study of the three-level cascade and lambda atomic system driven by probe and coupling field. The steady-state solution of the density matrix equations is obtained for both atomic systems. Also, the probe absorption without coupling field and at coupling strength of 5 MHz has been plotted. Further, the phenomenon of Electromagnetic Induced Transparency (EIT) and its dependency on probe decay rate has been discussed. Since at high values of probe decay, the EIT window becomes very less significant for both the cascade model, therefore one should not work in the regime where the decay rate of atomic states is too high. Thus, our findings can be utilized in deciding which atomic systems to be used for producing proper EIT peaks. This would help find the optimum results that would be beneficial for atom-based sensing and metrology.

**Acknowledgements** S. S. Nande is thankful to ISERC Department, Visva-Bharati University, Shantiniketan, West Bengal. S. S. Nande is also thankful to the Electromagnetic Metrology section of CSIR-National Physical Laboratory, New Delhi for supporting his visit. Monika is thankful to Union Grant Commission for Junior Research Fellowship. H. S. Rawat is thankful to Science and Engineering Research Board (SERB), Department of Science and Technology (DST), Government of India, Confederation of Indian Industry (CII), and Rohde & Schwarz India Pvt. Ltd. for the Prime Minister Fellowship for Doctoral Research. Monika and H. S. Rawat are also thankful to AcSIR for allowing to work at CSIR- National Physical Laboratory, New Delhi through their Ph.D. Program. The authors are also thankful to the CSIR for FTT project MLP191132 to established Rydberg atom-based sensing at CSIR-NPL.

## References


1. Raimond JM (2002) Quantum information processing with atoms and cavities. Superlattices Microstruct 32(4–6):187–193. https://doi.org/10.1016/S0749-6036(03)00019-3
2. Rawat HS, Dubey SK, Ojha VN (2019) Polarization dependence of interferences inside rubidium atomic vapor governing microwave vector E-field metrology. J Opt Soc Am B 36(12):3547. https://doi.org/10.1364/josab.36.003547
3. Monika, Rawat HS, Dubey SK (2020) RF E-field sensing using rydberg atom-based microwave electrometry. Mapan J Metrol Soc. https://doi.org/10.1007/s12647-020-00404-2
4. Liu X, Jiang Z, Qu J, Hou D, Sun F (2020) Rabi resonances in buffer-gas-filled Cs-vapor cells for SI-traceable microwave magnetic field detection. IEEE Trans Instrum Meas 69(4):1302–1307. https://doi.org/10.1109/TIM.2019.2918514
5. Song Z et al (2019) Rydberg-atom-based digital communication using a continuously tunable radio-frequency carrier. Opt Express 27(6):8848. https://doi.org/10.1364/oe.27.008848
6. Arora P et al (2014) Atomic clocks: a brief history and current status of research in India. Pramana J Phys 82(2):173–183. https://doi.org/10.1007/s12043-014-0709-6
7. Hall JL (2006) Nobel lecture: defining and measuring optical frequencies. Rev Mod Phys 78(4):1279–1295. https://doi.org/10.1103/RevModPhys.78.1279
8. Hanneke D, Fogwell S, Gabrielse G (2008) New measurement of the electron magnetic moment and the fine structure constant. Phys Rev Lett 100(12):1–4. https://doi.org/10.1103/PhysRevLett.100.120801
9. Holloway CL et al (2014) Broadband Rydberg atom-based electric-field probe for SI-traceable, self-calibrated measurements. IEEE Trans Antennas Propag 62(12):6169–6182. https://doi.org/10.1109/TAP.2014.2360208
10. Jenkins R (2019) Measuring magnetic field gradients using electromagnetically induced transparency in Rb vapor. Undergraduate Honors Theses, Paper 1370. https://scholarworks.wm.edu/honorstheses/1370
11. Autler SH, Townes CH (1955) Stark effect in rapidly varying fields. Phys Rev. https://doi.org/10.1103/PhysRev.100.703
12. Rawat HS, Dubey SK, Ojha VN (2018) Distinction between double electromagnetically induced transparency and double Autler-Townes splitting in RF-driven four-level ladder 87Rb atomic vapor. J Phys B At Mol Opt Phys 51(15):155401. https://doi.org/10.1088/1361-6455/aacdd9
13. Kumar S, Fan H, Kübler H, Sheng J, Shaffer JP (2017) Atom-based sensing of weak radio frequency electric fields using homodyne readout. Sci Rep. https://doi.org/10.1038/srep42981
14. Harris SE (1997) Electromagnetically induced transparency. Phys Today. https://doi.org/10.1063/1.881806
15. Marangos JP (1998) Electromagnetically induced transparency. J Mod Opt. https://doi.org/10.1080/09500349808231909
16. Lukin MD, Imamoğlu A (2001) Controlling photons using electromagnetically induced transparency. Nature. https://doi.org/10.1038/35095000
17. Petrosyan D, Otterbach J, Fleischhauer M (2011) Electromagnetically induced transparency with Rydberg atoms. Phys Rev Lett. https://doi.org/10.1103/PhysRevLett.107.213601
18. Harris SE, Field JE, Imamoğlu A (1990) Nonlinear optical processes using electromagnetically induced transparency. Phys Rev Lett 64(10):1107–1110. https://doi.org/10.1103/PhysRevLett.64.1107
19. Fleischhauer M, Imamoglu A, Marangos PJ (2005) Electromagnetically induced transparency. Rev Mod Phys 77(2):633–673. https://doi.org/10.1103/RevModPhys.77.633
20. Boyd R (1992) Nonlinear optics. Academic Press, Boston
21. Manzano D (2020) A short introduction to the Lindblad master equation. AIP Adv. https://doi.org/10.1063/1.5115323
22. Blaauboer M (1997) Steady state behaviour in atomic three-level $\Lambda$ and ladder systems with incoherent population pumping. Phys Rev A 55(3):2459–2462

23. Gavryusev V et al (2016) Density matrix reconstruction of three-level atoms via Rydberg electromagnetically induced transparency. J Phys B At Mol Opt Phys 49(16). https://doi.org/10.1088/0953-4075/49/16/164002.
24. Alam S (1999) Lasers without inversion and electromagnetically induced transparency. SPIE Optical Engineering Press, Bellingham, Wash.
25. Su JJ, Yu IA (2003) The study of coherence-induced phenomena using double-sided feynman diagrams. Chin J Phys 41(6):627–642
26. Boller KJ, Imamolu A, Harris SE (1991) Observation of electromagnetically induced transparency. Phys Rev Lett. https://doi.org/10.1103/PhysRevLett.66.2593
27. Krishna A, Pandey K, Wasan A, Natarajan V (2005) High-resolution hyperfine spectroscopy of excited states using electromagnetically induced transparency. Europhys Lett 72(2):221–227. https://doi.org/10.1209/epl/i2005-10228-6
28. Harris SE, Luo ZF (1995) Preparation energy for electromagnetically induced transparency. Phys Rev A 52(2):928–931. https://doi.org/10.1103/PhysRevA.52.R928
29. Ham BS (2002) Dark resonance based optical switching. J Mod Opt 49(14–15):2477–2484. https://doi.org/10.1080/09500340210000011301
30. Figueroa Barragan EV, Vewinger F, Appel J, Lvovsky AI (2013) Adiabatic transfer of quantum optical information in atomic vapor 32(19):PDP_FC3. https://doi.org/10.1364/ope.2006.pdp_fc3
31. Vestergaard Hau L, Harris SE, Dutton Z, Behroozi CH (1999) Light speed reduction to 17 metres per second in an ultracold atomic gas. Nature 397(6720):594–598, 1999. https://doi.org/10.1038/17561
32. Phillips DF, Fleischhauer A, Mair A, Walsworth RL, Lukin MD (2001) Storage of light in atomic vapor. Phys Rev Lett 86(5):783–786. https://doi.org/10.1103/PhysRevLett.86.783

# Reliability Measurement of Complex Industrial Redundant Systems Using Semi-Markov Process and Regenerative Point Technique

**Gitanjali**

**Abstract** In this paper, a repair-replacement problem is set out that improves the efficiency of the renovation of the manufacturing system. Using the semi-Markov process and RPT, we estimate the reliability measurements for instance availability, mean time to system failure, and busy periods of servers. For this, we recommend a non-identical unit redundant complex system of two inventive units and one duplicate unit. The system works in upstate if two out of three units are operational. For power supply in industries, electric generators are used as main units and solar panels can be taken as standby to fulfill the load requirement. Here repairing of inventive unit is done initially by common server and then by expert server. While attaining dissatisfaction in repairing in given time, main concern is given to inspecting the inventive unit, whereas duplicate unit directly undergoes for the replacement.

**Keywords** Reliability measurements · Industrial redundant system · Semi-Markov process · Specialist server · Priority of renovating the inventive unit

## 1 Introduction

In the emerging of digitalization, the latest scientific systems have emerged with complex configurations and high features. And, for that reason, these devices have taken place to execute singularly the task of old ones. Various researchers and scientists have designed a variety of complex industrial models and assert their strength under reliability procurement of standby system models [1–7]. Also sustaining the performance of the industrial systems adequately throughout their projected time seems to be a big challenge for almost all manufacturing systems. Despite this, it is the eminent fact that failure of the inventive unit may degrade the efficiency of the system and also cause damage to the entire system. Thus, authors suggested that after specific period of the failure of a repairable system, main concern should be given to inspection of inventive units which reveals whether the unit is further repairable or

Gitanjali (✉)
Department of Applied Sciences, Maharaja Surajmal Institute of Technology, GGSIPU, Delhi, India
e-mail: drgitanjali@msit.in

 
S. Yadav et al. (eds.), *Recent Advances in Metrology*, Lecture Notes in Electrical Engineering 906, https://doi.org/10.1007/978-981-19-2468-2_25

needs to be substituted [8–11]. Keeping all the views here, we designed a stochastic model of three non-identical units, in which two are inventive and other one is kept in cold standby. Repairing of inventive unit is done initially by common server and then by expert server. While attaining dissatisfaction in repairing in given time, duplicate unit directly undergoes for the replacement.

## 2 Notations

| | |
|---|---|
| $I_0/D_0$ | The inventive unit is in an operational state. |
| $D_{cs}$ | Duplicate unit is taken as cold standby. |
| $\alpha_0$ | Extreme rate of mending time of the inventive unit. |
| $\lambda_1/\lambda_2$ | The constant failure rate of the inventive unit/duplicate unit. |
| $a/b$ | The probability that the inventive failed unit is not repairable/repairable by some specialist server. |
| $f_0(t)/g_0(t)/g_e(t)$/h(t) | PDF of substitution time/repair time taken by common server/repair time taken by expert server/inspection time of the inventive unit. |
| $I_{ur}/I_{ui}/I_{urp}$ | The inventive unit and duplicate unit are failed and under repair/under inspection/under replacement to be done by a common server. |
| $I_{uR}/I_{uRe}$ | The inventive unit is failed and under repair to be done by a common server/specialist server from the previous state. |
| $I_{uI}/I_{uRp}$ | The inventive unit is under inspection/under replacement from the previous state to be done by a common server. |
| $D_{wr}/D_{ur}/D_{urp}$ | The duplicate unit is in a failure state and waiting for repair/under repair/under replacement to be done by a common server. |
| $D_{wR}/D_{uR}/D_{uRp}$ | The duplicate unit is in a failure state and waiting for repair/under repair/under replacement from the previous state to be done by a common server. |
| $\sim/*$ | Symbol for Laplace–Stieltjes transform (L.S.T)/Laplace transform. |
| ®/© | Symbols for Stieltjes convolution/Laplace convolution. |

The feasible state transition diagram for the system model is shown in Fig. 1.

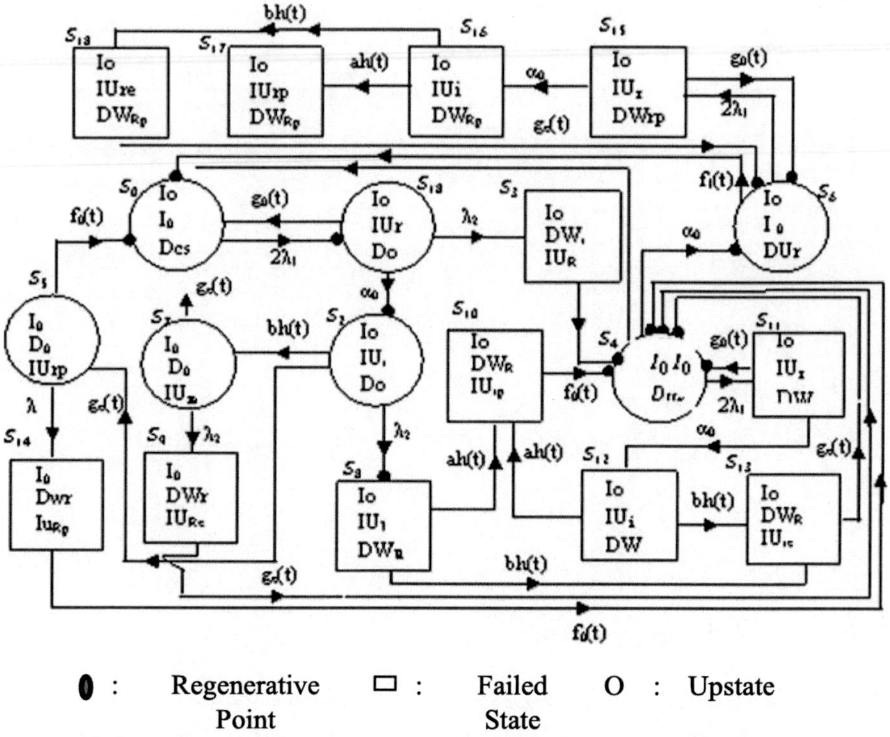


**Fig. 1** State transition diagram

## 3 Transition Probabilities and Mean Sojourn Times

Simple probabilistic considerations yield the following expressions for transition probabilities from state $i$ to $j$ is p $= p_{ij} = Q_{ij}(\infty) = Q(\infty)$. By probabilistic arguments, we have

$$p_{01}=\frac{2\lambda_1}{2\lambda_1},\ p_{10}=\frac{\theta_0}{\theta_0+\lambda_2+\alpha_0},\ p_{13}=p_{14.3}=\frac{\lambda_2}{\theta_0+\lambda_2+\alpha_0},\ p_{12}=\frac{\alpha_0}{\theta_0+\lambda_2+\alpha_0},$$
$$p_{25}=\frac{a\eta}{\eta+\lambda_2},\ p_{27}=\frac{b\eta}{\eta+\lambda_2},\ p_{28}=\frac{\lambda_2}{\eta+\lambda_2},\ p_{34}=\frac{\theta_0}{\theta_0},\ p_{40}=\frac{\theta_1}{\theta_1+2\lambda_1+\alpha_0},$$
$$p_{46}=\frac{\alpha_0}{\theta_1+2\lambda_1+\alpha_0},\ p_{4,11}=\frac{2\lambda_1}{\theta_1+2\lambda_1+\alpha_0},\ p_{50}=\frac{\beta_0}{\beta_0+\lambda_2},\ \ p_{5,14}=\frac{\lambda_2}{\beta_0+\lambda_2},$$
$$p_{60}=\frac{\beta_1}{\beta_1+2\lambda_1},\ \ p_{6,15}=\frac{2\lambda_1}{\beta_1+2\lambda_1},\ \ p_{70}=\frac{2\lambda_1}{\theta_1+\lambda_2},$$
$$p_{79}=\frac{\lambda_2}{\theta_2+\lambda_2},\ \ p_{8,10}=p_{12,10}=p_{16,17}=\frac{a\eta}{\eta},\ p_{8,13}=p_{12,13}=p_{16,18}=\frac{b\eta}{\eta},$$
$$p_{94}=p_{13,14}=p_{18,6}=\frac{\theta_2}{\theta_2},\ \ p_{10,4}=p_{14,4}=p_{17,6}=\frac{\beta_0}{\beta_0},$$
$$p_{11,4}=p_{15,6}=\frac{\theta_0}{\theta_0+\alpha_0},\ \ p_{11,12}=p_{15,16}=\frac{\alpha_0}{\theta_0+\alpha_0},\ \ p_{24.8,10}=\frac{\lambda_2}{\eta+\lambda_2}\frac{a\eta}{\eta}\frac{\beta_0}{\beta_0},$$
$$p_{24.8,13}=\frac{\lambda_2}{\eta+\lambda_2}\frac{b\eta}{\eta}\frac{\theta_2}{\theta_2},\ p_{44.11}=\frac{2\lambda_1}{\theta_1+2\lambda_1+\alpha_0}\frac{\theta_0}{\theta_0+\alpha_0},\ p_{54.14}=\frac{\lambda_2}{\beta_0+\lambda_2}\frac{\beta_0}{\beta_0},$$
$$p_{66.15}=\frac{2\lambda_1}{\beta_1+2\lambda_1}\frac{\theta_0}{\theta_0+\alpha_0},\ p_{11,4.12,10}=\frac{\alpha_0}{\theta_0+\alpha_0}\frac{a\eta}{\eta}\frac{\beta_0}{\beta_0},\ p_{11,4.12,13}=\frac{\alpha_0}{\theta_0+\alpha_0}\frac{b\eta}{\eta}\frac{\theta_2}{\theta_2},$$
$$p_{15,6.16,17}=\frac{\alpha_0}{\theta_0+\alpha_0}\frac{a\eta}{\eta}\frac{\beta_0}{\beta_0},\ p_{15,6.16,18}=\frac{\alpha_0}{\theta_0+\alpha_0}\frac{b\eta}{\eta}\frac{\theta_2}{\theta_2},\ p_{74.9}=\frac{\lambda_2}{\theta_2+\lambda_2}\frac{\theta_2}{\theta_2}.$$

It can simply be demonstrated to

$$p_{01}=p_{10}+p_{12}+p_{13}=p_{10}+p_{12}+p_{14.3}=p_{25}+p_{27}+p_{28}=p_{25}+p_{27}+$$
$$p_{24.8,10}+p_{24.8,13}=p_{34}=p_{40}+p_{46}+p_{4,11}=p_{40}+p_{46}+p_{44,11}=p_{50}+p_{5,14}=$$
$$p_{50}+p_{54,14}=p_{60}+p_{6,15}=p_{70}+p_{79}=p_{70}+p_{74.9}=p_{8,10}+p_{8,13}=p_{94}=$$
$$p_{10,4}=p_{11,4}+p_{11,4.12,10}+p_{11,4.12,13}=p_{12,10}+p_{12,13}=p_{14,4}=p_{15,6}+p_{15,6.16,17}+$$
$$p_{15,6.16,18}=p_{16,17}=p_{16,18}=p_{17.6}=p_{18,6}=1$$

The model has the following mutually exclusive states of the system:
Regenerative states:

$$S_0=(I_o,I_o,D_{cs}),\quad S_1=(I_o,I_{ur},D_o),\quad S_2=(I_o,I_{ui},D_o),\quad S_4=(I_o,I_o,D_{ur}),$$
$$S_5=\left(I_o,I_{urp},D_o\right),\quad S_6=\left(I_o,I_o,D_{urp}\right),\quad S_7=(I_o,I_{ure},D_o)$$

Non-regenerative states:

$$S_3=(I_o,I_{uR},D_{wr}),\quad S_8=(I_o,I_{uI},D_{wR}),\quad S_9=(I_o,I_{uI},D_{wR}),$$
$$S_{10}=\left(I_o,I_{urp},D_{wR}\right),S_{12}=(I_o,I_{ui},D_{wR}),S_{13}=(I_o,I_{ure},D_{wR}),$$
$$S_{14}=\left(I_o,I_{uRp},D_{wr}\right),\quad S_{15}=\left(I_o,I_{ui},D_{wrp}\right),S_{16}=\left(I_o,I_{uI},D_{wRp}\right),$$
$$S_{17}=\left(I_o,I_{urp},D_{wRp}\right),\quad S_{18}=\left(I_o,I_{ure},D_{wRp}\right),$$

The mean sojourn times $\mu_i$ in state $S_i$ is given by

$$\mu_0 = \text{m}_{01} = \frac{1}{2\lambda}, \mu_1 = \text{m}_{10} + \text{m}_{12} + \text{m}_{13} = \frac{1}{\theta_0 + \lambda_2 + \alpha_0},$$
$$\mu_1' = \frac{1}{\theta_0 + \lambda_2 + \alpha_0}\left(1 + \frac{\lambda_2}{\theta_0}\right),$$
$$\mu_2 = \frac{1}{\eta + \lambda_2}, \mu_2' = \frac{1}{\eta + \lambda_2}\left[1 + \lambda_2\left(\frac{a}{\beta_0} + \frac{1}{\eta} + \frac{b}{\theta_2}\right)\right],$$
$$\mu_3 = \frac{1}{\theta_0}, \mu_4 = \frac{1}{\theta_1 + 2\lambda_1 + \alpha_0}, \mu_5 = \frac{1}{\beta_0 + \lambda_2},$$
$$\mu_5' = \frac{1}{\beta_0 + \lambda_2}\left(1 + \frac{\lambda_2}{\beta_0}\right), \mu_6 = \frac{1}{\beta_1 + 2\lambda_1},$$
$$\mu_7 = \frac{1}{\theta_2 + \lambda_2}, \mu_7' = \frac{1}{\theta_2 + \lambda_2}\left(1 + \frac{\lambda_2}{\theta_2}\right).$$

# 4 Reliability and MTSF

Let $M_i(t)$ be the cumulative distribution function of the initial passage period from regenerative state $i$ to an abortive state (failed state), we have the following recursive relation for $M_i(t), i = 0, 1, 2, 4, 5, 6, 7$:

$$\text{M}_i(t) = \sum_j Q_{i,jj}(t) \boxed{S} \text{M}_j(t) + \sum_k Q_{i,jk}(t) \tag{1}$$

where Sj is an unfailed regenerative state to which the given regenerative state Si can transit and Sk is a failed state to which the state I can transit directly.

By making use of $L.S.T$ for Eq. (1) and resolving for $\tilde{M}_0(s)$ we get

$$\text{MSTF}(M_0) = \lim_{s\to 0} \frac{1 - \tilde{M}_0(s)}{s} = \frac{n_I}{d_I} \tag{2}$$

where $N_I = \mu_0 + \mu_1 + p_{12}(\mu_2 + p_{25}\mu_5 + p_{27}\mu_7)$ and $D_I = 1 - p_{10} - p_{12}(p_{25}p_{50} + p_{27}p_{70})$. For numerical analysis, we take

$$\begin{aligned} &g_0(t) = \theta_0 e^{-\theta_0 t}, g_1(t) = \theta_1 e^{-\theta_1 t}, g_e(t) = \theta_2 e^{-\theta_2 t}, \\ &h(t) = \eta e^{-\eta t}, f_0(t) = \beta_0 e^{-\beta_0 t} \text{ and } f_1(t) = \beta_1 e^{-\beta_1 t} \end{aligned} \tag{3}$$

From Eq. (2), we get

$$N_I = \frac{1}{2\lambda_1} + \frac{1}{(\theta_0 + \lambda_2 + \alpha_0)}\left(1 + \left(\frac{\alpha_0}{(\lambda_2 + \eta)}\left(\frac{1}{\eta} + \left(\frac{a}{(\lambda_2 + \beta_0)} + \frac{b}{(\lambda_2 + \theta_2)}\right)\right)\right)\right),$$

$$D_I = \left[1 - \frac{1}{(\theta_0 + \lambda_2 + \alpha_0)}\left(\frac{\alpha_0\eta}{(\lambda_2 + \eta)}\left(\left(\frac{a\beta_0}{(\lambda_2 + \beta_0)} + \frac{b\theta_2}{(\lambda_2 + \theta_2)}\right) + \theta_0\right)\right)\right] \quad (4)$$

## 5 Availability Analysis

At any instant '$t$', let $\Lambda_i(t)$ be the possibility of the system in upstate assuming that the system arrived in regenerative state $i$ at $t = 0$. The recursive relation for $\Lambda_i(t), i = 0, 1, 2, 4, 5, 6, 7$ is given as

$$\Lambda_i(t) = T_i(t) + \sum_j q_{i,j}^{(n)}(t) \copyright \Lambda_j(t) \quad (5)$$

where $T_i(t)$ is the probability that the system is active at first in regenerative state Si $\in E$ at time '$t$' deprived of visiting any further regenerative state.

By making use of Laplace transformation to (5) and resolve them for getting steady-state availability $\Lambda_0^*(s)$ as

$$\Lambda_0 = \lim_{s \to 0} s\Lambda_0^*(s) = \frac{n_{II}}{d_{II}} \quad (6)$$

where $n_{II} = p_{60}(p_{40} + p_{46})(\mu_0 + \mu_1 + p_{12}(\mu_2 + p_{25}\mu_5 + p_{27}\mu_7) + (p_{14.3} - p_{12}(p_{24.8,10} + p_{24.8,13} + p_{25}p_{54.14} - p_{27}p_{74.9}))(p_{46}\mu_6 + p_{60}\mu_4)$

$$\begin{aligned} n_{II} = {} & p_{60}(p_{40} + p_{46})(\mu_0 + \mu_1 + p_{12}(\mu_2 + p_{25}\mu_5 + p_{27}\mu_7) \\ & + \big(p_{14.3} - p_{12}\big(p_{24.8,10} + p_{24.8,13} + p_{25}p_{54.14} - p_{27}p_{74.9}\big)\big)(p_{46}\mu_6 + p_{60}\mu_4) \end{aligned}$$

and

$$\begin{aligned} d_{II} = {} & p_{60}(p_{40} + p_{46})(p_{10} + p_{12}(p_{25}p_{50} + p_{27}p_{70}) \\ & + ((p_{60} + p_{60}p_{46})\ \big(p_{14.3} - p_{12}\big(p_{24.8,10} + p_{24.8,13} + p_{25}p_{54.14} - p_{27}p_{74.9}\big)\big)) \\ & \mu_0 + \mu_1' + p_{12}\big(\mu_2' + p_{25}\mu_5' + p_{27}\mu_7'\big) \\ & + \big(p_{14.3} + p_{12}\big(p_{24.8,10} + p_{24.8,13} + p_{25}p_{54.14} + p_{27}p_{74.9}\big)\big)\ \big(p_{46}\mu_6' + p_{60}\mu_4'\big)\big), \\ & T_0(t) = e^{-2\lambda_1 t},\ T_1(t) = e^{-(\lambda_2+\alpha_0)t}\overline{G_0}(t),\ T_2(t) = e^{-\lambda_2 t}\overline{H}(t), \\ & T_4(t) = e^{-(2\lambda_1+\alpha_0)t}\overline{G_1}(t), T_5(t) = e^{-\lambda_2 t}\overline{F_0}(t), \\ & T_6(t) = e^{-2\lambda_1 t}\overline{F_1}(t),\ dT_7(t) = e^{-\lambda_2 t}\overline{G_e}(t). \end{aligned}$$

Applying distributions mentioned in (3), we get

$$N_{II} = \frac{(\theta_1+\alpha_0)}{(\theta_1+2\lambda_1+\alpha_0)}\frac{\beta_1}{(2\lambda_1+\beta_1)}\frac{1}{(\theta_0+\lambda_2+\alpha_0)}1$$
$$+\frac{(\theta_0+\lambda_2+\alpha_0)}{2\lambda_1}+\frac{\alpha_0\eta}{(\lambda_2+\eta)}\left(\frac{a}{(\lambda_2+\beta_0)}+\frac{b}{(\lambda_2+\theta_2)}+\frac{1}{\eta}\right)$$
$$+\frac{\lambda_2}{(\theta_1+\alpha_0)}1-\frac{\alpha_0\eta}{(\lambda_2+\eta)}\left(\frac{a}{(\lambda_2+\beta_0)}+\frac{b}{(\lambda_2+\theta_2)}+\frac{1}{\eta}\right)\left(\frac{\alpha_0}{\beta_1}+1\right),$$

$$D_{II} = \frac{(\theta_1+\alpha_0)}{(\theta_1+2\lambda_1+\alpha_0)}\frac{\beta_1}{(2\lambda_1+\beta_1)}\frac{1}{(\theta_0+\lambda_2+\alpha_0)}$$
$$\left[\theta_0+\frac{\alpha_0\eta}{(\lambda_2+\eta)}\left(\frac{a}{(\lambda_2+\beta_0)}+\frac{b}{(\lambda_2+\theta_2)}+\frac{1}{\eta}\right)+\right.$$
$$\left(\frac{(\theta_0+2\lambda_1+\alpha_0)}{(\theta_1+\alpha_0)}+\frac{\alpha_0}{(\theta_1+\alpha_0)}\right)\frac{\lambda_2}{2\lambda_1}$$
$$\left(1-\frac{\alpha_0\eta}{(\lambda_2+\eta)}\left(\frac{a}{(\lambda_2+\beta_0)}+\frac{b}{(\lambda_2+\theta_2)}+\frac{1}{\eta}\right)\right) \tag{8}$$
$$+\left(\frac{\lambda_2}{\theta_0}+1\right)+\frac{\alpha_0}{(\lambda_2+\eta)}\left(\left(1+\lambda_2\eta\left(\frac{a}{\beta_0}+\frac{b}{\theta_2}+\frac{1}{\eta}\right)\right.\right.$$
$$\left.\left.+\frac{a\eta}{(\lambda_2+\beta_0)}\left(\frac{\lambda_2}{\beta_0}+1\right)+\frac{b\eta}{(\lambda_2+\theta_2)}\left(\frac{\lambda_2}{\theta_2}+1\right)\right)\right)$$

## 6 Analysis of Busy Period of Servers

At any instant '$t$', let $C_i^r(t)$ be the probability that the common server is busy in repairing the failed unit assuming that the system entered regenerative state $S_i$ at $t = 0$. Here are the recursive relations for $C_i^r(t)$ are given as

$$C_i^r(t) = P_i(t) + \sum_j q_{i,j}^{(n)}(t) \copyright C_j^r(t) \tag{9}$$

where regenerative state i can transit through n $\geq$ 1 (natural number) transitions with any consecutive regenerative state j. By making use of Eq. (5) for Laplace transforms and resolving them for getting the busy period of common server $C_0^{r*}(s)$ as

$$C_0^r = \lim_{s\to 0} sC_0^{r*}(s) = \frac{n_{III}}{d_{II}} \tag{10}$$

where

$$n_{III} = p_{60}(p_{40} + p_{46})\big(P_1^*(0) + p_{12}\big(P_2^*(0) + p_{25}P_5^*(0)\big) + p_{14.3}+$$
$$p_{12}\big(p_{24.8,10} + p_{24.8,13} + p_{25}p_{54.14} + p_{27}p_{74.9}\big)\big(p_{60}P_4^*(0) + p_{46}P_6^*(0)\big)+$$
$$p_{60}p_{12}p_{27}(p_{40} + p_{46})P_7^*(0)$$

and $d_{II}$ is already specified,

Here,

$$P_1(t) = e^{-(\lambda_2+\alpha_0)t}\overline{G_0}(t) + \big(\lambda_2 e^{-(\lambda_2+\alpha_0)t}©1\big)\overline{G_0}(t),\ P_2(t) = e^{-\lambda_2 t}\overline{H}(t)+$$
$$\big(\lambda_2 e^{-\lambda_2 t}©1\big)\overline{H}(t) + \big(\lambda_2 e^{-\lambda_2 t}\overline{H}(t)©\overline{ah}(t)©1\big)©\overline{F_0}(t)+$$
$$\big(\lambda_2 e^{-\lambda_2 t}\overline{H}(t)©\overline{bh}(t)©1\big)\overline{G_e}(t),\ P_4(t) = e^{-(2\lambda_1+\alpha_0)t}\overline{G_1}(t),\ P_5(t) = e^{-\lambda_2 t}\overline{F_0}(t)+$$
$$\big(\lambda_2 e^{-\lambda_2 t}(1)\overline{F_0}(t),\ P_6(t) = e^{-2\lambda_1 t}\overline{F}_1(t) + \big(2\lambda_1 e^{-2\lambda_1 t}©\big)\overline{F}_1(t) \text{ and}$$
$$P_7(t) = e^{-(\lambda_2+\alpha_0)t}\overline{G_0}(t) + \big(\lambda_2 e^{-(\lambda_2+\alpha_0)t}©1\big)\overline{G_0}(t). \tag{11}$$

Using (3), we have

$$N_{III} = \frac{(\theta_1+\alpha_0)}{(\theta_1+2\lambda_1+\alpha_0)}\frac{\beta_1}{(2\lambda_1+\beta_1)}\frac{1}{(\theta_0+\lambda_2+\alpha_0)}$$
$$[(1+\lambda_2) + \frac{\alpha_0}{(\lambda_2+\eta)}(\left(1+\lambda_2\eta\left(\frac{a}{\beta_0}+\frac{b}{\theta_2}+\frac{1}{\eta}\right)\right)+$$
$$\frac{a\eta(1+\lambda_2)}{(\lambda_2+\eta)(\lambda_2+\beta_0)}) + \frac{\lambda_2}{(\theta_1+\alpha_0)}$$
$$\left(1+\frac{\alpha_0\eta}{(\lambda_2+\eta)}\left(\frac{b}{(\lambda_2+\theta_2)}+\frac{a}{(\lambda_2+\beta_0)}+\frac{1}{\eta}\right)\left(\frac{\alpha_0}{\beta_1}+1\right)\right)+\frac{\alpha_0 b\eta}{(\lambda_2+\eta)}\frac{(1+\lambda_2)}{(\lambda_2+\theta_2)}\Bigg]$$

## 7 Analysis of Numbers of Repairs and Replacements of the Failed Unit

Let $V_i^c(t)$ be the probable number of visits by the common server in (0, $t$] assuming that the system entered the regenerative state $S_i$ at $t = 0$. We get the recursive relation for $V_i^c(t)$ as

$$V_i^c(t) = \sum_j Q_{i,j}(t)\boxed{S}\left[\delta_j + V_j^c(t)\right] \tag{12}$$

where $j$ is any regenerative state to which the given regenerative state $i$ transits and $\delta_i. = 1$, if $j$ is the regenerative state wherever server ensures his work anew, else $\delta_i. =$ 0. By making use of $L.S.T$ for Eq. (12) and resolving them for getting an projected number of repairs of the inventive unit by a common server $\widetilde{V_0^c}(s)$ as

$$V_0^c = \lim_{s\to 0} s\widetilde{V_0^c}(s) = \frac{n_{IV}}{d_{II}} \tag{13}$$

$n_{IV} = ((p_{12}(p_{27}p_{74.9} + p_{24.8,13}) + 1)\,(p_{40} + p_{46})p_{60} + (p_{40} + p_{46})p_{12}p_{60}(p_{27} + p_{24.8,13})$ and $d_{II}$ is already specified.

$$N_V = \frac{(\theta_1 + \alpha_0)}{(\theta_1 + 2\lambda_1 + \alpha_0)} \frac{\beta_1}{(2\lambda_1 + \beta_1)} \frac{1}{(\theta_0 + \lambda_2 + \alpha_0)}$$
$$\left(\frac{\alpha_0\lambda_2\eta b}{(\lambda_2 + \eta)}\left(\frac{1}{(\lambda_2 + \theta_2)} + \frac{1}{\eta}\right) + (\theta_0 + \alpha_0 + \lambda_2) + \alpha_0 b\right)$$

## 8 Revenue Attainment

Profit attained to the system model in steady state can be analyzed as

$$P = C_1\Lambda_0 - C_2C_0^r - C_3E_0^r - C_4V_0^c - C_5V_0^e \tag{14}$$

where

| | |
|---|---|
| $C_1$ | Returns for each unit uptime of the system. |
| $C_2 =$ | Rate intended for each unit by which the common server does repair, inspection, and replacement of the unit. |
| $C_3$ | Rate intended for each unit by which the specialist server does repair of the unit. |
| $C_4$ | Rate intended for every visit of a common server for doing his job. |
| $C_5$ | Rate intended for every visit of a specialist server for repairing the inventive unit. |

## 9 Case Study

From above numerical results for particular values $C1 = 5000$, $C2 = 600$, $C3 = 900$, $C4 = 100$, and $C5 = 150$, we effectively obtained expected revenue of the system. By using regenerative point technique (RPT) and arbitrary distributions for inspection, repair, and replacement times, we get the graphical results as shown in Figs. 2, 3 and 4. It may be observed that all reliability measures increased for higher values of repair rates of common server for the main unit. However, their values decreased for the improved values of the failure rate of the main as well as duplicate units.

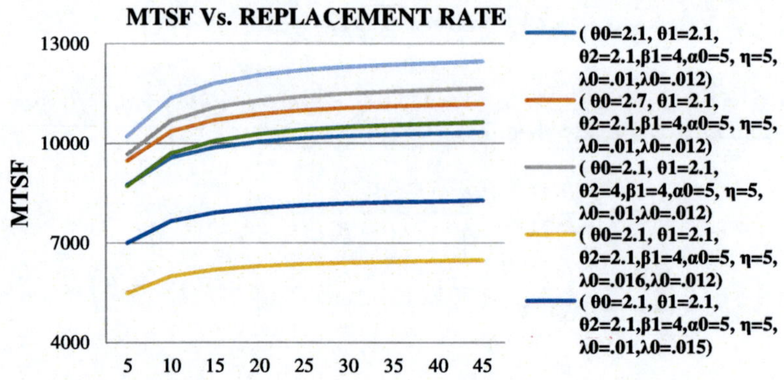


**Fig. 2** MTSF versus replacement rate

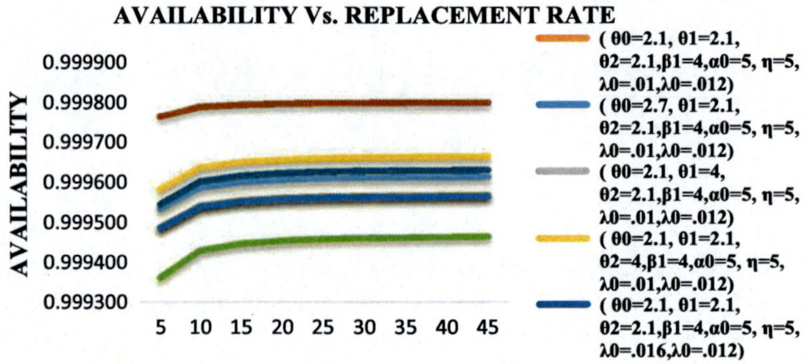


**Fig. 3** Availability versus replacement rate

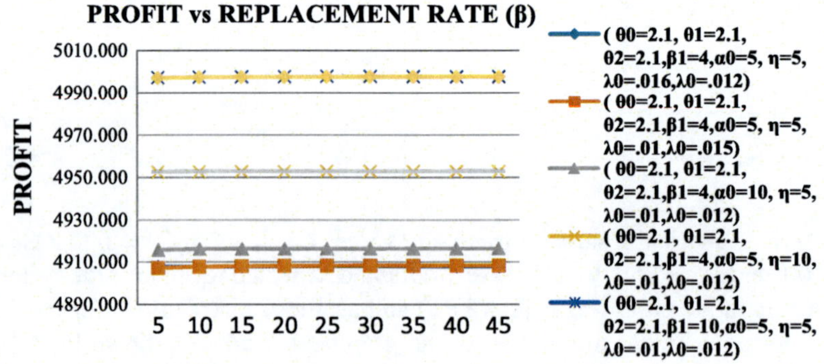


**Fig. 4** Profit versus replacement rate

## 10 Conclusion

Any complex redundant system of two non-identical inventive units and one duplicate unit where priority is given to inspection of inventive units for checking the feasibility of further repairing of the failed unit is beneficial when there will be a constantly increased in the repair rate of the expert server.

## References


1. Singhal N, Sharma SP (2019) Availability analysis of industrial systems using Markov process and generalized fuzzy numbers. Mapan 34:79–91
2. Lado A, Singh VV (2019) Cost assessment of complex repairable system consisting two subsystems in series configuration using Gumbel Hougaard family copula. Int J Qual Reliab Manag 36(10):1683–1698
3. Gitanjali, Malik SC (2019) Cost-benefit evaluation of a repairable system with alternate repair and Weibull distribution. Int J Agricult Stat Sci 15(2):741–749
4. Kadyan S, Malik SC, Gitanjali (2020) stochastic analysis of a three-unit non-identical repairable system with simultaneous working of cold st and by units. J Reliab Stat Stud 13(2–4):385–400
5 . Malik SC, Bhardwaj RK (2010) Stochastic analysis of 2-out-of-3 redundant systems with random appearance and disappearance of the repair facility. J Comb Inf Syst Sci (JCISS) 35(1–2):43–59
6. Gitanjali, Kumari,S.(2018) Reliability modeling of a 2- out-of-3 redundant system with alternate repair. Int J Stat Reliab Eng 5(2):101–107
7. Kumar, Pawar D, Malik SC (2019) Profit analysis of a warm standby non-identical unit system with single server performing in normal/abnormal environment. Life Cycle Reliab Saf Eng 8:219–226
8. Pundir PS, Patawa R, Gupta PK (2018) Stochastic outlook of two non-identical unit parallel system with priority in repair. Cogent Math Stat 5(1):1467208
9. Kumar A, Pawar D, Malik SC (2019) Weathering server system with non-identical units and priority to repair of the main unit. J Adv Res Dyn Control Syst 11(10)
10. Kadyan S, Barak MSG (2020) Stochastic analysis of a non-identical repairable system of three units with priority for operation and simultaneous working of cold standby units. Int J Stat Reliab Eng 7(2):269–274
11. Malik SC, Gitanjali (2014) Reliability measures of a parallel system with repair by an expert server subject to maximum repair and inspection times of ordinary server. Int J Math Arch 5(1):162–169

# Comparison of Different Feature Selection Approaches on Breast Cancer Dataset

**Parul Grover, Anita Choudhary, and Megha Khatri**

**Abstract** Breast cancer was found to be the second most common type of cancer which occurred around the world in the year 2018. It takes a huge amount of effort from the healthcare team to diagnose the type of breast cancer. It is crucial to correctly classify the tumor type, as misclassification might lead to the loss of life of the patient. This work aims at investigating different nuclear features present in the dataset and using this domain knowledge, determines the features which could assist in achieving the model with the highest accuracy. Also, select features using other feature selection algorithms such as Boruta, exhaustive search methods, and choose the best among them. Further, compare numerous model performances using domain knowledge and the selected feature selection approach, respectively.

**Keywords** FNA · FNAC · PCA · SMOTe

## 1 Introduction

Due to the advancements in science, it has been possible to treat patients who are diagnosed at an early stage of breast cancer. The most common methods of sampling are mammography or fine-needle aspiration. If a benign case is wrongly detected as malignant and vice versa, it might lead to incorrect surgery or loss of life of the patient. It is therefore quite crucial to accurately detect the cancer type.

P. Grover (✉)
Guru Gobind Singh Indraprastha University, Delhi, India
e-mail: groverparul91@gmail.com

A. Choudhary
Guru Tegh Bahadur Institute of Technology, Guru Gobind Singh Indraprastha University, Delhi, India

M. Khatri
School of Electronics and Electrical Engineering, Lovely Professional University, Phagwara, Punjab, India
e-mail: megha.25035@lpu.co.in

 
S. Yadav et al. (eds.), *Recent Advances in Metrology*, Lecture Notes in Electrical Engineering 906, https://doi.org/10.1007/978-981-19-2468-2_26

**Table 1** represents numerical features obtained after the process of FNAC [2]

| Nuclear feature | Explanation |
|---|---|
| Radius | Average of radial length from the center to each point on the nuclear border |
| Texture | It is measured by calculating the grayscale values of the pixels |
| Perimeter | The distance around the nuclear border |
| Area | It can be calculated by adding the pixels that are inside the border and half of the pixels on the perimeter |
| Compactness | It can be calculated using perimeter and area features using the following formula: $\text{perimeter}^2/\text{area}$ |
| Smoothness | Calculated by measuring the variation between the length of the radius and mean length of the adjacent radii |
| Concavity | It is measured using indentation size in the nuclei |
| Concave points | The number of indentations on the nucleus is known as concave points |
| Symmetry | The difference in length between line segments. These segments need to be perpendicular to the axis |
| Fractal dimension | The measure of nuclear border irregularities |

Machine learning can be used to assist the doctors and physicians in the process of classification of the tumor as benign or malignant which would reduce the chances of misclassification of tumor.

## *1.1 About the Dataset*

The dataset used during this research is provided by researchers from the University of Wisconsin Hospital. It consists of 33 variables which are fetched using image from the technique named fine-needle aspirate (FNA) and taken from samples of 569 patients [1]. It contains one categorical variable (diagnosis) with two levels 'B' (Benignant) and 'M' (Malignant) and 30 numerical columns.[1] These 30 numerical features present in the dataset are calculated using mean, standard error, and worst of each variable present in Table 1.

Fine-needle aspiration cytology (FNAC) is the technique used commonly by healthcare teams to diagnose the presence of cancer. It became quite popular in the 1950s and 1960s, where a 10-ml syringe along with 22-to-23-gauge needles is used to extract the matter from the breast and is studied. The study of nuclear features present in the tumor cells is known as morphometry.

[1] UCI (1995).

### 1.2 *Distribution of Variables*

In order to understand the distribution of the variables present in the dataset, different graphs were plotted such as density plots, swarm plots, and correlation plots. These plots helped in understanding which features can be used to distinguish the malignant from benign cases and vice versa (Fig. 1).

## 2 Data Preprocessing

After the process of data analysis using various plots, the first step in data preparation is to select a subset of variables that be used in diagnosing the type of tumor. This process of selecting a subset from the original set of variables is known as feature selection.

In this work, three different feature selection methods are applied. The first method involves the use of domain knowledge using Pearson correlation matrices and swarm plots. The second method uses Boruta technique for the selection of features, and the third method involves the use of the exhaustive search algorithm. Later, PCA for feature reduction and SMOTe for balancing of the dataset are implemented to each selected subset individually. Feature selection can be of three types: filter method, wrapper method, and embedded method.

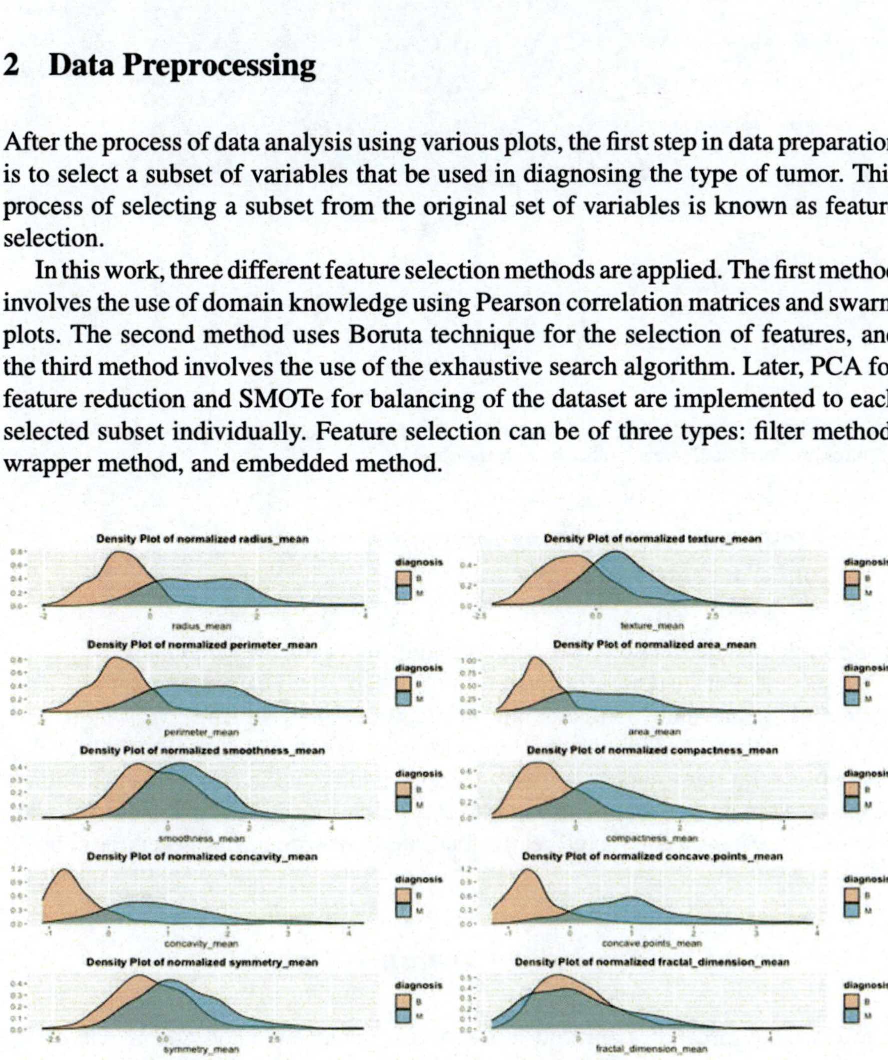


**Fig. 1** Distribution of variables using density plot. This plot is used while selecting variables using domain knowledge

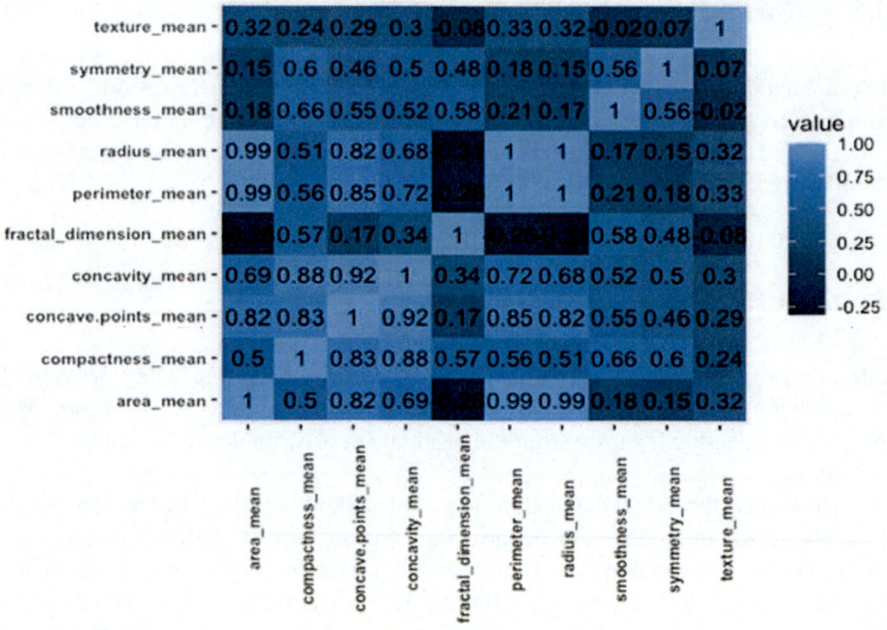


**Fig. 2** Correlation matrix for the mean subset. This shows a figure with values of how different features are correlated to each other in a given subset

## 2.1 *Feature Selection Using Domain Knowledge and Correlation Matrix*

The correlation matrices represent the correlation between the variables present in the same subsets. For instance, for feature 'area_mean', the Pearson correlation values of 'radius_mean' and 'perimeter_mean' are above 0.90 which is very high. This implies that they are highly correlated with each other, and hence, the removal of two among the three should be performed to reduce duplication. This selection is based on domain knowledge. Similarly, this process is again carried out for other remaining variables, and therefore, the final subset of variables is generated (Fig. 2).

## 2.2 *Feature Selection Using Exhaustive Search*

The exhaustive search algorithm is a type of wrapper method and is one of the greediest algorithms used for feature selection. It requires high computational power and is only suggested to be used when the dataset is small. It runs $2^{n-1}$ times to find the best set of variables with the highest accuracy. It was found that the adjusted $R$

square value was maximum when the selected number of variables was 13. Hence, from the 30 variables present in the dataset, a subset of 13 features was created.

## 2.3 Feature Selection Using Boruta

Boruta algorithm is another type of wrapper method which is based around random forest classification. The first step involves the creation of random variables called shadow variables in the dataset. Then, it trains a random forest classifier on the extended dataset where it checks whether a real feature has higher importance as compared to its shadow features. This process is repeated until all the features are either confirmed or rejected. After the application of Boruta, it was found that 3 variables could be confirmed as not important, while 27 were considered important variables. Hence, the three unimportant variables were eliminated from the dataset. It is computationally less expensive and is also recommended if the dataset used is large.

## 2.4 Code Used for Feature Selection

The code for feature selection using Boruta and exhaustive search is provided in this section.

```
library (Boruta)
breastCancer.train.norm.boruta$diagnosis<- factor(ifelse(breastCancer.train.norm.
   boruta$diagnosis== "B", 0, 1))
boruta<- Boruta(diagnosis~., data=breastCancer.train.norm.boruta[,−1])
plot(boruta, las = 2, cex.axis = 0.5) plotImpHistory(boruta)
TentativeRoughFix(Boruta)
attStats(Boruta)
getNonRejectedFormula(boruta) getConfirmedFormula(boruta)
```

The plot generated using the above code shows the 27 important variables, and 3 variables are found as not important. Hence, these not important variables can be removed while selecting the final subset using Boruta.

```
library(leaps)
breastCancer.train.norm$diagnosis<- factor(ifelse(breastCancer.train.norm
   $diagnosis == "B", 0, 1))
search<- regsubsets(diagnosis~., data = breastCancer.train.norm[,−1], nbest =
1, nvmax = dim(breastCancer)[2], method = "exhaustive")
sum <- summary(search)
sum$adjr2
```

The above code is used for finding the best subset using exhaustive search method for feature selection.

## 3 Results and Discussions

After the application of feature selection methods, three different subsets were created using each method. The selected variables further underwent feature reduction using PCA, which was then followed by the balancing of the dataset. Then, synthetic minority oversampling technique (SMOTe) is applied to balance the class variables.

It was found that among 30 variables, 17 features can be used for diagnosis of the tumor. Further, two other feature selection methods, Boruta and exhaustive search, were implemented on the standardized dataset. A subset of 27 variables and 13 variables was created using Boruta and exhaustive search, respectively (Fig. 3).

Further, six different algorithms were implemented on the cleaned and processed dataset to evaluate the performance. The six algorithms used were KNN, Naïve Bayes, boosted trees, logistic regression, neural network, and linear discriminant analysis. The selected features using different feature selection algorithms were compared using numerous model performances. Among the different matrices, confusion matrix was drawn for each algorithm which provided false and true values.

The total count of the false negative and false positive must be reduced; however, it is more important to minimize the number of false-negative cases as it is far more dangerous for patients to be wrongly detected as healthy when they are sick (False negative). Whereas, in the case of predicting non-cancerous as cancerous (False positive) is still an acceptable loss, as the patient can get a second opinion before proceeding with the treatment. Figure 4 shows the count of false negative obtained when different algorithms were implemented.

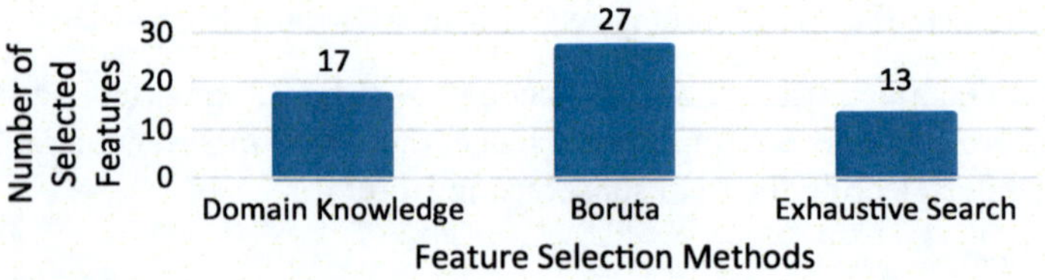


**Fig. 3** Count of selected features using three feature selection approaches. This figure shows the different number of selected subsets for each approach

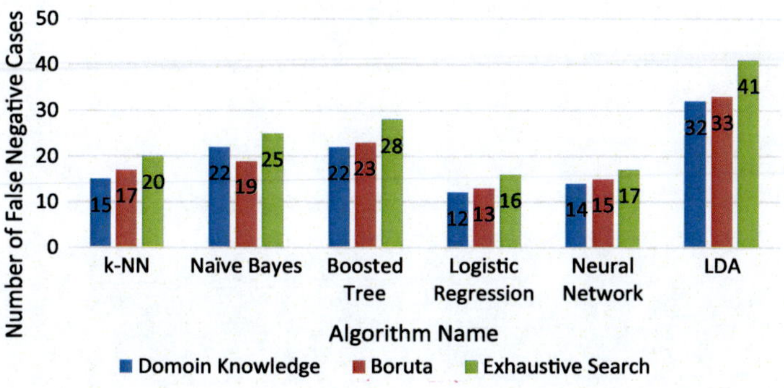


**Fig. 4** Count of false-negative cases using three feature selection approaches. The above figure shows the number of false-negative cases for each approach. Logistic regression had the minimum number of false-negative cases

## 4 Conclusions

On comparing incorrectly detected cases and majorly the false negatives, it was analyzed that models performed better with the features selected using domain knowledge. It was found that the results when using Boruta were also comparable. Therefore, in similar problems, with an extended dataset, if the users are not having the expertise or are not able to attain enough domain knowledge to select the best features, Boruta could be used as a good alternative for feature selection. Among the six implemented algorithms, logistic regression algorithm performed the best with the least number of false cases and the highest accuracy of 97.27%. This was also tested using the test set which acted as the real-world data. It was found that logistic regression passed the final evaluation having 96% accuracy with four misclassified diagnoses.

## References

1. Wolberg WHW, Street WN, Mangasarian OL (1995) UCI machine learning repository: breast cancer Wisconsin (diagnostic) data set. https://archive.ics.uci.edu/ml/datasets/Breast+Cancer+Wisconsin+(Diagnostic)
2. Laishram S (2017) Nuclear morphometric application in the quantitative description of breast lesions, J Med Res 3:255–257
3. Florentine BD, Nancy CW, Shanez W, Carol C, William H, Anwar R (1999) Fine needle aspiration (FNA) biopsy of palpable breast masses: comparison of conventional smears with the Cyto-Tek MonoPrep system. Cancer Cytopathol: Interdisc Int J Am Cancer Soc 87:278–285

# Double-Layer Secured Automatic Home Access System with Gas Detector Sensor Using Raspberry Pi and Python

**Deepshikha Yadav and Richa Gupta**

**Abstract** Automation of premises has evolved swiftly in all the domains like homes and offices. Integration automation with Internet capabilities has started registering its existence in various hybrid applications. Along with automation, remote monitoring of a network working devices has led to the development of surveillance systems for household appliances these days. Apart from being accessible and affordable, these technologies have proved promising security solutions. Robust security systems for restricted access are seen as a prime concern in recent times. In view of above concerns, we have proposed and tested a system which provides a double-layer secured access to the premises where the system is installed. Face recognition is implemented along with RFID tag scanning to unlock the door for authorized persons only. An additional safety feature of gas leakage detection is incorporated using gas sensor to avoid any accidental fires.

**Keywords** Raspberry Pi 4 · MQ2 sensor · RFID · Servomotor

## 1 Introduction

Latest advances in technology are helping people to achieve an improved and easy life day by day. To alleviate the concerns of safety and security, continuous efforts have been made by the scientist and researchers worldwide. Progress in the field of hardware and software is ensuring the integration of intelligent systems and devices in our ordinary life. As demands on our work front have increased, we tend to rely on more reliable devices to keep our home safe and secure from accidents and thefts. Even at offices where a high level of security and confidentiality is desired, restricted access is a challenging task. During last few years, numerous solutions have been designed to prevent access to unauthorized user. This makes biometric technology

D. Yadav (✉) · R. Gupta
MSIT, Janakpuri, New Delhi, India
e-mail: deepshikha@msit.in

R. Gupta
e-mail: richagupta@msit.in

 
S. Yadav et al. (eds.), *Recent Advances in Metrology*, Lecture Notes in Electrical Engineering 906, https://doi.org/10.1007/978-981-19-2468-2_27

a promising field to investigate upon. Conventional methods of access control like password protection are unreliable as they can be easily compromised or forgotten.

Fingerprint scanning was the first attempt toward securing the premises using biometric identification. Apart from finger print and retina scanning, face recognition is an important criterion for authorized personnel access to the premises. This motivated the researchers for further research in this area leading to present day improvements.

Automation of sensors and devices led to its integration with software and Internet of Things. Also, introduction of user-friendly network-based sensors and digital technology has progressed many folds in last few decades. Various techniques are RFID [1], finger print access [2], face recognition and voice recognition [3], etc. Face recognition using various algorithms has been seen as a promising alternative for access control for various applications including industrial uses [4–6]. Besides secure access safety, avoiding fires due to accidental gas leakage has been an integral part of home automation [7, 8]. Smart homes not only allow users to control various appliances but also ensure a fully reliable feedback and display mechanism for easy reference. Light control from a simple touch has been an interesting and convenient feature for end users lately [9]. Rapid development in major communication methodologies like Wi-Fi, Bluetooth, GSM, etc., has made remote monitoring and control of, the sensors and devices more convenient and reliable [10]. Security solutions for home automation are centered on robust and easy to operate controllers for easy amalgamation of wireless communication and technology implementation. Various computing and controlling modules have been tested and implemented from time to time to increase the efficiency and cost-effectiveness of complete system under testing [11–15].

In this work, an integrated automatic security module for home environment is proposed and designed. The automatic door, RFID tag verification, and the face recognition technologies are combined with leakage detection to build more secure system than usual. Proposed design focuses on building a home security system based on Python and Raspberry Pi having double layer of security for authorized access. Easily available physical components like camera, Raspberry Pi, RFID module, etc., are used to make design simpler and efficient. A person can be detected using face detection setup installed at the door and the RFID double verifies the visitor's identity and owner will get the information regarding it. Information of the person entered will be updated automatically. A display is setup at the door to guide the visitor in verification.

RFID system integrated with python would detect the person's face and update the information accordingly. Once the access is granted, the servomotor-controlled automatic door will let the visitor inside the premises. Also, a gas detection system is tested for the presence of flammable gases like LPG, Methane, etc.

## 2 Design of Proposed System

Internet Gateway Device Raspberry Pi enroutes the data coming from the sensor network to Internet and vice versa. It implements TCP/IP protocols and control and monitor device status through the webpage served by embedded web server (EWS). Figure 1 shows the architecture of the proposed system. The major components are Camera module, RFID module, Gas detector sensor, and Raspberry Pi unit.

**A. Raspberry Pi 4 Model B with 4 GB RAM**
Raspberry Pi 4 with Raspbian Operating system installed on it is the main controlling unit here. The used OS runs applications like PyCharm and codes written in different programming languages like python. We are using Python as programming language here.

**B. MQ2 Flammable Gas and Smoke Module**
MQ2, working on 5 V input, detects the LPG or methane gas, fire, or smoke in the unit under test. It will send the analog reading of the sensed gas by a Wi-Fi module to the Pi 4 unit. Script running on Pi 4 will monitor and inform about the danger mark of the gas level (Fig. 2).

**C. RC522 RFID Module**
The RC522 is a RF Module operating on 13.56 MHz industrial (ISM) band which do not require a license to operate. It consists of a RFID reader and RFID card. 3.3 V

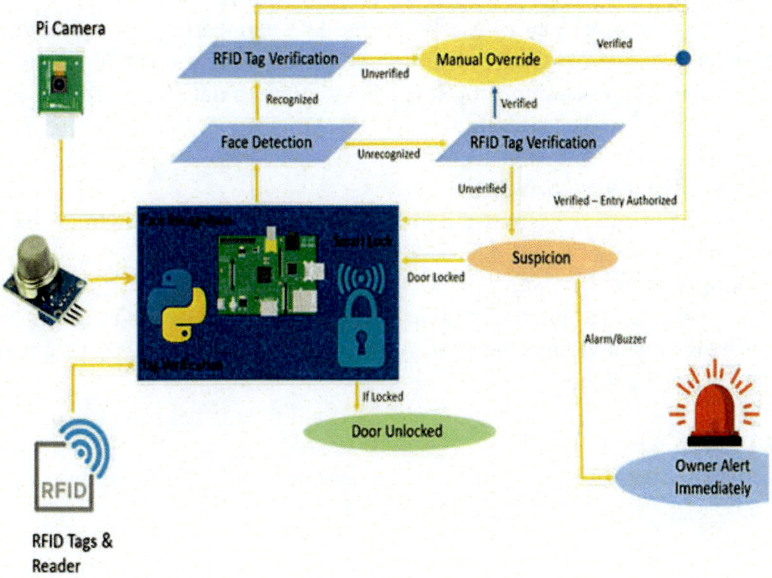


**Fig. 1** Block diagram of proposed system

**Fig. 2** MQ2 gas sensor

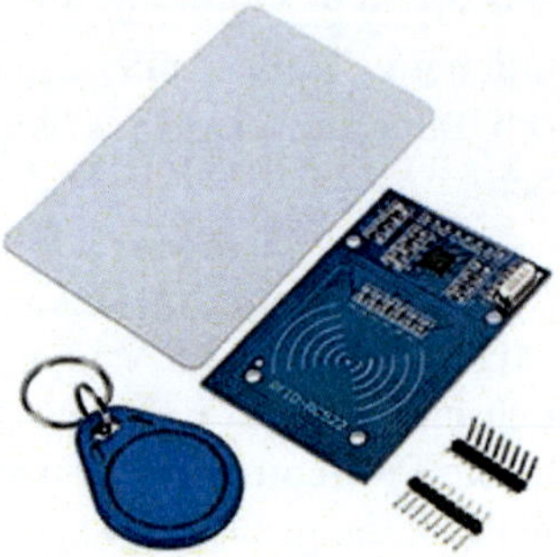

**Fig. 3** RC522 RFID module

is enough to operate this unit making our system power efficient. It is required in applications where a unique ID is desired for reliable operations (Fig. 3).

**D. ESP-01 ESP8266 Serial Wi-Fi Transceiver Module**

This is a cost-effective easy to use Wi-Fi module connecting the devices/module, i.e., Raspberry pi and MQ2 sensor on remote location. It is capable of working as access point as well as station spot. It is used to retrieve data from sensors and send it to Pi board for further processing.

**E. SanDisk Micro SD Card 32G**

The Raspbian Operating system and other application are stored in this.

**F. Servomotors**

The door locking and unlocking system is automatically controlled by the servomotors. Once the system verifies the identity of the visitor, the motors rotate in particular angle to unlock the door.

**G. TFT 2.4-in. Display**

Display shows the ongoing processes and current state of the system like "face verified," "swipe tag," and "Gas leakage."

## 3 Implementation and Results

For processing the images, we have used OpenCV which an open-source software library for real-time image and video processing imbibing machine learning capabilities. Python face_recognition package is used to compute the bounding box around each face, analyzing facial embedding and thereby comparing front face with the encoded dataset. Imutils, a set of functions, is used to integrate OpenCV computation on the Raspberry Pi 4. Following steps are necessary for successful installation of main unit:

First, the board is connected to all the necessary accessories for proper installation and operating system is loaded.

Step 1: Connect your USB mouse to one of the USB ports on the Pi.

Step 2: Connect your USB Keyboard to the other USB port on the Pi.

Step 3: Connect the micro-USB charger to the Pi but do not connect it to the power supply yet.

Step 4: Flash the SD Card with the Raspian OS.

For facial recognition with Pi, we have used OpenCV, face_recognition, and imutils packages to train our Raspberry Pi based on a set of collected real-time images. This works as dataset for training purpose giving us the freedom to test the reliability of system on every data rather an already existing dataset. Function train_model.py will create a mapping between names and faces in the file, encodings.pickle by analyzing the images in our dataset. Once training is done, facial_req.py is able to detect and identify faces. Once the above procedure is, complete, plug in your webcam into one of the USB ports of Raspberry Pi. Reboot the Pi board if required.

***Train the Model for Raspberry Pi Facial Recognition*** Copy the files of Python code on new terminal Ctrl-T on your system. From Raspberry Pi Desktop Open file Manager and store the dataset for training Pi board. A new window showing the view of webcam will open. Raspberry Pi 4 will take approximately 10 s to open webcam window. Pressing spacebar will take a photo of the face where webcam is pointed. Every time a spacebar is pressed, a new photo will be clicked and stored. We took 10 photos of familiar face from different angles to increase the reliability of match. These photos worked well as a dataset for training our model effectively. Esc will mark the end of clicking the photos. This process was repeated with other members of the group to include additional dataset for granting access to multiple familiar peoples. Pi took 3–4 s for analyzing each photo of created dataset. In total, around 1.5 min were taken by Pi to analyze and remember the complete dataset. Encoded information of received dataset is stored in encodings.pickle file. Figure 4 shows the real-time training of system using images taken at the time of starting the module.

Figure 5 shows the actual demonstration model of the complete system having RFID unit, display unit, and Raspberry Pi 4 board with MQ2 gas sensor connected to its port.

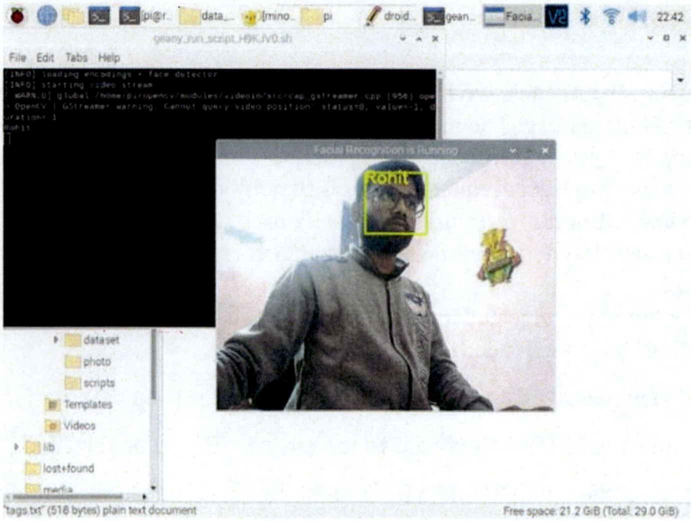


**Fig. 4** Face recognition model training

**Fig. 5** Actual system setup

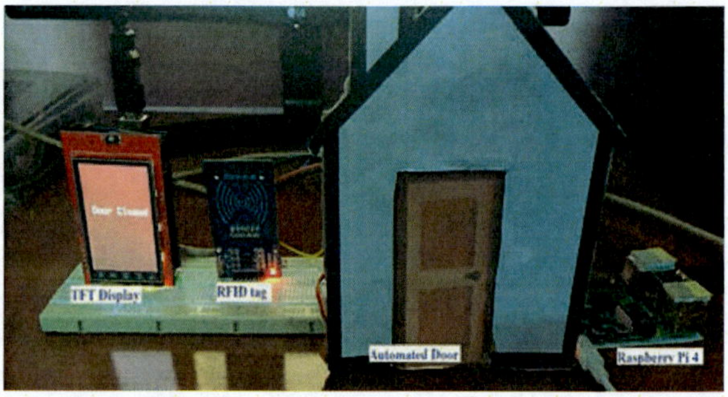


The automated door will open only after successful facial recognition (Fig. 6) and RFID ID matching process (Fig. 7). Servomotors are connected to demonstrate the locking and unlocking of the doors (Fig. 8).

## 4 Conclusion

A face recognition-based double-layer security automatic home access system is presented in this paper. Raspberry Pi 4 with Python functionalities is used to train and analyze the familiar faces. The proposed system can be used not only as an authentic access grant system for home security but for other applications also. Pi 4 board provides a faster computation, thereby reducing the overall time delay. Apart from

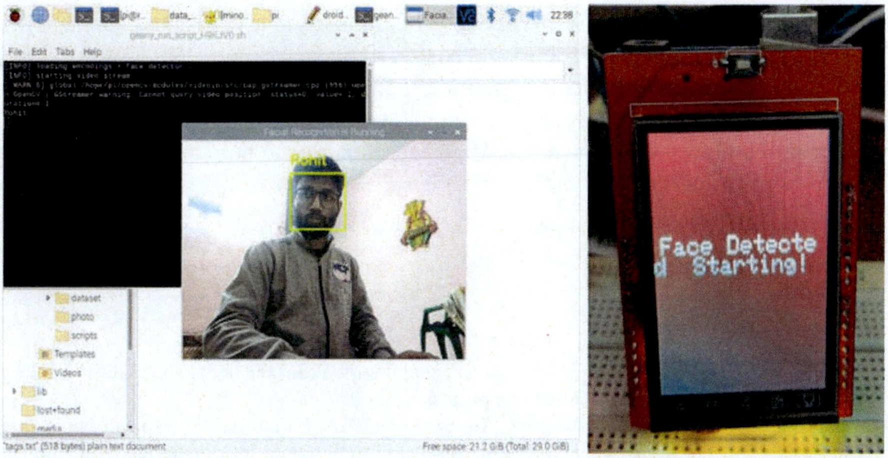


**Fig. 6** Face detection

**Fig. 7** RFID scanning

**Fig. 8** Door opened by servomotors

being cost-effective, system tested above is portable, user-friendly, and very, easy to install. Components used are robust and cost-effective making system more effective and efficient. Gas detection feature provides additional security from accidental gas leakage and subsequent fire threats. Although system itself is automatic and reliable, safe handling of the kit and incorrect keys operation due to human errors may cause system to malfunction temporarily.

## References

1. Park YT, Sthapit P, Pyun JY (2009) Smart digital door lock for the home automation. In: 2009 TENCON IEEE region 10 conference, pp 1–6
2. Komol MMR, Podder AK, Ali MN, Ansary SM (2018) RFID and finger print based dual security system: a robust secured control to access through door lock operation. Am J Embed Syst Appl 6(1):15–22
3. Kim T, Park H, Hong SH, Chung Y (2013) Integrated system of face recognition and sound localization for a smart door phone. IEEE Trans Consum Electron 59(3):598–603
4. Garg S, Yadav A, Jamloki S, Sadana A, Tharani K (2020) IoT based home automation. J Info Optim Sci 41(1):261–271
5. Wang S, Ye J (2010) Research and implementation of embedded face recognition system based on ARM9. In: 2010 International conference on mechanic automation and control engineering. IEEE, pp 2618–2621
6. Prakash R, Chithaluru P (2021) Active security by implementing intrusion detection and facial recognition. In: Nanoelectronics, circuits and communication systems. Springer, pp 1–7
7. Subramanian M A, Selvam N, Rajkumar S, Mahalakshmi R, Ramprabhakar J (2020). Gas Leakage detection system using IoT with integrated notifications using Pushbullet—A review. In: 2020 Fourth international conference on inventive systems and control. IEEE, pp 359–362
8. Alani S, Mahmood SN, Attaallah SZ, Mhmood HS, Khudhur ZA, Dhannoon AA (2021) IoT based implemented comparison analysis of two well-known network platforms for smart home automation. Int J Electr Comput 11(1):2088–8708
9. Gota DI, Puscasiu A, Fanca A, Miclea L, Valean H (2020) Smart home automation system using Arduino microcontrollers. In: 2020 IEEE international conference on automation, quality and testing, robotics. IEEE, pp 1–7
10. Zahra MMA, Mohsin MJ, Abdul-Rahaim LA (2020) Artificial intelligent smart home automation with secured camera management-based GSM, cloud computing, and arduino. Period Eng Nat Sci 8(4):2160–2168
11. Sinha A, Tatikonda R (2019) IoT based home automation using Raspberry Pi. Int J Adv Res 5:558–560
12. Arora Y, Pant H (2019) Home automation system with the use of Internet of Things and artificial intelligence. In: 2019 International conference on innovative sustainable computational technologies. IEEE, pp 1–4
13. Susany R, Rotar R (2020) Remote control android-based applications for a home automation implemented with Arduino Mega 2560 and ESP 32. Technium 2(2):1–8
14. Chekired F, Houtti S, Bouroussis C A, Rahmani A, Tilmatine A, Canale L (2020) Low cost automation system for smart houses based on PIC microcontrollers. In: 2020 IEEE international conference on environment and electrical engineering and 2020 IEEE industrial and commercial power systems Europe. IEEE, pp 1–5
15. Hanif M, Mohammad N, Harun B (2019) An effective combination of microcontroller and PLC for home automation system. In: 2019 1st international conference on advances in science, engineering and robotics technology. IEEE, pp 1–6

# FIR Filter Design Using Grasshopper Optimization Algorithm

**Sandeep Singh, Gagan Singh, Sourav Bose, and Shiva**

**Abstract** In this paper, digital finite impulse response (FIR) low-pass filter (LPF) and high-pass filter (HPF) are designed using a novel meta-heuristic algorithm named grasshopper optimization algorithm (GOA). The GOA is meta-heuristic population-based optimization algorithm, which mimics the food searching behaviour of the grasshopper. The filter design aims to evaluate the optimal filter parameters and find the minimum objective function value so that the output of the designed filter matches with the output response of the ideal filter. Mean square error (MSE) is taken as the error objective function. The results obtained using GOA are compared with the other two algorithms, namely particle swarm optimization (PSO) algorithm and grey wolf optimization (GWO) algorithm. The simulated results reveal that GOA is best suited algorithm for FIR filter design problem.

**Keywords** FIR filter design · Mean Square Error · Particle Swarm Optimization Algorithm · Grey Wolf Optimization Algorithm · Grasshopper Optimization Algorithm

## 1 Introduction

In digital signal processing (DSP), filters have a significant role so that useful information can be processed. Digital filters are used for different applications in domains like signal processing, communication systems, channel equalization and noise reduction [1]. Digital filters are most used in signal separation and restoration. When noise or external signal is added to the information, then separation is needed and restoration is required when the receiving or transmitting signal gets deformed. Besides, these digital filters have a wide range of applications in image processing [2–4], system modelling [5–9], speech processing [10] and audio processing [11].

Digital filters are classified as finite impulse response (FIR) and infinite impulse response (IIR) filters. The present output of the FIR filters depend upon the present

S. Singh (✉) · G. Singh · S. Bose · Shiva
Department of Electronics and Communication Engineering, Maharaja Surajmal Institute of Technology, Delhi, India
e-mail: sandeep@msit.in



S. Yadav et al. (eds.), *Recent Advances in Metrology*, Lecture Notes in Electrical Engineering 906, https://doi.org/10.1007/978-981-19-2468-2_28

input samples only. FIR filters are designed recursively and have the characteristic of linear phase and stability, whereas the present output of IIR filters depend upon the present and past input and past output also. They have non-recursive nature and require fewer coefficients and less storage in comparison with FIR filters. Easy implementation, high stability and linear phase are the main attributes of FIR filters. In literature, FIR filters are designed using conventional methods such as windowing method and frequency sampling method. But due to lack of the control on cut-off frequency and transition width these conventional methods fails. This motivated the researchers to use the swarm-based optimization algorithms.

In this paper, a newly proposed optimization algorithm called grasshopper optimization algorithm (GOA) [12] is used for the design of FIR low-pass filter (LPF) and high-pass filter (HPF). The results obtained using GOA are compared with other two existing algorithms, particle swarm optimization (PSO) [13] and grey wolf optimization (GWO) [14] algorithms.

The exclusive features of GOA are as follows:

1. The GOA does not get trapped in local minima, and it does not concentrate towards the target too hastily as for exploration and exploitation, Different weights are used.
2. Unlike other algorithms, in GOA, updating of position of search agents depends on all the search agent positions. This makes GOA different from other algorithms.
3. It has less elapsed time and fast convergence.

The whole paper is organised as: Sect. 2 describes the problem formulation, Sect. 3 briefly explain the applied GOA. Section 4 discusses the results and analysis and Sect. 5 concludes the work done.

## 2 Problem Formulation

The system transfer function of FIR filter is given by

$$D(z) = \sum_{k=0}^{K-1} d(k)z^{-k} \tag{1}$$

where $d(k)$ denotes the filter coefficients and $K$ represents the length while $K - 1$ is order of the filter. Our purpose is to evaluate the optimal value of filter coefficients, using PSO, GWO and GOA, in such a manner that the designed filter response converges towards ideal filter response. Stop band ripples, lower pass band ripples, reduced transition bandwidth and higher stop band attenuation are some of the desiderate attributes of the filter. The filter coefficients $d(k)$ represent dimension of search agents in evolutionary algorithms under consideration while the position of search agents is updated until maximum iteration is met or minimum value is achieved by error fitness function.

The magnitude response of digital FIR filter can be written as:

$$D(\omega) = \sum_{k=0}^{K-1} d(k)\mathrm{e}^{-j\omega k} \tag{2}$$

$\omega$ is the digital frequency between $-\pi$ to $\pi$ in rad/s.

The fitness function plays an important role in process of designing a FIR filter. Different fitness functions might give different results for same problem. The fitness function defines the relative importance of a design. A higher fitness value implies a better design. The fitness function may be defined in several different ways. Therefore, appropriate choice of error objective function is crucial to get the bona fide output. Mean square error (MSE) is taken as error objective function given in Eq. 3. The minimization of MSE between the ideal and evaluated response of the filter in terms of magnitude is the primary objective of this paper.

$$E(\omega) = \frac{1}{\omega} \sum\nolimits_{\forall\omega} (|D_{id}(\omega)| - |D(\omega)|)^2 \tag{3}$$

## 3 Grasshopper Optimization Algorithm

Grasshoppers creates one of the largest group found in nature for food-seeking, even though they are individual in nature. Nymph and adulthood exhibit this swarming behaviour. Gravity force on the grasshopper, social interaction between grasshoppers and direction of the wind are the factors which affect process of seeking food. The strength of attractive forces and repulsive forces between grasshoppers constitute social interaction. Since nymph do not have wings, therefore, their movement is also affected by direction of wind. It is assumed that direction of wind is in the direction of target. Force of gravity is not taken into consideration while evaluating next position of agent. Next position which is denoted by $X_{t+1}$ of a search agent can be evaluated using following equation

$$X_i^d = c\left(\sum_{j=1}^{N} c\frac{ub_d - lb_d}{2}\, s\left(\left|x_d^j - x_i^d\right|\right)\frac{x_j - x_d}{d_{ij}}\right) + T_d \tag{4}$$

First $c$ behaves as the inertia weight in order to support exploration while another $c$ represents the diminishing factor which shrinks all the comfort zones to achieve the best exploitation. $S$ represents the social interaction force on grasshopper while target position is denoted by $T_d$.

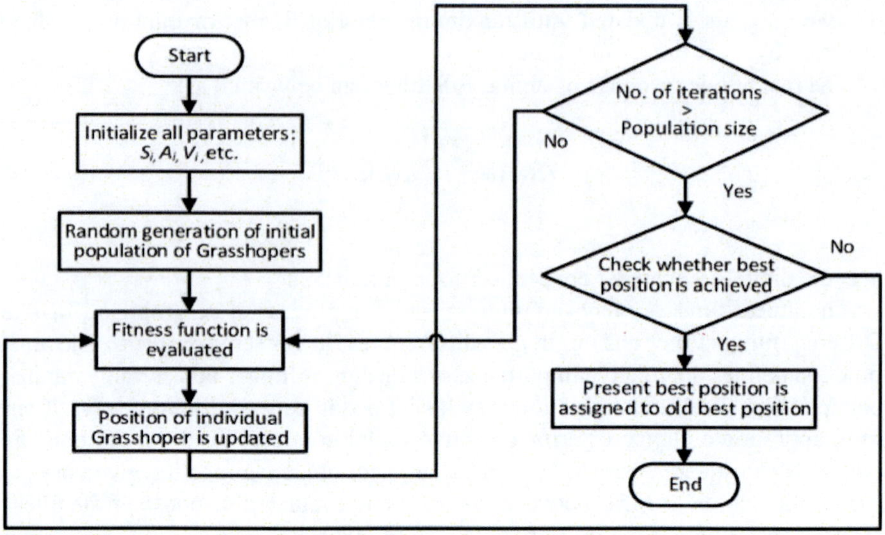


**Fig. 1** Flowchart of GOA

$$c = C_{\max} - l\frac{C_{\max} - C_{\min}}{L} \tag{5}$$

The value of $c$ gets updated in compliance with Eq. 5, which varies on the value of iteration $l$ that changes from one to maximum iteration $L$ (maximum iteration) (Fig. 1).

**Algorithm of Grasshopper Optimization**

1. Initialize swarm N, $C_{max}$, $C_{min}$ and highest iterations $L$
2. Initialize $\omega_n$, $k$, $\omega_l$, $\omega_u$
3. T = Target search agent (best)
4. ***While*** (l < L)
    a. Update c as given in equation
5. ***For*** each search agent
    a. Evaluate error fitness value using Eq. (3)
    b. Get T (best search agent)
    c. Current position of agent is updated using Eq. (4)
    d. Check if search agent is overflowing boundaries, if yes, relocate it
6. ***End for***
7. If there is a better solution, Update T
8. l = l + 1
9. ***End while***
10. Return T

## 4 Result Analysis

The results of FIR 30th order low-pass and high-pass filter designed using PSO, GWO and GOA are evaluated. A total of 31 optimal filter parameters and MSE are calculated of all the three algorithms. MATLAB version R2007b software is used for attaining the results on computer having i5 10th Generation, 3.20 GHz and 8 GB RAM.

### *4.1 Low Pass Filter*

GOA, PSO and GWO are used to build the 30th order LPF having cut-off frequency $0.5\pi$. To verify the performance of applied GOA optimal parameters, minimum MSE and magnitude profile are taken as performance measures. A total of 31 optimal filter parameters are calculated of GWO, PSO and GOA. These calculated coefficients are listed in Table 1. Statistical results of MSE are obtained and described in Table 2. Based on the evaluated results, it can be stated that GOA is the right choice for FIR filter design problem.

**Table 1** Optimal values of coefficient for low-pass filter

| Coefficients | | Algorithms | |
|---|---|---|---|
| $d$(k) | GOA | GWO | PSO |
| $d(1) = d(31)$ | 0.002974 | −0.009662 | −0.060535 |
| $d(2) = d(30)$ | 0.004206 | −0.052051 | −0.127387 |
| $d(3) = d(29)$ | 0.004598 | −0.052079 | −0.139994 |
| $d(4) = d(28)$ | −0.000082 | 0.000191 | −0.068167 |
| $d(5) = d(27)$ | −0.002873 | 0.002382 | 0.047525 |
| $d(6) = d(26)$ | −0.016523 | −0.081511 | 0.174482 |
| $d(7) = d(25)$ | −0.057546 | −0.106657 | 0.305179 |
| $d(8) = d(24)$ | −0.108592 | −0.018908 | 0.360952 |
| $d(9) = d(23)$ | −0.145199 | −0.007881 | 0.227013 |
| $d(10) = d(22)$ | −0.177904 | −0.236520 | −0.030196 |
| $d(11) = d(21)$ | −0.230559 | −0.465051 | −0.131969 |
| $d(12) = d(20)$ | −0.256672 | −0.359833 | 0.043604 |
| $d(13) = d(19)$ | −0.181992 | −0.003267 | 0.235016 |
| $d(14) = d(18)$ | −0.049464 | 0.210026 | 0.154931 |
| $d(15) = d(17)$ | 0.034123 | 0.140082 | −0.068761 |
| $d(16)$ | 0.069105 | 0.003435 | −0.110270 |

**Table 2** Statistical results of MSE for 30th order low-pass filter

| Algorithms | Mean | Median | Standard deviation | Best | Worst |
|---|---|---|---|---|---|
| PSO | 0.0006261 | 0.0003793 | 0.00057613 | 0.0000696 | 0.0012176 |
| GOA | 0.0001013 | 0.0001343 | 0.00008083 | 0.00000529 | 0.00018906 |
| GWO | 0.0013840 | 0.0011752 | 0.00063232 | 0.00088929 | 0.0024147 |

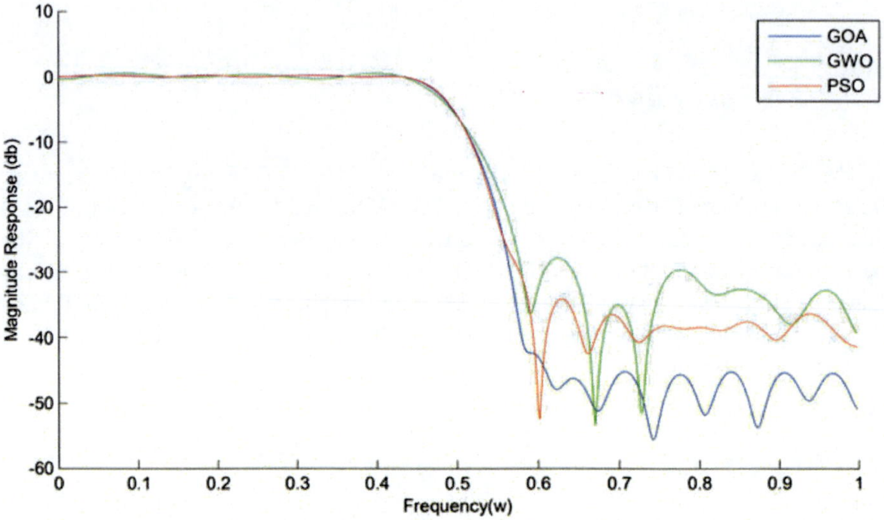


**Fig. 2** Magnitude response of 30th order FIR LPF

Figure 2 represents the graphical representation of the normalized magnitude response of low-pass filter of the algorithms GOA, PSO and GWO. Table 2 clearly specifies that the GOA is best among the three as it has the lowest value of the statistical results obtained as the best value obtained for GOA is 0.00000529 with standard deviation of 0.00008083.

## 4.2 *High Pass Filter*

GOA, GWO and PSO are used to build 30th order high-pass filter having cut-off frequency $0.5\pi$. The performance analysis of applied GOA in terms of optimal parameters, minimum MSE and magnitude profile is done. A total of 31 optimal filter parameters of PSO, GWO and GOA are calculated. These calculated coefficients are listed in Table 3. Statistical results of MSE in terms of best, worst and mean values are obtained and are described in Table 4. Based on the evaluated results, it can be observed that GOA is best suited for FIR filter design problem.

**Table 3** Optimal values of coefficient for high-pass filter

| Coefficients | | Algorithms | |
|---|---|---|---|
| $d$(k) | GOA | GWO | PSO |
| $d(1) = d(31)$ | −0.011044 | −0.000112 | −0.008881 |
| $d(2) = d(30)$ | 0.014477 | 0.003834 | 0.010540 |
| $d(3) = d(29)$ | −0.003506 | −0.000089 | 0.000966 |
| $d(4) = d(28)$ | −0.025647 | −0.054498 | −0.036362 |
| $d(5) = d(27)$ | 0.033194 | 0.075735 | 0.076116 |
| $d(6) = d(26)$ | 0.006496 | −0.000973 | −0.045934 |
| $d(7) = d(25)$ | −0.048536 | −0.086302 | −0.062135 |
| $d(8) = d(24)$ | 0.033809 | −0.000232 | 0.153233 |
| $d(9) = d(23)$ | −0.004093 | 0.122914 | −0.146555 |
| $d(10) = d(22)$ | 0.007811 | −0.010648 | 0.073185 |
| $d(11) = d(21)$ | 0.008518 | −0.326612 | 0.028163 |
| $d(12) = d(20)$ | −0.089737 | 0.493445 | −0.178168 |
| $d(13) = d(19)$ | 0.114505 | −0.301954 | 0.303609 |
| $d(14) = d(18)$ | 0.049272 | −0.001523 | −0.228606 |
| $d(15) = d(17)$ | −0.231102 | 0.075495 | −0.040295 |
| $d(16)$ | 0.124234 | 0.000491 | 0.214423 |

**Table 4** Statistical results of MSE for 30th order high-pass filter

| Algorithms | Mean | Median | Standard deviation | Best | Worst |
|---|---|---|---|---|---|
| GOA | 0.0001670 | 0.0001886 | 0.00006853 | 0.00007381 | 0.00024633 |
| GWO | 0.0025153 | 0.0016288 | 0.00185862 | 0.0011986 | 0.0056436 |
| PSO | 0.0003334 | 0.0002294 | 0.0006792 | 0.00013409 | 0.0016351 |

Figure 3 represents the graphical representation of the normalized magnitude response of high-pass filter of the algorithms GOA, PSO and GWO. Table 4 clearly specifies that the GOA is best among the three to as it has the lowest value of the statistical results obtained as the best value for GOA obtained is 0.00007381 with standard deviation of 0.00006853.

## 5 Conclusion

This work considers the design of FIR LPF and HPF, 30th order using PSO, GWO and GOA and set side by side the performance between the three in terms of their proficiency to advance towards the ideal filter response. Filter coefficients for LPF and HPF have been attained by minimizing the error function. GOA-based FIR filter was

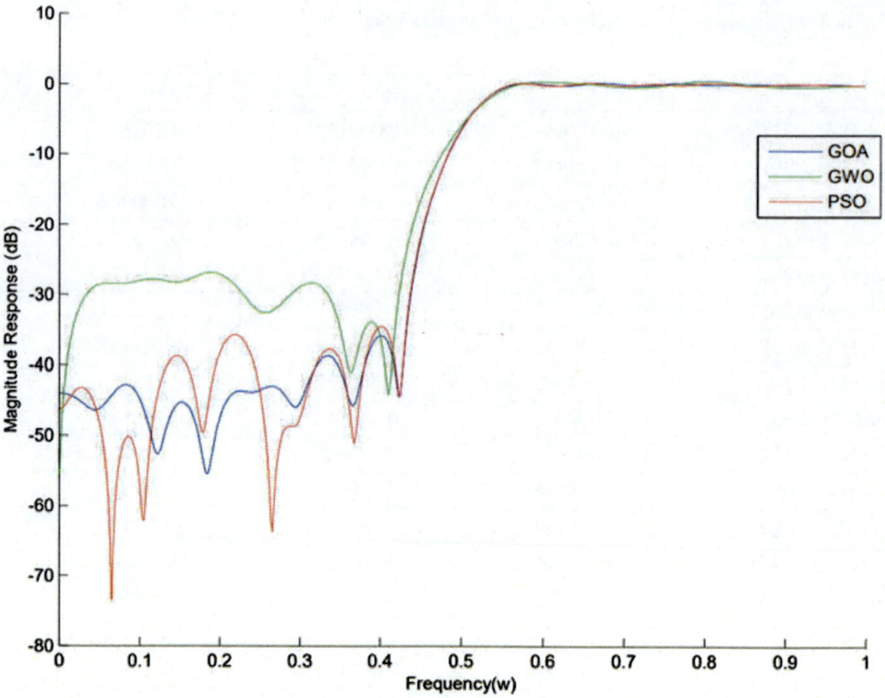


**Fig. 3** Magnitude response of 30th order FIR HPF

found to provide the best solution in comparison between the other two algorithms and provides least pass band ripple, stop band ripple and greater attenuation stop band.

## References

1. Vaidyanathan PP (2001) Filter banks in digital communications. IEEE Circ Syst Mag 1(2):4–25
2. Kumar A, Komaragiri R, Kumar M (2018) From pacemaker to wearable: techniques for ECG detection systems. J Med Syst 42(2):1–17
3. Kumar A, Komaragiri R, Kumar M (2018) Design of wavelet transform based electrocardiogram monitoring system. ISA Trans 80:381–398
4. Kumar A, Komaragiri R, Kumar M (2018) Heart rate monitoring and therapeutic devices: a wavelet transform based approach for the modeling and classification of congestive heart failure. ISA Trans 79:239–250
5. Kadam G, Bhaskar PC (2012) Reduction of power line interference in ECG signal using FIR filter. Int J Comput Eng Res 2(2):314–319
6. Latifoğlu F (2013) A novel approach to speckle noise filtering based on artificial bee colony algorithm: an ultrasound image application. Comput Methods Programs Biomed 111(3):561–569
7. Zhang S, Salari E (2003) Reducing artifacts in coded images using a neural network aided adaptive FIR filter. Neurocomputing 50:249–269

8. Zhang S (2011) Image denoising using FIR filters designed with evolution strategies. In: 2011 3rd International workshop on intelligent systems and applications. IEEE, pp 1–4
9. Torbati N, Ayatollahi A, Kermani A (2013) Ultrasound image segmentation by using a FIR neural network. In: 2013 21st Iranian conference on electrical engineering (ICEE). IEEE, pp 1–5
10. Zahorian S, Gordy P (1983). Finite impulse response (FIR) filters for speech analysis and synthesis. In: ICASSP'83. IEEE international conference on acoustics, speech, and signal processing, vol 8. IEEE, pp 808–811
11. Ponce H, Ponce P, Molina A (2014) Adaptive noise filtering based on artificial hydrocarbon networks: an application to audio signals. Expert Syst Appl 41(14):6512–6523
12. Mirjalili SZ, Mirjalili S, Saremi S, Faris H, Aljarah I (2018) Grasshopper optimization algorithm for multi-objective optimization problems. Appl Intell 48(4):805–820
13. Aggarwal A, Rawat TK, Upadhyay DK (2016) Design of optimal digital FIR filters using evolutionary and swarm optimization techniques. AEU-Int J Electron Commun 70(4):373–385
14. Mirjalili S, Mirjalili SM, Lewis A (2014) Grey wolf optimizer. Adv Eng Softw 69:46–61

# Optimization of Flux for Bead Geometry and Weld Dilution in SAW Using Gray Analysis

**Brijpal Singh and Narender Singh**

**Abstract** This study has been conducted to find out and optimal composition of flux for low width, reinforcement, dilution, and high penetration of the weld. The weld geometry decides the load-carrying capacity of the weld, and dilution effects describe the elements transfer to the welds. Both of these factors affect the load-bearing capacity of the welded material. The fluxes were designed using RSM and were made by agglomeration technique. The optimized flux contains $CaF_2$, FeMn, and NiO each as 8, 2, and 2%.

**Keywords** Shielded metal arc welding · Weld width · Reinforcement · Weld reinforcement factor · Depth of penetration

## 1 Introduction

The submerged arc welding is very useful and old process, and it was made operational in 1920s in USA. In SAW, the electrode which is in the form of wire is non consumable, and it travels to generate the arc. The arc heats the weld metal and melts it for making weld. In this, welding current is passed through the electrode and the workpiece. The arc is generated between the base metal and electrode. The heat produced by the arc melts the electrode, base plate, and the adjacent flux. This welding is called submerged arc welding because in this welding the arc is buried in the molten flux, and in thus, double protection from the atmospheric gases and from other sources is obtained. Very high welding current is used in SAW as it is often used for welding of thicker metal per unit area of the electrode is larger in comparison to shielded metal arc welding, and it can use 5–6 times higher current density. In this welding, due to high current used, the melting rate of electrode and work piece is

B. Singh (✉) · N. Singh
Department of Mechanical Engineering, Maharaja Surajmal Insititute of Technology, Janak-Puri, New-Delhi 110058, India
e-mail: brijpal.singh@msit.in

N. Singh
e-mail: narender_singh@msit.in



S. Yadav et al. (eds.), *Recent Advances in Metrology*, Lecture Notes in Electrical Engineering 906, https://doi.org/10.1007/978-981-19-2468-2_29

very high. The electrodes used for SAW are either bare, and usually, they are coated with copper in order to avoid corrosion. The coating of electrodes helps in increasing the electrical conductivity of the electrode wire. There are varieties of combinations of electrode wire and fluxes, which are used in this welding.

Submerged arc welding is widely used for making high-quality joints as it gives the double protection from atmospheric gases (Houldcroft [1]). In this welding process, the arc is submerged and hidden as it is not visible to the operator. The flux is supplied with an arrangement, and a hopper is used for this. This hopper moves ahead with the torch. There is no need of any shielding gas for the protection of the weld, and the unused flux gives additional shield from the atmospheric gases. Very high currents up to 2000 amp can be used. This welding is suitable for both type of currents, AC and DC. The welds made by the SAW process have high strength and ductility with low hydrogen and nitrogen content. The various applications of SAW are shipbuilding; pressure vessels are made of low-alloy steels, mild steels, low-carbon steels. This is also used for welding nickel-based alloys, stainless steels, and other non-ferrous metals.

## 2 Saw Fluxes

The fluxes are used to reduce d the oxidation of the weld metal and clean the weld from impurities. The SAW fluxes contain calcium fluoride, ferromanganese, manganese oxide, titanium oxide, lime, and other compounds. In SAW, the weld pool is submerged under the layer of molten flux, and this gives the protection from the atmospheric gases. The flux which is melted becomes a conductive medium and serves as a path for carrying the current from electrode to the work piece. Fluxes can be classified as bonded fluxes, fused fluxes, and agglomerated fluxes. They can also be classified based on the value of basicity index value. The weld metal properties and the composition of the weld are decided by the flux and wire composition.

## 3 Bead Width

The width of the weld bead may be defined as the width of the metal deposited on base plate. Bead width and reinforcement both are supposed to be responsible for the strength of a welded joint. The weld chemistry is affected by these two-bead geometry parameters (Srihari [2]). Tregelsky (1968) [3], in his research, found that the reinforcement should usually be 20% of plate thickness. The excessive reinforcement in weld bead is not desirable, and it does not improve the strength of the weld. It has been found that the flux consumption is increased with increasing reinforcement of the welds.

## 4 Weld Penetration

The weld penetration may be defined as the distance between the base plate top surface and depth of the weld up to which the fusion has taken place. The penetration is an important parameter in bead geometry as it increases the strength of the weld. If the penetration is high, then the strength of the weld will be large. In order to join thick plates, less no. of passes are required to complete the weld if the weld penetration is large. The production rate is also high if the penetration is large. There are various factors which influence the depth of penetration in the weld. Some of the factors are flux composition, the welding speed, type of polarity, travel speed, electrode stick out, basicity index, and physical properties of the flux. The penetration mainly is decided by the heat input by the arc. The depth of penetration also depends upon the diameter of the electrode. The thickness of the electrode also affects the weld bead geometry and shape (Caddell [4])

## 5 Weld Dilution

The weld dilution may be defined as the ratio of the area of penetration and the total area of the weld metal. Numerically, it may be written as the ratio of $\frac{A_P}{A_P+A_R}$ where $A_P$ is area of penetration and $A_R$ is area of reinforcement. The loss or gain of the elements by the weld depends upon the dilution. It has a significant effect on weld bead shape and geometry.

## 6 Agglomerated Fluxes

For making agglomerated fluxes, the various ingredients in the raw state are converted in to powder form, and they are dry mixed. After that a binder is used. Potassium silicate or sodium silicate is added to the dry mixture for providing the strength. This binder increases the strength of the flux. These fluxes are also called as bonded fluxes. After this, the wet mixture is pelletized and baked to some temperature which is lower than the melting point of the ingredients. The pellets are formed after drying. After this, the pellets are broken up, and the flux is passed through a sieve to screen to size, and packaging is done. The agglomerated fluxes are consumed in low quantity as these are having low bulk density. The agglomerated fluxes are preferred over the other fluxes as the alloying and deoxidizer's elements can be added in the dry state. The main limitation of using agglomerated flux is that they easily absorb the moisture from the atmosphere and forms the gas or pores in the welding.

## 7 Optimization and Flux Formulation

The scientific method DOE should be used in place of hit and trial methods for developing the commercial fluxes. The optimization techniques decide that which flux is best suited for the desirable outputs. So, multi-objective optimization should be used for getting desirable output. The multi-objective optimization for impact strength and hardness was reported for input factors such as NiO, MnO, and MgO. With this technique, a single MRPI factor was calculated for $SiO_2$ base fluxes by Kumar et al. [5]. Jindal [6] used the composite desirability approach for optimization of UTS, impact strength, and hardness for HSLA steel. They suggested the flux mixture for maximum UTS, maximum impact strength, and a targeted value of microhardness.

The bead geometry is characterized by its weld width (*W*), reinforcement (*R*), and weld penetration. For weld bead geometry and shape relationship, the weld width (*W*), depth of penetration (*P*), height of reinforcement (*R*), dilution (D), weld penetration shape factor (WPSF) or $\frac{W}{P}$ and weld reinforcement factor (WRFF) or $\frac{W}{R}$ are measured. The photograph shown in Fig. 1 depicts bead width, reinforcement, and penetration. The weld joint properties are decided by these parameters and affect the mechanical properties of the welding joints. The important process parameters voltage, current, travel speed, nozzle to plate distance, the flux composition also decide the joint strength and other properties. Various researchers have used various techniques for the optimization of the bead geometry and shape relationship. As the load-carrying capacity of a weld joint also depends on the shape and size of the weld, so, the study of bead geometry and shape relationship was carried out. The polished weld beads and a photo of weld bead parameters are given in Fig. 2.

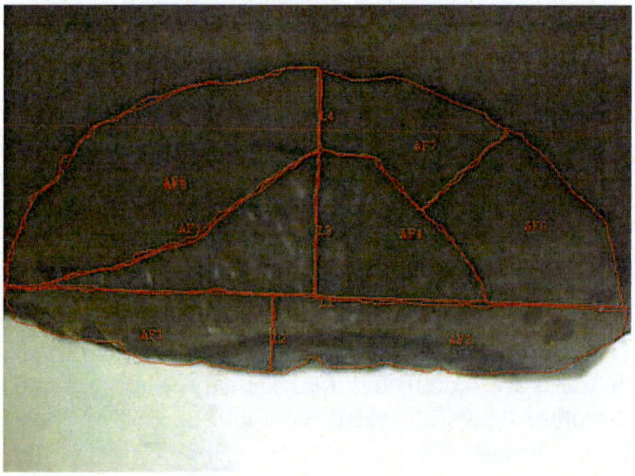

**Fig. 1** Polished weld beads and a photo of weld bead parameters

**Fig. 2** Polished weld beads and a photo of weld bead parameters

## 8 Experimental Procedure

For correlating the input factors with the output, following procedure for experimentation was followed.

1. The fluxes were designed by using RSM methodology. CCD design has been used as it is compact and effective. The designed matrix has been shown in Table 1.
2. Three fluxes CaO, $SiO_2$, and $Al_2O_3$ were selected as base and were mixed in the ratio of 7:10:2. This was decided with the help of r binary and ternary phase diagrams. The range of additives was decided as (2–8) %.
3. In this experiment, as per the design, twenty fluxes were made d by agglomeration technique. The base flux constituents has been given in Table 2
4. The out factors such as e width, reinforcement, penetration, and weld dilution have been shown in Table 2.

## 9 Gray Relational Analysis

The gray relational process was developed by Deng [7]. This is called gray theory, and it is used in decision-making. In this theory, it is assumed that there are two types of data: One is called black, and other is called white. In this system, it is assumed that the black data show the unknown information, while white data show the known information. In gray system, it is assumed that there is some known information, and rest information about the system is unknown. There may be some incomplete information about the system. This incomplete information is known as a gray system. In this study, two responses were considered. One is weld penetration

**Table 1** Design matrix for input factors

| No. of experiment | $CaF_2$ wt % A | FeMn wt% B | NiO wt % C |
|---|---|---|---|
| 1 | +1 | −1 | −1 |
| 2 | 0 | +1 | 0 |
| 3 | +1 | −1 | +1 |
| 4 | −1 | −1 | −1 |
| 5 | 0 | 0 | 0 |
| 6 | 0 | 0 | 0 |
| 7 | +1 | +1 | +1 |
| 8 | 0 | 0 | 0 |
| 9 | 0 | −1 | 0 |
| 10 | +1 | 0 | 0 |
| 11 | 0 | 0 | +1 |
| 12 | −1 | −1 | +1 |
| 13 | 0 | 0 | 0 |
| 14 | 0 | 0 | 0 |
| 15 | +1 | +1 | −1 |
| 16 | −1 | 0 | 0 |
| 17 | 0 | 0 | 0 |
| 18 | 0 | 0 | −1 |
| 19 | −1 | +1 | +1 |
| 20 | −1 | +1 | −1 |

which is to be maximized for joint strength, while the weld dilution was considered to be minimum as it has a significant effect on weld chemistry. As the units and ranges of various input and output factors may differ so, the normalization of data is very much essential. The original data must be normalized to make it comparable. The first step in GRA is to normalize the data. The normal procedure for applying this analysis is as follows Vijayan and Rao [8].

1. Firstly, the data are normalized as the various factors are in different units.
2. After normalization of data, next step is to calculate the gray relational coefficients for all the twenty experiments.
3. After the calculation of gray relational coefficient, gray relational grade was calculated.

**Table 2** Output responses

| Flux | $CaF_2$ (%) | FeMn (%) | NiO (%) | Width (mm) | Reinforcement (mm) | Penetration (mm) | Dilution $\frac{A_P}{A_P+A_R}$ |
|---|---|---|---|---|---|---|---|
| 1 | 8 | 2 | 2 | 34 | 5.53 | 7.19 | 0.387 |
| 2 | 5 | 8 | 5 | 48.73 | 11.71 | 9.31 | 0.179 |
| 3 | 8 | 2 | 8 | 41.7 | 8.01 | 9.18 | 0.454 |
| 4 | 2 | 2 | 2 | 42.1 | 9.53 | 8.01 | 0.366 |
| 5 | 5 | 5 | 5 | 51.47 | 9.4 | 8.83 | 0.467 |
| 6 | 5 | 5 | 5 | 49.6 | 9.5 | 9.42 | 0.281 |
| 7 | 8 | 8 | 8 | 45.51 | 11.71 | 6.5 | 0.257 |
| 8 | 5 | 5 | 5 | 42.42 | 9.42 | 9.53 | 0.555 |
| 9 | 5 | 2 | 5 | 43.51 | 8.89 | 10 | 0.399 |
| 10 | 8 | 5 | 5 | 42.42 | 7.02 | 6.25 | 0.388 |
| 11 | 5 | 5 | 8 | 46.53 | 8.96 | 9.25 | 0.251 |
| 12 | 2 | 2 | 8 | 44.23 | 8.59 | 9.76 | 0.351 |
| 13 | 5 | 5 | 5 | 51.47 | 8.02 | 8.83 | 0.473 |
| 14 | 5 | 5 | 5 | 51.5 | 10.13 | 9 | 0.411 |
| 15 | 8 | 8 | 2 | 36.8 | 8.38 | 6.04 | 0.338 |
| 16 | 2 | 5 | 5 | 44.02 | 7.44 | 7.33 | 0.398 |
| 17 | 5 | 5 | 5 | 48.14 | 8.69 | 9.51 | 0.482 |
| 18 | 5 | 5 | 2 | 39.5 | 8.17 | 8.69 | 0.405 |
| 19 | 2 | 8 | 8 | 40.6 | 7.6 | 9.24 | 0.454 |
| 20 | 2 | 8 | 2 | 35.53 | 9.59 | 9 | 0.566 |

Data processing is required to make the original sequences into a comparable sequence. After normalization, each response is given or assigned a value between 0 and 1.The standard equations are used to normalize the data. The following Eq. (1) is used for maximizing the output responses Vijayan and Rao [8].

$$X_i^*(k) = \frac{X_i(k) - \min X_i(k)}{\max X_i - \min X_i(k)} \tag{1}$$

In order to get the nominal or a set value, the normalization is done according to the given below in Eq. 2.

$$X_i^*(k) = 1 - \frac{\left|X_i^0(k) - X^0\right|}{\max X_i^0(k) - X^0} \tag{2}$$

where $X^0$ the original sequence, min is $X_i^0(k)$ the minimum $X_i^0(k)$ value, max $X_i^0(k)$ is the largest $X_i^0(k)$ value, and $X^0$ is the desired value.

In order to get lower the better, following procedure may be used for the normalization. This has been depicted in Eq. 3.

$$X_i^*(k) = \frac{\max X_i(k) - X_i(k)}{\max X_i - \min X_i(k)} \tag{3}$$

After getting the above values, the next step is to calculate the gray relational coefficient. The purpose of the calculating the gray relational coefficients is to express the relationship between the idle and the actual experimental result. This is generally represented by $\tau_i(k)$ and can be calculated from the below given Eq. 4.

$$\tau_i(k) = \frac{\Delta_{\min} + \tau\Delta_{\max}}{\Delta_i(k) + \tau\Delta_{\max}} \tag{4}$$

In the above equation, $(\tau)$ is the distinguishing coefficient. It is usually taken as 0.5. The last step in this process is to calculate gray relational grade. The formula for this is given below in Eq. 5 (Table 3)

$$X_i = \frac{1}{n}\sum_{k=1}^{n} \tau_i(k) \tag{5}$$

## 10 Conclusion(s)

The gray relational analysis shows that experiment no 1 gives the optimum results. The composition of flux no 1 is $CaF_2$ as 8%, FeMn as 2%, and NiO as 2%. The experiment was repeated with the above values, and it was validated from the experiment that the weld bead geometry was good, and dilution was also low. So, it may be used for design of fluxes for optimized bead geometry and weld dilution.

**Table 3** Rank calculated for fluxes

| Flux No | Penetration (P) mm | Nor. value | Dev. P | GRC-penetration | Dilution (D) | Nor. value | Dev. D | GRC-dilution | GRG | Rank |
|---|---|---|---|---|---|---|---|---|---|---|
| 1 | 7.19 | 0.29 | 0.71 | 0.413 | 0.387 | 0.46 | 0.54 | 0.482 | 0.448 | 18 |
| 2 | 9.31 | 0.83 | 0.17 | 0.746 | 0.179 | 1.00 | 0.00 | 1.000 | 0.873 | 1 |
| 3 | 9.18 | 0.79 | 0.21 | 0.704 | 0.454 | 0.29 | 0.71 | 0.413 | 0.559 | 10 |
| 4 | 8.01 | 0.50 | 0.5 | 0.500 | 0.366 | 0.52 | 0.48 | 0.509 | 0.504 | 15 |
| 5 | 8.83 | 0.70 | 0.3 | 0.625 | 0.467 | 0.26 | 0.74 | 0.402 | 0.513 | 13 |
| 6 | 9.42 | 0.85 | 0.15 | 0.769 | 0.281 | 0.74 | 0.26 | 0.655 | 0.712 | 4 |
| 7 | 6.5 | 0.12 | 0.88 | 0.362 | 0.257 | 0.80 | 0.20 | 0.713 | 0.538 | 11 |
| 8 | 9.53 | 0.88 | 0.12 | 0.806 | 0.555 | 0.03 | 0.97 | 0.340 | 0.573 | 7 |
| 9 | 10 | 1.00 | 0 | 1.000 | 0.399 | 0.43 | 0.57 | 0.468 | 0.734 | 2 |
| 10 | 6.25 | 0.05 | 0.95 | 0.345 | 0.388 | 0.46 | 0.54 | 0.481 | 0.413 | 20 |
| 11 | 9.25 | 0.81 | 0.19 | 0.725 | 0.251 | 0.81 | 0.19 | 0.729 | 0.727 | 3 |
| 12 | 9.76 | 0.94 | 0.06 | 0.893 | 0.351 | 0.56 | 0.44 | 0.529 | 0.711 | 5 |
| 13 | 8.83 | 0.70 | 0.3 | 0.625 | 0.473 | 0.24 | 0.76 | 0.397 | 0.511 | 14 |
| 14 | 9 | 0.75 | 0.25 | 0.667 | 0.411 | 0.40 | 0.60 | 0.455 | 0.561 | 9 |
| 15 | 6.04 | 0.00 | 1 | 0.333 | 0.338 | 0.59 | 0.41 | 0.549 | 0.441 | 19 |
| 16 | 7.33 | 0.33 | 0.67 | 0.427 | 0.398 | 0.43 | 0.57 | 0.469 | 0.448 | 17 |
| 17 | 9.51 | 0.88 | 0.12 | 0.806 | 0.482 | 0.22 | 0.78 | 0.390 | 0.598 | 6 |
| 18 | 8.69 | 0.67 | 0.33 | 0.602 | 0.405 | 0.42 | 0.58 | 0.461 | 0.532 | 12 |
| 19 | 9.24 | 0.81 | 0.19 | 0.725 | 0.454 | 0.29 | 0.71 | 0.413 | 0.569 | 8 |
| 20 | 9 | 0.75 | 0.25 | 0.667 | 0.566 | 0.00 | 1.00 | 0.333 | 0.500 | 16 |

## References

1. Hould Croft PT (1989) Submerged arc welding, 2nd edn. Abington Publishing, Cambridge, England. Srihari T (1992) Submerged arc welding of high strength low alloy steels. Ph.D. thesis, I.I.T, Delhi
2. Srihari T (1992) Submerged arc welding of high strength low alloy steels. Ph.D. thesis, I.I.T, Delhi
3. Trindade VB, Payao JC, Souza LFG, Paranhos RR (2007) The role of addition of Ni on the microstructure and mechanical behaviour of C–Mn weld metals. Exacta 5(1):177–183
4. Caddell RM (1967) The influence of physical properties on penetration in arc welding. Trans ASME J Eng Ind 37(5):328–332
5. Kumar A, Singh H, Maheshwari S (2012) Modeling and analysis by RSM of hardness for welded joints using developed agglomerated fluxes. Ind J Eng Mater Sci 19:379–385
6. Jindal S (2013) Development of submerged arc welding fluxes for welding of structural steels. Ph.D. thesis, MMU, Ambala, Haryana, India
7. Deng J (1989) Introduction to grey system. J Grey System 1(1):1–24
8. Vijayan D, Rao VS (2014) C Fricton stir welding of age hardenable aluminium alloys: a parametric approach using RSM base GRA coupled with PCA. J Inst Eng India Sec C 95(2):127–141

# Uncertainty Estimation for Liquid Dielectric Measurements Using Open-Ended Coaxial Probe Method

**Asheesh Kumar Sharma and Satya Kesh Dubey**

**Abstract** Evaluation of uncertainties associated with liquid dielectric properties is a necessary part of SAR assessment and associated uncertainty estimation, used for certification of mobile and wireless devices. The paper presents the uncertainty estimation and budget preparation for dielectric parameters of tissue equivalent liquids using an open-ended coaxial probe system in accordance with IEEE-1528 standard document. The method uses the open-short-load (O-S-L) calibration for measurement probe using a reference material (i.e., DI water, Ethane-di-ol, etc.) on a Vector network analyzer (VNA). Tissue equivalent liquid have been prepared by following a standardized process to achieve frequency-specific target values for TEL dielectric properties (dielectric constant and conductivity simultaneously).

**Keywords** Specific Absorption rate (SAR) · Tissue Equivalent Liquid (TEL) · Open-ended coaxial probe · IEEE-1528

## 1 Introduction

This is the age of technology, connectivity, and innovation. With the technology advancing and newer methods of connectivity being established each day, the numbers and type of communication devices are increasing at very high rate. Humans are getting more proximity to advance wireless communication devices (including mobiles, tablets, Bluetooth, and wearable devices). Such massive increase in wireless device usage has increased awareness and concerns toward the existence of RF exposure as a potential concern among the general public and media. This aspect, however, has been known to manufacturers and RF design community from start [1], and the regulations have been in place [2–4] to ensure human safety. The exposure limit of such wireless devices is measured in terms of their SAR (Specific Absorption rate) value, which is the rate of energy absorbed by human body per unit mass, when

A. K. Sharma · S. K. Dubey (✉)
Academy of Scientific and Innovative Research (AcSIR), Ghaziabad 201002, India
e-mail: dubeysk@nplindia.org

National Physical Laboratory, CSIR, New Delhi 110012, India



S. Yadav et al. (eds.), *Recent Advances in Metrology*, Lecture Notes in Electrical Engineering 906, https://doi.org/10.1007/978-981-19-2468-2_30

exposed to a type of electromagnetic (EM) radiation. The SAR value of a wireless device can be evaluated by measurement of E-field in a tissue equivalent liquid (TEL) medium, as represented in following equation:

$$\text{SAR} = \frac{\sigma E^2}{\rho} \tag{1}$$

where $|E|$ is the electric field strength magnitude measured, $\sigma$ is the conductivity of medium (TEL) and $\rho$ is the mass density of medium. As SAR value is directly linked with dielectric conductivity, it is hence very important to estimate accurate values of the dielectric properties.

The procedure of SAR evaluation, measurement techniques to be used, and the uncertainty evaluation have been laid out by IEEE and are updated time to time to incorporate the evolving technology and measurement techniques as well [2, 5, 6]. One of the key parameter in the SAR evaluation is the dielectric conductivity of the TEL. The TELs are artificial/synthesized material (chemical composites), which mimic the electric properties of human body tissues at their specific frequency. The SAR evaluation of various wireless devices are carried out in accordance with methods provided in standards [5, 6] by making use of phantom models and these TELs. Hence, it is very important to assess the dielectric property of the TELs accurately and evaluate the uncertainty associated.

The uncertainties associated with dielectric property measurement are generally dependent on the technique used. There are three techniques identified in the IEEE standard for dielectric measurement, namely Slotted coaxial transmission line, Contact probe (open-ended coaxial transmission-line probe/sensor), and TEM line. All these methods have their own limitations and advantages. However, the contact probe method or the open-ended Coaxial Probe method is most commonly used due to its ease of measurement and simplicity.

This paper discusses the details of open-ended coaxial probe method-based Liquid Dielectric assessment facility available at CSIR-National Physical Laboratory (NPL), which utilizes a commercial dielectric assessment system comprising DAK 3.5 sensor probe (Make: Speag) and a vector network analyzer (ZNB 8, Make: R&S) along with commercial software package DAK 2.4.0.814. The theoretical background of the open-ended probe method is discussed in next section followed by detailed discussion on measurement procedure. The uncertainty evaluation is also carried out for TELs for 3G (2.1, 2.3 GHz) and Wi-Fi (2.45 GHz) frequencies.

### *1.1 Open-Ended Coaxial Probe Method*

The open-ended coaxial probes are essentially coaxial transmission-line sections, usually with an end flange, which serve as a ground plane extension for fringing fields. Connector end of the probe is connected to one port of the network analyzer. The material/sample under test (MUT) is kept in direct contact with the open end of probe. Under such condition, when an EM wave propagates, reflection of EM energy occurs at interface. Hence, the dielectric properties can be obtained from the experimentally measured reflectivity at the interface. A representation along with a simplified capacitive-equivalent model [7] for the open-ended probe in contact with MUT is shown below in Fig. 1a, b.

The capacitive-equivalent model consists of two components: (i) $C(\varepsilon_r)$ which is related to dielectric properties of sample (MUT) and (ii) $C_f$ which is independent of dielectric properties of MUT. When a probe is in contact with a sample, the reflection coefficient at the interface is given by [7]:

$$\Gamma^* = \Gamma e^{j\phi} = \frac{1 - j\omega Z_0[C(\varepsilon_r) + C_f]}{1 + j\omega Z_0[C(\varepsilon_r) + C_f]} \quad (2)$$

where $\omega$ is the measurement frequency and $Z_0$ is the impedance of the coaxial line.

The above Eq. 2 can be rearranged and the dielectric properties of MUT ($\varepsilon_r$) can be obtained as:

$$\varepsilon_r = \varepsilon' + j\varepsilon'' = \frac{1 - \Gamma^*}{j\omega Z_0 C_0[1 + \Gamma^*]} - \frac{C_f}{C_0} \quad (3)$$

The coefficient $C_f$ and $C_0$ can be obtained by calibrating the probe against a standard sample with known dielectric property such as de-ionized water (DI water) [8] and ethane-di-ol. The materials to be used as reference material for calibration of probes and their dielectric properties have been studied in detail by multiple researchers [8–10]. The coefficients $C_f$ and $C_0$ are automatically calculated and corrections applied in the DAK system. This has been elaborated in next section.

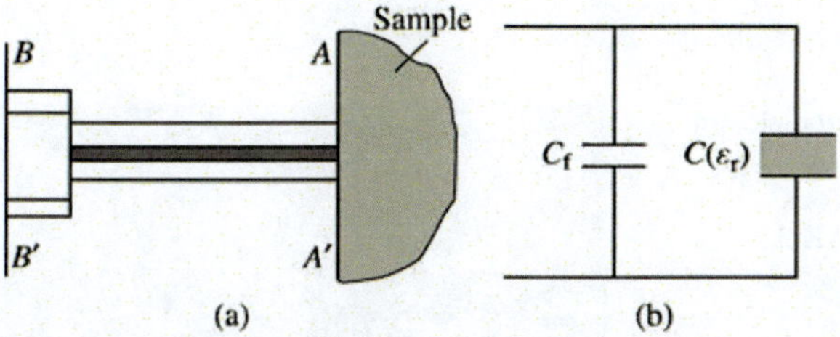


**Fig. 1** **a** Systematic of open-ended coaxial probe system in contact with a sample (MUT), The plane BB' represents the VNA port and plane A-A′ represents the interface between the sample and probe at which reflectivity is measured. **b** Simplified capacitance-equivalent circuit at plane A-A′ for open-ended probe system shown

### *1.2 Calibration of DAK System, DAK Probe, and Further Validation*

Figure 2 shows the commercial DAK system (with DAK 3.5 probe) used for dielectric measurement of liquids at CSIR-NPL. The liquid dielectric assessment method employs standard O-S-L method for calibration of the probe. In this, the probe is calibrated at "open" and "short" conditions and DI water at known temperature is used as "Load." The calibration is verified by subsequent measurement against reference liquid (DI water).

The measurement of reference liquid (DI water) after O-S-L calibration along with the deviation from target value and standard deviation is shown in Fig. 3a, b.

## 2 Uncertainty Evaluation for Dielectric Measurement

The Liquid Dielectric measurement system (DAK) at CSIR-NPL uses a vector network analyzer (VNA) connected through an automatic dielectric evaluation software, which measures reflectivity at probe interface and provides dielectric measurement results (dielectric constant/relative permittivity and conductivity) for MUT, using method explained in Sect. 1.1. The uncertainty budget for dielectric measurement essentially derives from measurement inaccuracies in the calibration data (i.e., reference), analyzer drift, and random errors [5]. Certain other uncertainty contributors may need to be considered for specific test setups. These include sample-to-probe air gaps/bubbles, frequency interpolations, numerical analysis/data extraction artifacts, sensor dimensional/positioning considerations, coaxial probe finite flange effects, etc.

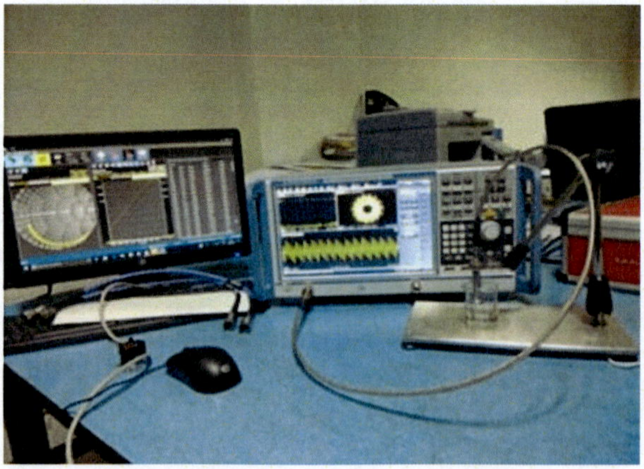

**Fig. 2** DAK liquid dielectric assessment system showing measurement on DAK software using DAK 3.5 probe and ZNB 8 VNA

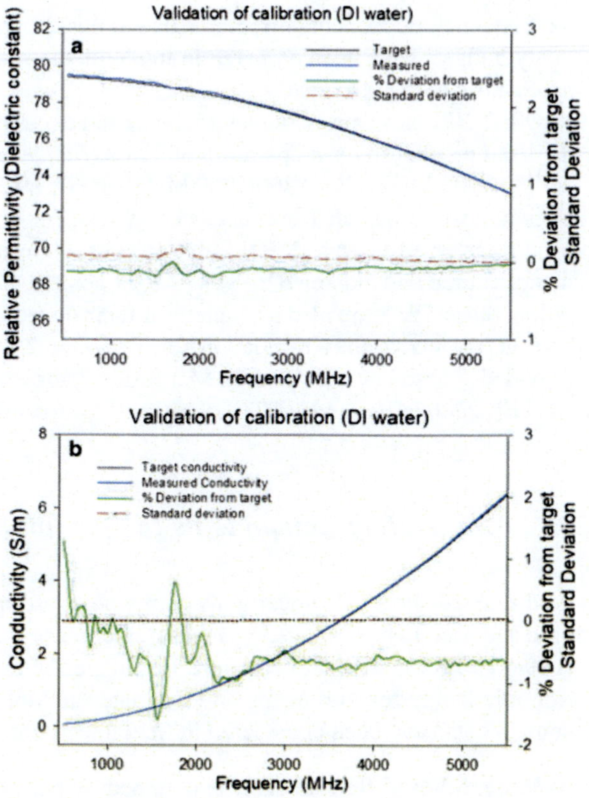


**Fig. 3** **a** Measured versus target relative permittivity of reference liquid (DI water at 21.99 ± 0.5 °C) along with standard deviation and % deviation from target. **b** Measured versus target conductivity of reference liquid (DI water at 21.99 ± 0.5 °C) along with standard deviation and % deviation from target

Literature has provided insights on how to evaluate the contributing factors and also how to reduce uncertainty contributors by modifying the test setup. The calibration data required for probe calibration for reference materials has been widely studied [9, 10]. Evans and Michelson [11] have shown the application of least square analysis to bilinear model (as discussed in Sect. 1.1) for probe coefficient correction upon calibration. Berube et al. [8] have studied four different geometries for open-ended coaxial probes and analysis of flange effects, dimensional effects for analysis and correction of these uncertainties. Through a comparative analysis, Jenkins et al. [12] have suggested computation for systematic uncertainties for calculable geometries of open-ended probe sensors for automatic network analyzers. Gregory et al. [13] have carried out detailed study for evaluating systematic uncertainties for high-frequency coaxial cell measurement techniques. They have incorporated Monte Carlo modeling for evaluation of uncertainties incorporating behavior of systematic measurement errors. IEEE-1528 (2013) (Annexure E.3) [5] also has provided an example uncertainty evaluation table for dielectric analysis.

The DAK system at CSIR-NPL employs abovementioned correction in the automatic evaluation software for specific probe geometry (e.g., DAK 3.5, DAK 12, etc.). The DAK system utilizes an iterative NRW-based algorithm and least square fitting for frequency-dependent corrections due to phase part in measurement [14]. Uncertainty budgets have been prepared in line with the standard document [5] and VNA application note [14].

CSIR-NPL has carried out another detailed experimental study to develop a standardized procedure for preparation of TELs (frequency specific) for frequencies for 2G (834/900 MHz), 3G (1800/1900 MHz), 4G (2.1/2.3 GHz), and Wi-Fi (2.45 GHz) communication bands. The process know-how for development of these TELs has been transferred to M/s FARE Labs. Pvt. Ltd., Gurugram, India. These TELs are manufactured and marketed by M/s FARE Labs Pvt. Ltd. commercially. After preparation, these TELs are also evaluated for their dielectric properties. In the following section, the uncertainty evaluation for TELs for 2.1, 2.3, and 2.45 GHz has been provided. These are marketed by M/s FARE Labs Pvt. Ltd. as HDTEL 2100 MHz, HDTEL 2300 MHz, and HDTEL 2450 MHz, respectively.

## 2.1 *Uncertainty Components Identification and Evaluation*

As discussed above, the uncertainty components are specific to measurement system used and need to be evaluated by measurement over long period of time and analysis. A standard measurement procedure is finalized after long-term study and analysis to minimize systematic errors and evaluate random errors. Following uncertainty components have been considered in final uncertainty estimation:

**(i) Repeatability** Repeatability is obtained by performing minimum five measurements per sample TEL (MUT), over a period of five days at same temperature (within ±0.5 °C). Standard deviation in measured values is obtained for the set of data at particular frequency and used as repeatability component of uncertainty.

**(ii) Deviation from Target** The target values for reference liquids (DI water is used as reference liquid for the budget shown here) are available in DAK software at pre-defined temperatures (15, 20 25, 30, 35 °C, etc.) and comparable with standard reference data provided in literature. The target values at specific measurement temperatures are obtained by interpolation methods. This feature is readily available in DAK software.

**(iii) Network Analyzer Drift** VNA drift uncertainty has been evaluated according to latest VNA calibration guidelines [15]. This includes uncertainties for test port linearity, test port mismatch, connector repeatability, noise floor, etc.

**(iv) Test Port Cable Variation** Test port cables are sensitive to temperature changes, movement, and other mechanical influences. As recommended in [15], cable used in dielectric measurement is individually characterized for errors due to bending and movement within the constraints of measurement procedure specified in Annexure-I.

**Table 1** Measured dielectric properties for different TELs and reference liquid (DI Water) at designated frequencies at 22 ± 1 °C

| Sample/MUT | Frequency (GHz) | Measured value | | Standard deviation | |
|---|---|---|---|---|---|
| | | $\varepsilon_r'$ | $\sigma$ (S/m) | $\varepsilon_r'$ | $\sigma$ (S/m) |
| DI water | 2.1 | 78.5881 | 1.008576 | 0.076752 | 0.011303 |
| | 2.3 | 78.3523 | 1.213328 | 0.069419 | 0.00687 |
| | 2.45 | 78.17108 | 1.374803 | 0.088579 | 0.008084 |
| HDTEL 2100 MHz | 2.1 | 40.84634 | 1.448022 | 0.021509 | 0.007803 |
| HDTEL 2300 MHz | 2.3 | 40.31826 | 1.645454 | 0.029831 | 0.008922 |
| HDTEL 2450 MHz | 2.4 | 38.9216 | 1.781078 | 0.024558 | 0.0007 |

**(v) Ambient Condition** The standard laboratory temperature for electrical measurements is 23 °C. Also, the measurement temperature of TELs as required for SAR evaluation is between 18 and 25 °C (within ±1 °C). Presented here are results for temperature (22 ± 1) °C.

# 3 Results and Discussions

## 3.1 *Measurement of Dielectric Properties for TELs Using DAK System*

The dielectric properties for DI water and Tissue Equivalent Liquids sample (MUTs) (namely HDTEL 2100 MHz, HDTEL 2300 MHz, and HDTEL 2450 MHz) have been measured using the standard procedure and tabulated below (Table 1). The measurements are carried out for the entire calibration range for the DAK 3.5 probe using DAK system, i.e., 300 MHz to 5 GHz. Measurements are carried out over one-week time (05 times minimum) by maintaining the MUTs at temperature 22 ± 1 °C. On each day of measurement during the one-week period, 05 measurement values are recorded. The mean and standard deviation for the combined data set are calculated and tabulated in Table 1.

## 3.2 *Uncertainty Evaluation*

Table 2 shows the combined uncertainty budget for dielectric parameters (relative permittivity and conductivity) prepared as per procedure discussed in previous sections. Table 2 also includes experimental uncertainty estimates for relative permittivity of one specimen TEL (MUT) at frequency 2.45 GHz. The TEL used as MUT is frequency-specific TEL for Wi-Fi frequency having target values within tolerance range (±5%) for frequency 2.45 GHz at 22 ± 1 °C.

**Table 2** Combined uncertainty budget for relative permittivity ($\varepsilon_r'$) and conductivity ($\sigma$) measurement for TELs using open-ended Coaxial Probe method (showing measured values for relative permittivity ($\varepsilon_r'$) for TEL for Wi-Fi frequency, i.e., 2.45 GHz)

| Contribution | Estimate | Uncertainty contribution | Distribution | Divisor | Std. Unc |
|---|---|---|---|---|---|
| **UA** Repeatability | 0.024558 | 0.024558 | Normal | 1 | 0.024558 |
| **UB1** Deviation from target (reference) | −0.07587 | 0.075875 | Rectangular | 1.732 | 0.043808 |
| VNA Linearity | 0.006884 | 0.006884 | Rectangular | 1.732 | 0.003975 |
| Test port Mismatch | 0.004251 | 0.004251 | U-shaped | 1.414 | 0.003006 |
| Noise floor | 0.004 | 0.004 | Gaussian | 2 | 0.002 |
| Connector repeatability | 0.001255 | 0.001255 | Gaussian | 2 | 0.000628 |
| **UB2** Combined VNA Drift | | | | | 0.005406 |
| **UB3** Cable Fixture Variation | 0.000895 | 0.000895 | Gaussian | 2 | 0.000448 |
| **UB4** Ambient Condition | 0.002 | 0.002 | Rectangular | 1.732 | 0.001155 |
| **Combined standard uncertainty** | | | | | **0.050527** |
| **Expanded uncertainty** | | | | | **0.101559** |

UA denotes "Type-A Uncertainty"
UB denotes "Type-B Uncertainty", the suffix after UB (e.g. UB1, UB2,...) denote the ith uncertainty component
This is the general method to write and identify the uncertainty components in uncertainty budgets. The components "VNA Linearity, test port mismatch, Noise floor, Connector repeatability", are combined to form "Combined VNA Drift" which is denoted by "UB2" i.e. 2nd component of Type-B Uncertainty

A detailed uncertainty analysis for TELs (namely HDTEL 2100MHz, HDTEL 2300 MHz, and HDTEL 2450 MHz) has been carried out. Uncertainty budgets for dielectric constant (relative permittivity) and conductivity have been separately evaluated for each of these TELs. The expanded uncertainties along with measured mean values for these TELs have been summarized and presented in Table 3 at their respective frequencies. The expanded uncertainties shown are for a coverage factor, $k = 2$, which corresponds to a coverage probability of approximately 95% for a normal distribution.

**Table 3** Measured Dielectric properties for Tissue Equivalent Liquids at 22 ± 1 °C at their designated frequencies along with measurement uncertainties showing percentage deviation from target values

| Sample (MUT) (MHz) | Designated frequency (GHz) | Measured value | | Expanded Uncertainty | | % Deviation from target | |
|---|---|---|---|---|---|---|---|
| | | $\varepsilon_r'$ | $\sigma$ (S/m) | $\varepsilon_r'$ | $\sigma$ (S/m) | $\varepsilon_r'$ | $\sigma$ (S/m) |
| HDTEL 2100 | 2.1 | 40.85 | 1.45 | ±0.159 | ± 0.023 | −2.63 | 2.82 |
| HDTEL 2300 | 2.3 | 40.32 | 1.65 | ±0.127 | ± 0.027 | −2.07 | 1.47 |
| HDTEL 2450 | 2.45 | 38.92 | 1.78 | ±0.102 | ± 0.019 | 0.71 | 1.05 |

## 4 Conclusion

The uncertainty evaluation for Tissue Equivalent Liquids for frequencies 2.1, 2.3, and 2.45 GHz (namely HDTEL 2100 MHz, HDTEL 2300 MHz, and HDTEL2450 MHz, respectively, as marketed by M/s FARE Labs Pvt. Ltd.) has been carried out using commercial Liquid dielectric assessment system (DAK) at CSIR-NPL, employing an open-ended coaxial probe (DAK 3.5). The uncertainty budget has been prepared, following the guidelines and methods as per the IEEE standards. The estimated dielectric properties along with their measurement uncertainties have been presented. The deviation from target values (as specified in IEEE-1528 standard) has also been estimated and the dielectric values of Tissue Equivalent Liquids, marketed by M/s FARE Labs Pvt. Ltd., have been found to be within tolerance range (±5%) at specified temperature, within the experimental errors and limitations.

**Acknowledgements** Authors like to express their gratitude to the Director, CSIR-National Physical Laboratory for providing necessary facilities to carry out the work done for this paper. Authors also like to thank Council of Scientific and Industrial Research (CSIR) for the monetary help via direct-SRF award to the one of the author. Authors also like to thank Academy of Scientific and Innovative Research (AcSIR) to provide opportunity to one of the author to work at CSIR-NPL through their Ph.D. (Physical Science) Program.

## References

1. Cleveland RF, Fields RE, Ulcek JL (1999) Questions and answers about biological effects and potential hazards of radiofrequency electromagnetic fields. FCC OET Bulletin vol 56, 4th edn.
2. IEEE Std. C95.3TM (1991), IEEE recommended practice for the measurement of potentially hazardous electromagnetic fields—RF and microwave
3. IEEE Std. C95.6TM (2002) IEEE standard for safety levels with respect to human exposure to electromagnetic fields, 0–3 kHz
4. ICNIRP (1998) Guidelines for limiting exposure to time-varying electric, magnetic and electromagnetic fields (up to 300 GHz). Health Phys 74:494–522

5. IEEE 1528 (2013) IEEE recommended practice for determining the peak spatial-average specific absorption rate (SAR) in the human head from wireless communications devices: measurement techniques
6. IEC/IEEE 62209-1528 (2020) Measurement procedure for the assessment of specific absorption rate of human exposure to radio frequency fields from hand-held and body-mounted wireless communication devices—Part 1528: human models, instrumentation, and procedures (frequency range of 4 MHz to 10 GHz). IEC/IEEE 62209-1528: 1.0 edn., 2020-10
7. Chen LF, Ong CK, Neo CP, Varadan VV, Varadan VK (2004) Microwave electronics: measurement and material characterization. Wiley
8. Berube D, Ghannouchi FM, Savard P (1996) A comparative study of four open-ended coaxial probe models for permittivity measurements of lossy dielectric biological materials at microwave frequencies. IEEE Trans Microw Theory Tech 44(10):1928–1934
9. Gregory AP, Clarke 1 RN (2001) Tables of the complex permittivity of dielectric reference liquids at frequencies up to 5 GHz'. NPL report CETM 33. ISSN 1467-3932
10. Gregory AP, Clarke RN (2005) Traceable measurements of the static permittivity of dielectric reference liquids over the temperature range 5–50 °C. Meas Sci Technol 16:1506–1516
11. Evans S, Michelson SC (1995) Intercomparison of dielectric reference materials available for the calibration of an open-ended probe at different temperatures. Meas Sci Technol 6(12):1721–1732
12. Jenkins S et al (1990) Dielectric measurements on reference liquids using automatic network analysers and calculable geometries. Meas Sci Technol 1(7):691–702
13. Gregory AP, Clarke RN, Cox MG (2009) Traceable measurement of dielectric reference liquids over the temperature interval 10–50 °C using coaxial-line methods. Meas Sci Technol 20:075106, 19 p
14. Measurement of dielectric material properties. Application note, RAC0607–0019_1_4E. Rohde & Schwarz GmbH & Co. KG
15. Guidelines on the evaluation of vector network analyzers (VNA). EURAMET calibration guide no. 12 version 3.0, Mar 2018

# Biosensors: Remediation Tool for Indoor Air Pollution

**Rakhi Tyagi and Archana Chaudhary**

**Abstract** Nowadays, the majority of people spend their lot of time indoors as compared to outdoors. In recent times, indoor air quality is poor or even worse than outdoor air quality. Some reasons for this bad air quality are the use of cleaning chemicals (releases volatile organic compounds), microbial flora, and the circulation of the same polluted air, especially for kitchens. To lessen the effects of indoor air pollutants on health and the environment, there is a need for developing rapid, cost-effective, and sensitive technology to detect these pollutants and helps with remediation. Biosensors proved themselves ideal to overcome the problem of indoor air pollution as these provide real-time data and simple to use. The biosensors are devices containing biological elements that can be enzyme, antibody, or microorganism; this biological element helps in the recognition of the pollutant present in the environment. This paper provides a review of the current status of various types of biosensors developed for indoor air pollution with the future outlook of biosensor technology.

**Keywords** Biosensors · Indoor air pollution · Bioreceptors · Health · Environment

## 1 Introduction

Today, air pollution is an upcoming and major global issue which creates so much health issue such as heart disease, lung cancer, and respiratory infection. It is always advisable to stay indoors when the AQI (Air Quality Index) is in the unhealthy range and people believe that the air they breathe inside the buildings is clean and pure [1]. But according to the Environmental Protection Agency (EPA), indoor air quality is worse than outdoor air and can cause major health issue [2]. However, the government and researchers have shown little concern about indoor air pollution.

R. Tyagi · A. Chaudhary (✉)
Faculty of Science, Shree Guru Gobind Singh Tricentenary University, Gurugram, India
e-mail: dr.archana57@gmail.com; archanac_fosc@sgtuniversity.org

R. Tyagi
e-mail: tyagirakhi0@gmail.com



S. Yadav et al. (eds.), *Recent Advances in Metrology*, Lecture Notes in Electrical Engineering 906, https://doi.org/10.1007/978-981-19-2468-2_31

One can easily recognize indoor air pollution due to the presence of some signs which include poor air circulation (the major cause of indoor air pollution), unusual odors, musty and stifling air, humid indoor environment, poor hygienic condition (dusty carpets), and excessive microbial flora (mold and bacteria on stale food and other items), and most important is feeling healthier outside the house rather than inside. This paper reviews the source and health effects of indoor air pollutants and highlights the biosensor techniques to monitor indoor air toxicity.

Cooking and heating are considered as a major anthropogenic source of indoor air pollution, additionally chemicals from house cleaning products (adds VOCs), mold, and other toxins from building materials also some sources of indoor air pollution [3]. Some other pollutants which characterize the indoor air quality are CO, $SO_2$, P.M2.5, $NO_x$, $NH_3$, PAHs, and volatile organic compounds (VOCs) [4]. Energy-efficient buildings are highly acceptable nowadays, and it requires buildings to be more compact which have an unavoidable impact on (IAQ) indoor air quality [5]. During the last 30 years, no study has been done to study the relation of energy efficiency and energy use of buildings, indoor air pollutants, and outdoor air quality [6]. Indoor air pollution rises mostly in winters due to poor ventilation and high level of humidity [7]. SBS (Sick building syndrome) is a major health issue that occurs due to poor IAQ in addition to fatigue, eye irritation, headache, etc. SBS is the situation in which residents of the building have some comfort issues during the time they spend in the building but no specific illness is diagnosed. According to WHO, nearly two million people die each year from indoor air pollution that harms women and young children around the world [8, 9].

To analyze and evaluate the effects of indoor air pollutants, reliable techniques or device that would continuously monitor indoor air pollution in a real-time manner is required. For pollutant monitoring, two approaches have been used. Traditional/conventional methods such as fluorometry and gas chromatography provide accurate and precise detection of environmental samples, but these techniques require costly instruments and require well-skilled manpower. One major drawback of conventional methods is that only a single compound can be identified at once and they do not provide bioavailability of pollutants and their impact on biological systems. A complementary solution to resolve the limitations of the conventional approach is being tested in which biological units such as algae, bacteria, and enzymes are used for analyses. Bacterial systems due to their rapid growth rate and low maintenance cost are particularly used for bioassay of environmental samples. Biosensors that are composed of whole-cell systems seem to be a desirable alternative in this context. Biosensors are not an independent unit but it is a component of instrumentation that uses any biological mechanism for pollutant detection [10]. They can be used for on-site monitoring as field devices due to their portable nature [11]. A biosensor is an integrated device containing different elements, i.e., analyte, recognition element (enzyme, receptors, microorganisms), signal transducer, electric signal, and display [12] (Fig. 1).

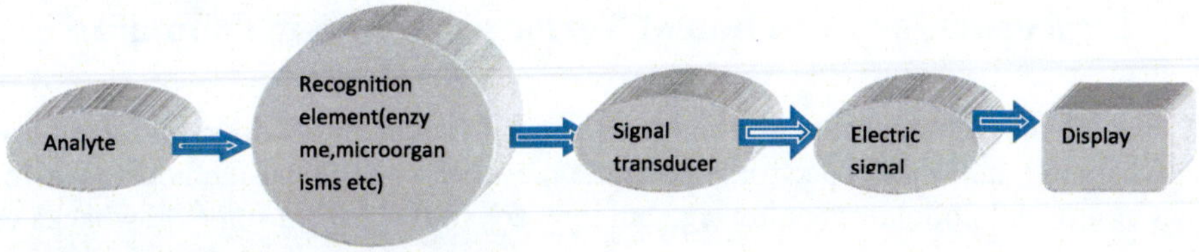


**Fig. 1** Different elements of biosensor unit

## 2 Indoor Sources and Health Effects

Indoor air pollution causes a severe problem due to a lack of knowledge of sources of indoor air pollutants and their impacts on health. Given below are some of the sources of indoor pollutant and the health impacts of these pollutants (Table 1).

**Table 1** Sources of indoor air pollution, a pollutant released, and their health effect

| S. No. | Household/office equipment | Pollutant | Health impact | Reference |
|---|---|---|---|---|
| 1 | Office PCs (after 3 months of use) | Phenol, toluene, 2-ethyl hexanol, formaldehyde, and styrene | Sick building symptoms | Bako-Biro et al. [13] |
| 2 | Office occupant (a by-product of metabolism) | Carbon dioxide ($CO_2$) | Drowsiness, Headaches | Hong et al. [14] |
| 3 | Cleaning products and ineffective ventilation rate, paints, building material | Volatile organic compounds—Formaldehyde | Eye, nose, and throat irritation | Suzuki et al. [15] |
| 4 | Moisture-prone buildings dust, carpets, and air conditioning | Microbial communities (including bacteria and fungi) | Legionnaire's disease, allergic reactions, hypersensitivity pneumonitis, humidifier fever, and tuberculosis | Hui et al. [16] Becher et al. [17] Eltzov et al. [18] |
| 5 | Structural defects in floors and walls, drains and piping, electrical connections, and cellars with earth floors | Radon | Lung cancer | Kim et al. [19] |

# 3 Biosensors for Detection of Various Indoor Air Pollutants

Conventional techniques such as high-performance liquid chromatography, gas chromatography, and mass spectrometry are some of the analytical methods used for environmental pollutant monitoring, and they are suffered from some drawbacks as the requirement of portable devices, expensive chemicals, reagents, time-consuming methodology, etc. [20]. Biosensors prove themselves as ideal instruments as they overcome the limitations of conventional techniques.

Biosensors are classified based on the recognition element used and the physicochemical nature of transducers. Recognition elements can be enzymes (enzymatic biosensors), plants, microbial fuel cells, antibodies (immunosensors), nucleic acids (genosensors), transducers can be piezoelectrical (used quartz crystal microbalance biosensors), electrochemical (includes amperometric biosensors), optical(surface plasmon resonance biosensors), calorimetric, etc. [21].

## 3.1 *Whole-Cell Sensors*

Whole-cell biosensor is a microbial biosensor, in this type of biosensor microbes are immobilized on a transducer, and microbes are used as biorecognition part on a transducer for monitoring of biochemical process [22]. Whole-cell sensors in which bacteria are used can detect various pollutants along them also provide information about the biological effects of detected pollutants [23]. It is very important to know the toxic effects of compounds rather than the chemical composition of the air pollutants. Valdman et al. [24] detected the naphthalene by using a strain of *Pseudomonas* in whole cell and a biosensor as a recognition material. The bioluminescent nature of bacterial cells helps in the recognition of pollutants as the inhibition of bacterial light shows the presence of air pollutants.

*Klebsormidium* a filamentous green alga was used for the detection of methanol and formaldehyde in the vapor [25]. This sensor showed good sensitivity (75 ppm) to the tested pollutant, and it is concluded from that study that VOCs can be detected with the help of algal-based biosensors. Seo et al. [26] also studied that Fungi (e.g., *Alternaria alternata, Eurotium herbariorum*, and *Aspergillus penicillioides*) can also be used for the detection of formaldehyde which is also a major indoor air pollutant (Fig. 2).

## 3.2 *Enzyme-Based Sensors*

Enzymes due to their availability and specificity are considered as an appropriate material to be used as a recognition element. Enzyme-based sensors are the type

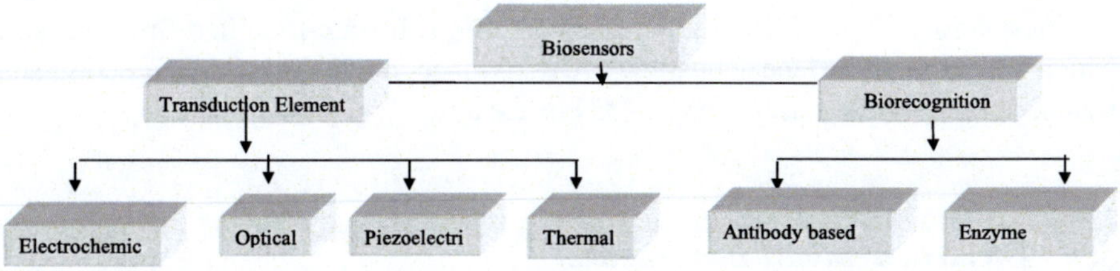


**Fig. 2** Classification of biosensors

of electrochemical biosensors which binds enzymes to the electrode via immobilization technique to enable high sensitivity and selectivity with short response time measurement. Enzyme-based biosensors worked either in direct mode or in indirect mode, and this classification is based on the mode of analyte monitoring. The analyte concentration is evaluated based on a percentage of the inhibited enzyme [27]. Mitsubayashi et al. [28] studied the detection of formaldehyde which is emitted from timber materials with formaldehyde dehydrogenase (FALDH) which is immobilized on a Pt-electrode coated with hydrophilic PTFE.

### 3.3 *Protein-Based Sensors*

Nowadays, innovative technologies which are based on protein engineering and synthetic biology by using microorganisms with specific signal outputs, selectvity, and sensitivity are most widely used in environmental monitoring and bioremediation. Ray et al. [29] showed that the protein-based enzymes detect benzene and its derivatives and also have the ability to differentiate between alkyl-substituted benzene derivatives such as toluene, *m*-xylene, and mesitylene. Raja et al. [30] studied green fluorescent protein-based Escherichia coli bacterial biosensor constructed based on green fluorescent protein expression under the control cadR gene of Pseudomonas aeruginosa BC15 is useful and applicable in determining the availability of heavy metals in soil and wastewater.

### 3.4 *Piezoelectric Sensor*

Piezo means the application of pressure or squeeze. Due to mechanical stress in some materials, a potential difference is created, which is called piezoelectricity. Piezoelectric biosensors are sensitive to changes in mass, density, or viscosity of samples in contact with their active surface. These biosensors work on the theoretical principle, i.e., the relation between mass deposited on the crystal surface and the resonant frequency of an oscillating piezoelectric [31].

Constantinoiu et al. [32] studied piezoelectric substrate-modified $SnO_2$ sensors which are based on surface acoustic wave (SAW) technique. This SAW-based sensor was used to detect organic volatiles [33] in the air.

### 3.5 *Electrochemical Biosensors*

Electrochemical biosensors carry some advantages like construction simplicity, low cost, and easy to use, these sensors use a variety of recognition systems but enzymes are mostly used in designing biosensors [22]. Indoor air pollutants (HCHO and $NO_2$) were detected by A $SnO_2$-based gas sensor fabricated with radio-frequency induction plasma deposition (IPD) [34].

Herschkovitz et al. [35] reported the new detection method in which flow injection system and biosensor measuring device are coupled by using formaldehyde dehydrogenase enzyme an $Os(bpy)_2$-poly(vinyl pyridine) (POs-EA) chemically modified screen-printed electrode. This novel technique can detect 30 ng $ml^{-1}$ of formaldehyde in an aqueous solution. The developed sensor is selective, inexpensive, stable over several days, and disposable, as well as simple, to manufacture and operate.

### 3.6 *Optical Biosensor*

The optical transducer works on the principle of use of optical transducer to detect changes that occur in optical properties, and these sensors use electromagnetic waves for analyte detection and could immobilize bio-element interactions which are known by the name of surface Plasmon resonance (SPR) [36]. These biosensors utilize the illumination released by certain microorganisms and they are compact and flexible but can measure generated electrical noise [37]. The utilization of optical biosensors technology is an essential approach for exploring and evaluating biomedical research, pharmaceuticals, environmental monitoring, national security, and warfare.

## 4 Application of Biosensors in Other Environmental Monitoring Fields

Presently, various anthropogenic activities destroy the environmental resources which become a matter of concern for everyone. Apart from conventional techniques, research on the construction of biosensors for monitoring of environmental pollutants will provide a sustainable solution to environmental pollution. biosensors can detect the pollutant at a very low concentration which is the foremost condition for environmental protection [38]. Biosensors exhibit numerous promising applications in various fields such as environmental monitoring, molecular diagnostics, soil quality monitoring, food quality monitoring, disease detection, pathogen discovery, drug discovery, water monitoring, and toxin detection [39].

## 5 Conclusion

Due to the rising concern of indoor air pollution, proper monitoring of pollutants and their remediation is most important. Biosensors fulfill real-time detection and help in removing indoor air pollutants. Currently, many techniques are unable to analyze multiple pollutants even though health issues occur only when these pollutants present in a group. So, future research will focus on creating techniques that can detect multiple pollutants at a time such as sensors based on biological receptors (enzyme, protein, whole cell, etc.). Indoor air pollution problems can be solved if proper ventilation and environment-friendly products have been used in buildings. Recent advances in the development of the biosensor over the past few years have paved the way for future researchers to further modify these biosensing elements to enhance them to the extent that would be able to detect even the most dangerous diseases like most viral diseases (HCV, HIV, Ebola, Crimean-Congo Virus, Rabies, etc.) and not only in humans but also viral diseases related to plants.

## References


1. Hänninen O, Rumrich I, Asikainen A (2017) Challenges in estimating health effects of indoor exposures to outdoor particles: Considerations for regional differences. Sci Total Environ 589:130–135
2. US EPA O (2014) Introduction to indoor air quality [collections and lists]. US EPA
3. Zhang X, Chen B, Fan X (2015) Different fuel types and heating approaches impact on the indoor air quality of rural houses in Northern China. Procedia Eng 121:493–550
4. Challoner A, Gill L (2014) Indoor/outdoor air pollution relationships in ten commercial buildings: PM2.5 and $NO_2$. Build Environ 80:159–173
5. Bajcinovci B, Jerliu F (2016) Achieving energy efficiency in accordance with bioclimatic architecture principles. Environ Climate Technol 18(1):54–63
6. Tham KW (2016) Indoor air quality and its effects on humans—a review of challenges and developments in the last 30 years. Energy Build 130:637–650
7. Wang Y, Kuckelkorn J, Zhao FY, Spliethoff H, Lang W (2017) A state of art of review on interactions between energy performance and indoor environment quality in Passive House buildings. Renew Sustain Energy Rev 72:1303–1319
8. Gall ET, Carter EM, Matt Earnest C, Stephen B (2013) Indoor air pollution in developing countries: research and implementation needs for improvements in global public health. Am J Public Health 103(4):67–72
9. Martin WJ 2nd, Glass RI, Balbus JM, Collins FS (2011) Public health. A major environmental cause of death. Science 334(6053):180–181
10. Mulchandani P, Lei Y, Chen W, Wang J, Mulchandani A (2002) Microbial biosensor for p-nitrophenol using *Moraxella* sp. Anal Chim Acta 470(1):79–86

11. Acerini CL (2006) Endocrine disrupting chemicals: a new and emerging public health problem? Arch Dis Child 91(8):633–641
12. Sethi RS (1994) Transducer aspects of biosensors. Biosens Bioelectron 9(3):243–264
13. Bako-Biro Z, Wargocki P, Weschler CJ, Fanger PO (2004) Effects of pollution from personal computers on perceived air quality, SBS symptoms and productivity in offices. Indoor Air 14(3):178–187
14. Hong T, Kim J, Lee M (2018) Integrated task performance score for the building occupants based on the $CO_2$ concentration and indoor climate factors changes. Appl Energy 228:1707–1713
15. Suzuki Y, Nakano N, Suzuki K (2003) Portable sick house syndrome gas monitoring system based on novel colorimetric reagents for the highly selective and sensitive detection of formaldehyde. Environ Sci Technol 37(24):5695–5700
16. Hui PS, Wong LT, Mui KW, Law KY (2007) Survey of unsatisfactory levels of airborne bacteria in air-conditioned offices. Indoor and Built Environment 16(2):130–138
17. Becher R, Øvrevik J, Schwarze P, Nilsen S, Hongslo J, Bakke J (2018) Do carpets impair indoor air quality and cause adverse health outcomes: a review. Int J Environ Res Public Health 15(2):184
18. Eltzov E, De Cesarea AL, Low KA, Marks RS (2019) Indoor air pollution and the contribution of biosensors. EuroBiotech J 3(1):19–31
19. Kim SH, Hwang WJ, Cho JS, Kang DR (2016) Attributable risk of lung cancer deaths due to indoor radon exposure. Ann Occup Environ Med 28(1)
20. Lang Q, Han L, Hou C, Wang F, Liu A (2016) A sensitive acetylcholinesterase biosensor based on gold nanorods modified electrode for detection of organophosphate pesticide. Talanta 156–157:34–41
21. Medeiros A, Maria L, Farias A (2011) Biosensor for environmental applications. In: Environmental biosensors
22. Ronkainen NJ, Halsall HB, Heineman WR (2010) Electrochemical biosensors. Chem Soc Rev 39(5):1747
23. Roda A, Pasini P, Mirasoli M, Michelini E, Guardigli M (2004) Biotechnological applications of bioluminescence and chemiluminescence. Trends Biotechnol 22(6):295–303
24. Valdman E, Valdman B, Battaglini F, Leite SG (2004) On-line detection of low naphthalene concentrations with a bioluminescent sensor. Process Biochem 39(10):1217–1222
25. Podola B, Nowack EC, Melkonian M (2004) The use of multiple-strain algal sensor chips for the detection and identification of volatile organic compounds. Biosens Bioelectron 19(10):1253–1260
26. Seo J, Kato S, Tatsuma T, Chino S, Takada K, Notsu H (2008) Biosensing of an indoor volatile organic compound on the basis of fungal growth. Chemosphere 72(9):1286–1291
27. Rebollar-Pérez G, Campos-Terán J, Ornelas-Soto N, Méndez-Albores A, Torres E (2016) Biosensors based on oxidative enzymes for detection of environmental pollutants. Biocatalysis 1(1)
28. Mitsubayashi K, Nishio G, Sawai M, Kazawa E, Yoshida H, Saito T, Saito H (2007) A biochemical sniffer-chip for convenient analysis of gaseous formaldehyde from timber materials. Microchim Acta 160(4):427–433
29. Ray S, Panjikar S, Anand R (2018) Design of protein-based biosensors for selective detection of benzene groups of pollutants. ACS Sensors
30. Edward Raja C, Selvam GS (2011) Construction of green fluorescent protein based bacterial biosensor for heavy metal remediation. Int J Environ Sci Tech 8(4):793–798
31. Marrazza G (2014) Piezoelectric biosensors for organophosphate and carbamate pesticides: a review. Biosensors 4(3):301–317
32. Constantinoiu I, Miu D, Viespe C (2019) Surface acoustic wave sensors for ammonia detection at room temperature based on $SnO_2/Co_3O_4$ bilayers. J Sens, 1–6
33. Johnson S, Shanmuganantham DT (2014) Design and analysis of SAW based MEMS gas sensor for the detection of volatile organic gases. Carbon 4(3):5

34. Spinelle L, Gerboles M, Kok G, Persijn S, Sauerwald T (2017) Review of portable and low-cost sensors for the ambient air monitoring of benzene and other volatile organic compounds. Sensors 17(7):1520
35. Herschkovitz Y, Eshkenazi I, Campbell CE, Rishpon J (2000) An electrochemical biosensor for formaldehyde. J Electroanal Chem 491(1):182–187
36. Monošík R, Streďanský M, Šturdík E (2012) Biosensors—classification, characterization and new trends. Acta Chimica Slovaca 5(1):109–120
37. Riedel K, Renneberg R, Wollenberger U, Kaiser G, Scheller FW (1989) Microbial sensors: Fundamentals and application for process control. J Chem Technol Biotechnol 44(2):85–106
38. Rodriguez-Mozaz S, Lopez de Alda MJ, Barceló D (2006) Biosensors as useful tools for environmental analysis and monitoring. Anal Bioanal Chem 386(4):1025–1041
39. Singh S, Kumar V, Dhanjal DS, Datta S, Prasad R, Singh J (2020) Biological biosensors for monitoring and diagnosis. In: Singh J, Vyas A, Wang S, Prasad R (eds) Microbial biotechnology: basic research and applications, pp 317–335

# Steering Algorithm for Generation of a Stable Time Scale

**Deboshree Roy, Preeti Kandpal, and Ashish Agarwal**

**Abstract** An atomic clock drifts from UTC (Coordinated Universal Time) due to factors such as aging and frequency drift. Algorithms are designed to provide a stable timescale by predicting and steering clocks to standard reference time (UTC). Kalman filter is an optimal estimator that can estimate present and future behavior of an atomic clock with respect to a stable reference timescale. A clock in general comprises white phase noise, white frequency noise, random walk frequency noise, random walk frequency aging, and constant frequency aging. Kalman filter is required for removal of noises and improving clock performance. The role of Kalman filter in improving the long- and short-term stability of a free running clock is investigated. A state matrix is formulated using a phase difference and frequency of clock with respect to a reference signal in this algorithm. The estimated state from Kalman is further used in a control algorithm for steering of clocks. Steering removes the drift in atomic clock and generates an accurate clock output with respect to UTC. The level of steering is determined by user-defined parameters. The variation in steering coefficient leading to a change in clock performance is studied.

**Keywords** Kalman filter · Prediction model · Clock steering

## 1 Introduction

A stable, accurate, and robust timescale is realized from a master clock steered in frequency via a phase micro-stepper according to an algorithm which combines the frequency comparison data between the master clock and other clocks in the ensemble. The algorithm plays an integral role to overcome the natural phenomena of

D. Roy (✉) · P. Kandpal · A. Agarwal
Time and Frequency Metrology, Indian Standard Time Division, National Physical Laboratory, Dr K S Krishnan Marg, New Delhi 110012, India
e-mail: deboshree_roy22@yahoo.com

A. Agarwal
e-mail: ashish@nplindia.org

 
S. Yadav et al. (eds.), *Recent Advances in Metrology*, Lecture Notes in Electrical Engineering 906, https://doi.org/10.1007/978-981-19-2468-2_32

frequency drift in clocks, to synchronize the time to UTC and eventually design a low-noise and a stable continuous timescale. The principle aim of algorithm is to generate a local Coordinated Universal Time (UTC), denoted as UTC ($k$), where $k$ represents the specific laboratory. The UTC($k$) should follow the UTC with minimum deviation [1, 2]. The algorithm can be designed, with an atomic clock, with reference to UTC and a weighted average of other clocks in the ensemble. The timescale algorithm, thus, in general, comprises three parts—prediction, weighting, and steering algorithm [3]. The paper describes an automated algorithm with the ultimate goal of real-time state prediction and control of individual atomic clocks for generation of a stable timescale output.

Application of Kalman filter in removal of noises on atomic clocks dates back to 1980s [4, 5]. The Kalman filter algorithm is based on a clock prediction model. For a clock, the phase or frequency are predicted with a set of linear stochastic differential equation [6]. The prediction algorithm considers clock aging, frequency drift factor, and various sources of noises associated with the clock [7, 8]. Kalman filter plays a key role in prediction algorithm and removal of noises in clocks. It is used to estimate the phase, frequency, and frequency drift information of a clock. A quantitative investigation of Kalman filter in state prediction and improving overall stability of a clock is presented in this paper. Alan deviation is used for stability analysis and noise component measurement of a clock [9]. Alan deviation is also used herein as a comparison tool between clock (UTC–clock) and Kalman filter output.

In this paper, we use Kalman filter for steering of atomic clock. The aim of a steering algorithm in a timescale is to minimize the deviation of the atomic clock with respect to a standard reference. Different steering algorithms to improve overall stability of the timescale are explored and implemented by different research groups [10–15]. The paper presents a steering technique of atomic clock through established control algorithms. Linear Quadratic Gaussian (LQG) as a steering method for steering an active hydrogen maser was described by Koppang and Leland [16]. Application of LQG and role of different parameters in LQG algorithm and their effects on steering an atomic clock is explored.

### *1.1 Clock Prediction Model*

A clock is characterized by parameters such as phase deviation $X_1(t)$ and frequency deviation $X_2(t)$ with respect to a standard reference timescale. Theoretically, a clock is defined by a set of stochastic differential equations: [17]

$$\mathrm{d}X_1(t) = X_2(t)\mathrm{d}t + \sigma_1 \mathrm{d}W_1(t) \tag{1a}$$

$$\mathrm{d}X_2(t) = d\ \mathrm{d}t + \sigma_2 \mathrm{d}W_2(t) \tag{1b}$$

With $X_1(0) = x_0$ and $X_2(0) = y_0$ represents the initial conditions, d is the frequency drift term. $\sigma_1$ and $\sigma_2$ are the coefficients for two wiener noise $W_1(t)$ and $W_2(t)$, respectively. The $\sigma_1$ and $\sigma_2$ are obtained through Alan variance [17, 18]. The relation between Alan variance and the coefficients, $\sigma_1$ and $\sigma_2$ are

$$\sigma_y^{\mathrm{WFN}}(t) = \sqrt{\frac{\sigma_1^2}{\tau}}; \sigma_y^{\mathrm{RWFN}}(t) = \sqrt{\frac{\tau\sigma_2^2}{3}} \tag{2}$$

where $\sigma_{\mathrm{RWFN}}$ and $\sigma_{\mathrm{WFN}}$ represents Random Walk Frequency Noise (RWFN) and White Frequency Noise (WFN) noises in Alan variance, respectively. They are the two dominant noise components in atomic clock [14].

The solution to the differential Eq. (1) is described as

$$X_1(t) = x_0 + y_0 t + d \cdot \frac{t^2}{2} + \sigma_1 W_1(t) + \sigma_2 \int_0^t W_2(s)\,\mathrm{d}s \tag{3a}$$

$$X_2(t) = y_0 + d \cdot t + \sigma_2 W_2(t) \tag{3b}$$

The clock progression with time is defined by Eq. 3a and b. Phase deviation and frequency deviation at epoch $(t_{k+1})$ are represented by $X_1(t_{k+1})$ and $X_2(t_{k+1})$, respectively.

*J* herein represents the noise component that shows a normal behavior with zero mean and a covariance described by Eq. (5)

$$X_1(t_{k+1}) = X_1(t_k) + X_2(t_k)\tau + d\frac{\tau^2}{2} + J_{k,1} \tag{4a}$$

$$X_2(t_{k+1}) = X_2(t_k) + d \cdot \tau + J_{k,2} \tag{4b}$$

$$J_k \sim N\left(0, \begin{pmatrix} \sigma_1^2\tau + \sigma_2^2\frac{\tau^3}{3} & \sigma_2^2\frac{\tau^2}{2} \\ \sigma_2^2\frac{\tau^2}{2} & \sigma_2^2\tau \end{pmatrix}\right) \tag{5}$$

## *1.2 Kalman Filter Formulation in Atomic Clock*

In this paper, Kalman filter is used to study the phase and frequency deviation of a clock with respect to a standard timescale. A Kalman filter is formulated with two set of equations.

$$X(k+1|k) = \varphi X(k) + \alpha, \quad \alpha \sim N(0, Q) \tag{6a}$$

$$Z(k+1) = H \cdot X(k+1|k) + \beta, \quad \beta \sim N(0, R) \tag{6b}$$

Equation (6a) is matrix formulation of clock system model defined in Eq. (4). $X(k)$ is the estimated state vector of the system at epoch $k$, and $X(k+1 \mid k)$ is the predicted state vector of the system at epoch $k+1$. $\Phi$ is the transition matrix, which links $X(k)$ and $X(k+1 \mid k)$. In the present algorithm, the state matrix and transition matrix are defined in the Eq. (7) as

$$x = \begin{pmatrix} x_1 \\ x_2 \end{pmatrix}; \varphi = \begin{pmatrix} 1 & \Delta t \\ 0 & 1 \end{pmatrix} \tag{7}$$

where $x_1$ and $x_2$ represent phase difference and fractional frequency with respect to a reference. The Eq. (6b) represents the measurement model (Fig. 2). Where $H$ is the measurement matrix and is formulated as $H = \begin{pmatrix} 0 & 1 \end{pmatrix}$ in this paper. $Z(k)$ is the measured value at epoch k. $\beta$ is the measured noise, characterized by a covariance matrix, $R$, and zero mean. In a Kalman filter, an error covariance matrix ($P$) prior to $k$th measurement is the expectation value of the difference between predicted value and measurement data

$$P(k|k-1) = E\left[(x(k|k-1) - x(k))(x(k|k-1) - x(k))^T\right] \tag{8}$$

Kalman gain ($G$) determines a relative ratio between the measurements and the prediction state covariance and is defined by the equation:

$$G(k) = P(k|k-1)H(k)^T\left[H(k)P(k|k-1)H(k)^T + R(k)\right]^{-1} \tag{9}$$

The new predicted state parameter depends on previous predicted state, the present measurement, and Kalman gain and is calculated by:

$$x(k|k) = \varphi x(k-1|k-1) + G(k)[z(k) - H \cdot \varphi \cdot x(k-1|k-1)] \tag{10}$$

### 1.3 Linear Quadratic Gaussian Formulation for Steering of Atomic Clock

A linear quadratic Gaussian (LQG) is a combination of linear quadratic estimator (Kalman filter) and linear quadratic regulator. The estimated output of Kalman filter is fed into the linear quadratic regulator to obtain a steered output for a timescale. A state equation for a LQG comprises an additional steering control term

$$X(k+1|k) = \varphi X(k) + Bu(k) + \alpha, \quad \alpha \sim N(0, Q) \tag{11}$$

where $B$ is the steering control term defined as $(1 \quad \Delta t)$ and $u$ is the steering control term. The control term $u$ is defined such that it minimizes the cost function defined by Eq. (12). The optimal control term for minimizing the cost function is defined as $u(k) = -K_r x(k)$, where $K_r$ is the feedback gain obtained by solving Albert Riccati equation.

$$C = \sum_k x(k)^T W_Q x(k) + u(k)^T W_R u(k) \tag{12}$$

where $x(k)$ is the estimated state defined in Eq. (10), and $W_Q$ and $W_R$ are the user-defined terms useful in controlling the level of steering. A comparatively large value of $W_R$ will drive the output toward reference timescale rapidly. This in turn may affect the short-term stability of the clock. A low level of $W_R$ compared to $W_Q$ will lead to reduced accuracy of the designed steered timescale. Hence, a trade-off between the steering level and accuracy is to be decided using parameters $W_Q$ and $W_R$. The variation in clock stability with these two parameters are investigated in the paper.

## 2 Hardware Architecture

CSIR-NPL maintains all seven International Standards (SI) being National Metrology Institute, and under 'Time' standard, it maintains a primary timescale for India which consist of an ensemble of five cesium atomic clocks and two active hydrogen masers along with supporting equipment. The aim of maintaining timescale of India is to maintain international traceability with BIPM and to generate Universal coordinated time for India which is designated as UTC (NPLI). The output signal of clocks in the form of frequency and pulse per second are inter-compared to a standard reference UTC (NPLI). The frequency 10 MHz is compared with respect to UTC (NPLI) by using a multichannel phase comparator. The output of phase comparator is used to derive matrix $Z(k)$ and formulate the measurement matrix $H$ shown in Eq. (6) (Fig. 1).

## 3 Results

Application of Kalman filter on a cesium clock data is studied. Assuming the measurement and initial error covariance matrix to be 1, respectively, the results are depicted in Fig. 2.

Figure 2a, b shows the variation of Kalman gain and error covariance matrix with time. The Kalman gain monotonically reduces and eventually stabilizes at a low value. This is due to the fact that the Kalman filter is weighting more on the clock prediction model than the measured data as per Eqs. (9) and (10). Figure 2c depicts the measured cesium clock output and Kalman output. The Alan deviation of clock

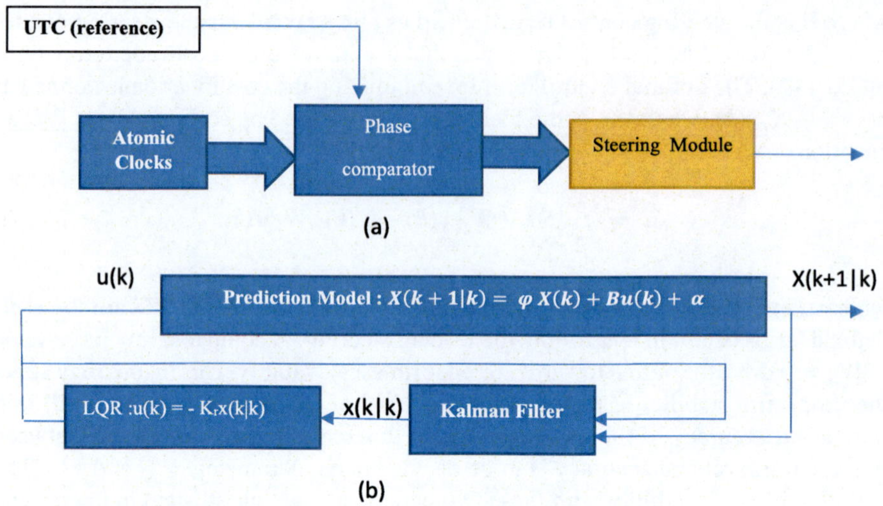


**Fig. 1** **a** Steering algorithm in timescale. **b** Steering module—LQG Algorithm

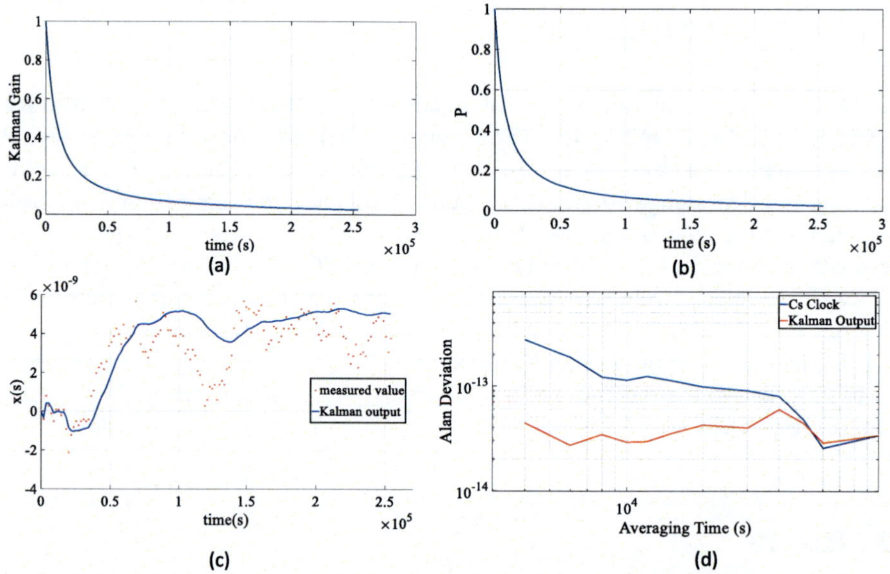


**Fig. 2** Variation of **a** Kalman gain and **b** covariance error matrix with time, **c** measured Cs clock data and Kalman filter output. **d** Alan deviation of Cs clock and Kalman filter output

**Table 1** Two cases with different user-defined parameter $W_Q$ in LQG algorithm leading to different feedback gain

| | $W_R$ | $W_Q$ | Feedback gain |
|---|---|---|---|
| Case I | 1 | $\begin{bmatrix} 10^{-11} & 0 \\ 0 & 10^{-11} \end{bmatrix}$ | $\begin{bmatrix} 5.5 \times 10^{-4} \\ 0.9 \end{bmatrix}$ |
| Case II | 1 | $\begin{bmatrix} 10^{-4} & 0 \\ 0 & 10^{-4} \end{bmatrix}$ | $\begin{bmatrix} 2.9 \times 10^{-6} \\ 0.1 \end{bmatrix}$ |

and Kalman filter output in Fig. 2d shows a minimization of noise particularly at a lower averaging time.

Further the Kalman filter output is used for steering of clock data. Table 1 defines two cases with different steering coefficient $W_Q$ with respect to $W_R$ that results in different feedback gain and, subsequently, different steered output. The steered output of the cesium clock for the two cases is described in Fig. 3.

Figure 3 shows that steering level of the cesium clock varies with a variation in $W_Q$. Further, the Alan deviation of the steered output shows a remarkable improvement when compared the clock output. Quantitatively, the noise components in Eq. (2) for cesium clock and steered output defined for two cases are described in Table 2. The

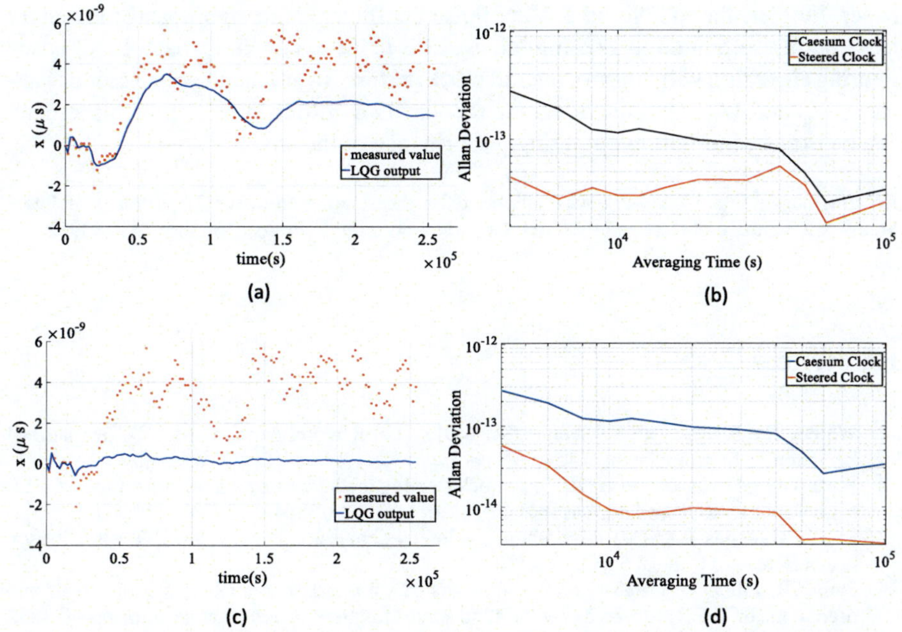


**Fig. 3** **a** Free running Cesium clock and output of LQG Algorithm with time for Case I. **b** Alan deviation for Case I. **c** Free running Cesium clock and output of LQG algorithm with time for Case II, and **d** Alan deviation for Case II. The cases I and II are defined in Table 1

**Table 2** Noise components (in Eq. 2) for free running clock and LQG output defined for cases I and II in Table 1

| | $\sigma_1$ | $\sigma_2$ |
|---|---|---|
| Cesium clock | $2.8 \times 10^{-13}$ | $5.4 \times 10^{-18}$ |
| Case I | $4.5 \times 10^{-14}$ | $3.4 \times 10^{-18}$ |
| Case II | $6.0 \times 10^{-14}$ | $5.6 \times 10^{-19}$ |

table shows that the noise components of steered output are lesser when compared with cesium clock.

## 4 Conclusion

The paper reports application of Kalman filter for estimation of states of a cesium clock. Kalman filter shows a general smoothening of measured clock data. The Alan deviation shows an improvement in short-term stability of Kalman filter output when compared to cesium clock. Kalman filter is further used in steering control algorithm for cesium clock. Variation in steering output with a change in user-defined parameters $W_Q$ and $W_R$ is studied. Alan deviation of LQG output shows that the long-term stability is poorer when steering coefficient $W_Q$ is low compared to $W_R$ in this paper. Further, the WFN and RWFN noise coefficients are significantly lower for LQG output compared to a cesium clock. This helps to conclude that a LQG-based steering algorithm will improve overall clock performance at appropriate user parameters, $W_Q$ and $W_R$. The role of standard control algorithm, such as pole placement, can be explored to improve steering performance of the clock.

**Acknowledgements** This work was supported by the funding from the Department of Legal Metrology, Ministry of Consumer Affairs, Government of India [Project Number TSP180932].

## References

1. Whibberley PB, Davis JA, Shemar SL (2011) Local representations of UTC in national laboratories. Metrologia 48(4):154
2. Wu Y, Gong H, Zhu X, Ou G (2015) A clock steering method: using a third-order type 3 DPLL equivalent to a Kalman filter with a delay. Metrologia 52(6):864
3. Panfilo G, Arias F (2009) Algorithms for international atomic time. IEEE Trans Ultrason Ferroelectr Freq Control 57(1):140–150
4. Stein SR, Filler RL (1988) Kalman filter analysis for real time applications of clocks and oscillators. In: IEEE proceedings of the 42nd annual frequency control symposium, pp 447–452
5. Yao J, Parker TE, Levine J (2017) JY1 time scale: a new Kalman-filter time scale designed at NIST. Meas Sci Technol 28(11):115004
6. Panfilo G, Harmegnies A, Tisserand L (2011) A new prediction algorithm for the generation of International Atomic Time. Metrologia 49(1):49
7. Formichella V, Camparo J, Tavella P (2017) Atomic clocks and the continuous-time random-walk. Eur Phys J B 90(11):1–9

8. Panfilo G, Tavella P (2008) Atomic clock prediction based on stochastic differential equations. Metrologia 45(6):108
9. Galleani L (2008) A tutorial on the two-state model of the atomic clock noise. Metrologia 45(6):175
10. Zhao S et al (2016) A new steering strategy for UTC (NTSC) based on hydrogen maser. In: 2016 IEEE international frequency control symposium (IFCS)
11. Bernier L-G, Dudle G (2004) Practical performance of the UTC (CH. R) real time realization of UTC (CH) and prospects for improvement, pp 170–174
12. Farina M, Galleani L, Tavella P, Bittanti S (2010) A control theory approach to clock steering techniques. IEEE Trans Ultrason Ferroelectr Freq Control 57(10):2257–2270
13. Matsakis D (2019) The effects of proportional steering strategies on the behavior of controlled clocks. Metrologia 56(2):025007
14. Gödel M, Schmidt TD, Furthner J (2019) Comparison between simulation and hardware realization for different clock steering techniques. Metrologia 56(3):035001
15. Ling J, Jiao D, Li L, Lin Y, Zhi-Jia L, Qi G, Jian X (2019) Research on GNSS receiver time and frequency steering algorithm based on linear quadratic Gaussian control. In: 2019 IEEE 2nd international conference on electronics technology (ICET), pp 99–103
16. Koppang P, Leland R (1999) Linear quadratic stochastic control of atomic hydrogen masers. IEEE Trans Ultrason Ferroelectr Freq Control 46(3):517–522
17. Zucca C, Tavella P (2005) The clock model and its relationship with the Allan and related variances. IEEE Trans Ultrason Ferroelectr Freq Control 52(2):289–296
18. Chaffee JW (1987) Relating the Allan variance to the diffusion coefficients of a linear stochastic differential equation model for precision oscillators. IEEE Trans Ultrason Ferroelectr Freq Control 34:655–658

# The Effect of Dust Charge Fluctuations on Modulational Instability of a Electrostatic Wave in a Complex Plasma

**Ajay Gahlot**

**Abstract** The effect of dust charge fluctuations (DCF) is investigated on modulational instability (MI) of a electrostatic lower hybrid wave in a complex plasma slab. A non-local theoretical model is developed for this process involving four wave interactions. Numerical calculations of the growth rate and unstable mode frequencies for this process have been carried out using the fluid equations and the dynamics of charged dust grain for the typical complex plasma parameters. The unstable mode frequency increases significantly in the presence of DCF while enhancement of growth rate is found proportional to pump amplitude.

**Keywords** Complex plasma · Dust charge fluctuations · Modulational instability

## 1 Introduction

The study of parametric instabilities of large amplitude electrostatic [1–5] and electromagnetic [6, 7] waves in a plasma has been an active field of research for the last few decades for their relevance to radio frequency current drive, heating of fusion plasmas [1, 8, 9], laser driven fusion [10], small scale laboratory experiments [11–13], and ionospheric modification experiments.

In recent years, considerable interest has been given in studying electrostatic waves in dusty plasmas [14–18]. The experimental enhancement of growth rate of electrostatic ion-cyclotron (EIC) wave in complex plasma was observed by Barkan et al. [16]. Chow et al. [17, 18] investigated EIC instability in collisionless plasma using kinetic theory. Their result proved that as parameter '$\delta$' (ion to electron density) enhances, critical drift velocity for wave excitation is decreased.

The presence of dust grains influences parametric process involving three wave interactions in unmagnetized [19–21] and magnetized plasmas [22]. Liu and Tripathi [8] considered MI of lower hybrid waves in infinite plasma. Konar et al. [23] investigated MI of a electrostatic lower hybrid wave in a plasma slab in the absence of

A. Gahlot (✉)
Department of Applied Sciences, Maharaja Surajmal Institute of Technology, C-4, Janakpuri, Delhi 110058, India
e-mail: ajaygahlot@msit.in

 
S. Yadav et al. (eds.), *Recent Advances in Metrology*, Lecture Notes in Electrical Engineering 906, https://doi.org/10.1007/978-981-19-2468-2_33

dust grains. In this paper, we examine the influence of DCF and dust dynamics on the MI of lower hybrid waves in a plasma slab.

In Sect. 2, the analysis is carried out using fluid approximation and first order perturbation theory. Dust charge fluctuations (DCF) are incorporated following Whipple et al. [14]. Results and conclusion part are mentioned in Sects. 3 and 4, respectively.

## 2 Instability Analysis

A homogenous plasma slab filled dust grains is bounded between $x = 0$ and $x = a_0$. The static uniform magnetic field is applied along z-direction, i.e., $\vec{B}_s = B_s\hat{k}$. The densities, temperature, mass, and charge of the three species dust grains, electrons, and ions are denoted by ($n_{d0}$, $T_d$, $m_d$, $-Q_{d0}$), ($n_{e0}$, $T_e$, $m_e$, $-e$), and ($n_{i0}$, $T_i$, $m_i$ $e$), respectively, in slab. We assume the potentials of the four waves of the form

$$\phi_0 = \phi_0(x)\exp\left[-i(\omega_0 t - k_{0z}z)\right],\ \phi_1 = \phi_1(x)\exp\left[-i(\omega_1 t - k_{1z}z)\right],$$

$$\phi_2 = \phi_2(x)\exp\left[-i(\omega_2 t - k_{2z}z)\right],\ \phi = \phi(x)\exp\left[-i(\omega t - k_z z)\right].$$

The dispersion relation describing electrostatic lower hybrid pump wave is given by

$$\frac{\partial^2\phi_0}{\partial x^2} + K_0^2\phi_0 = 0, \tag{1}$$

where
$K_0^2 = \frac{\omega_{LH}^2 m_i}{\omega_0^2 m_e}k_{0z}^2 - k_{0z}^2$, $\omega_{LH}^2 = \frac{\omega_{pi}^2}{1+\frac{\omega_{pe}^2}{\omega_{ce}^2}}$, $\omega_{pe}\left(= \sqrt{\frac{4\pi n_{e0}e^2}{m_e}}\right)$, $\omega_{pi}\left(= \sqrt{\frac{4\pi n_{i0}e^2}{m_i}}\right)$, and $\omega_{ce}\left(= \frac{eB_s}{m_e c}\right)$ are lower hybrid, electron plasma, ion plasma, and electron cyclotron frequency, respectively.

The parallel component of the ponderomotive force ($F_{pz}$) exerted by lower hybrid pump wave and the sidebands ($\phi_{1,2}$) on the electrons is given by

$$F_{pz} = -iek_z\phi_p \tag{2}$$

where

$$\phi_p = \frac{-ek_{0z}^2}{2m_e\omega_0^2}\left[\phi_0\phi_1 + \phi_0^*\phi_2\right] \tag{3}$$

Electron response to $\phi_p$ and self-consistent potential $\phi$ turns out to be

$$n_{e1} = \frac{-n_{e0} e k^2 (\phi + \phi_p)}{m_e \omega^2} \tag{4}$$

where $n_{e1}$ is the perturbed density of electrons.

The ion density perturbation at $(\omega, k)$ can be written as

$$n_{i1} = \frac{n_{i0} e k^2 \phi}{m_i \omega^2}. \tag{5}$$

Similarly, the dust density perturbation is given by

$$n_{d1} = -\frac{n_{d0} Q_{d0} k^2 \phi}{m_d \omega^2}. \tag{6}$$

The dust charge fluctuation is given by

$$Q_{d1} = \frac{|I_{e0}|}{i(\omega + i\eta)} \left( \frac{n_{i1}}{n_{i0}} - \frac{n_{e1}}{n_{e0}} \right). \tag{7}$$

where dust charging rate

$$\eta = 0.79a \left( \frac{\omega_{pi}}{\lambda_{Di}} \right) \left( \frac{1}{\delta} \right) \left( \frac{m_i}{m_e} \frac{T_i}{T_e} \right)^{\frac{1}{2}} \sim 10^{-2} \omega_{pe} \left( \frac{a}{\lambda_{De}} \right) \frac{1}{\delta}.$$

In Eq. (7), $(\omega^{-1})$ is approximately equal to $(\eta^{-1})$.

Using Eqs. (4) and (5) in Eq. (7), we get

$$Q_{d1} = \frac{|I_{e0}| e k^2}{i(\omega + i\eta)\omega^2} \left[ \frac{\phi}{m_i} + \frac{(\phi + \phi_p)}{m_e} \right] \tag{8}$$

In equilibrium, there is overall charge neutrality, i.e., $-e n_{i0} + e n_{e0} + Q_{d0} n_{d0} = 0$ $\frac{n_{i0}}{n_{e0}} = 1 + \frac{n_{d0}}{n_{e0}} \frac{Q_{d0}}{e}$ or $\frac{n_{d0}}{n_{e0}} = (\delta - 1) \frac{e}{Q_{d0}}$, where $\delta = n_{i0} / n_{e0}$.

Using Eqs. (4), (5), (6), and (8) in the Poisson's equation,

$$\nabla^2 \phi = 4\pi [n_{e1} e - n_{i1} e + n_{d0} Q_{d1} + Q_{d0} n_{d1}],$$

we get

$$\phi = \frac{-\chi_e \left[ 1 + \frac{i\beta}{(\omega + i\eta)} \right] \phi_p}{\varepsilon_d}, \tag{9}$$

where

$\varepsilon_d = 1 + \chi_e\left[1 + \frac{i\beta}{(\omega+i\eta)}\right] + \chi_i\left[1 + \frac{i\beta}{(\omega+i\eta)\delta}\right] + \chi_d$, $\chi_e = \frac{-\omega_{pe}^2}{\omega^2}$, $\chi_i = \frac{-\omega_{pi}^2}{\omega^2}$, $\chi_d = \frac{-\omega_{pd}^2}{\omega^2}$, $\omega_{pd}\left(= \sqrt{\frac{4\pi n_{d0} Q_{d0}^2}{m_d}}\right)$ and $\beta = \frac{|I_{e0}| n_{d0}}{e n_{e0}} = 0.397\left(1 - \frac{1}{\delta}\right)\left(\frac{a}{v_{te}}\right)\omega_{pi}^2\left(\frac{m_i}{m_e}\right)$, is the coupling parameter.

$\chi_e$ is electron susceptibility while $\omega_{pd}$, $\chi_i$, and $\chi_d$ denote dust plasma frequency, ion, and dust susceptibility, respectively.

Nonlinear electron density perturbation for both sidebands are

$$n_1^{nl} = \frac{\nabla.\left(n_{e1} V_0^*\right)}{2i\omega_1} \simeq -\frac{e k_{0z}^2 \phi_0^* n_{e1}}{2 m_e \omega_0^2} \tag{10}$$

and

$$n_2^{nl} = \frac{\nabla.(n_{e1} V_0)}{2i\omega_2} \simeq -\frac{e k_{0z}^2 \phi_0 n_{e1}}{2 m_e \omega_0^2}, \tag{11}$$

where $\omega_1 \simeq -\omega_0$ and $\omega_2 \simeq \omega_0$.

Equations representing lower and upper sidebands are obtained using Eqs. (10) and (11) in the Poisson's equation as

$$\begin{aligned} \frac{\partial^2 \phi_1}{\partial x^2} + K_{1d}^2 \phi_1 = {} & \frac{e^2 k_{0z}^4 k^2 \phi_0^* \chi_e}{4 m_e^2 \omega_0^2 \varepsilon_d M}\left[1 + \frac{i\beta}{(\omega + i\eta)}\right] \\ & \left\{1 + \chi_i\left[1 + \frac{i\beta}{(\omega + i\eta)\delta}\right] + \chi_d\right\}\left[\phi_0\phi_1 + \phi_0^*\phi_2\right] \end{aligned} \tag{12}$$

and

$$\begin{aligned} \frac{\partial^2 \phi_2}{\partial x^2} + K_{2d}^2 \phi_2 = {} & \frac{e^2 k_{0z}^4 k^2 \phi_0 \chi_e}{4 m_e^2 \omega_0^2 \varepsilon_d M}\left[1 + \frac{i\beta}{(\omega + i\eta)}\right] \\ & \left\{1 + \chi_i\left[1 + \frac{i\beta}{(\omega + i\eta)\delta}\right] + \chi_d\right\}\left[\phi_0\phi_1 + \phi_0^*\phi_2\right] \end{aligned} \tag{13}$$

where

$$K_{1d}^2 = \frac{\frac{\omega_{pi}^2}{\omega_1^2}\frac{m_i}{m_e}\left[1 + \frac{i\beta}{(\omega+i\eta)}\right]k_{1z}^2 - k_{1z}^2}{1 + \frac{\omega_{pe}^2}{\omega_{ce}^2}\left[1 + \frac{i\beta}{(\omega+i\eta)}\right] - \frac{\omega_{pi}^2}{\omega_1^2}\left[1 + \frac{i\beta}{(\omega+i\eta)\delta}\right] - \frac{\omega_{pd}^2}{\omega_1^2}}, \tag{14}$$

$$K_{2d}^2 = \frac{\frac{\omega_{pi}^2}{\omega_2^2}\frac{m_i}{m_e}\left[1+\frac{i\beta}{(\omega+i\eta)}\right]k_{2z}^2 - k_{2z}^2}{1+\frac{\omega_{pe}^2}{\omega_{ce}^2}\left[1+\frac{i\beta}{(\omega+i\eta)}\right] - \frac{\omega_{pi}^2}{\omega_2^2}\left[1+\frac{i\beta}{(\omega+i\eta)\delta}\right] - \frac{\omega_{pd}^2}{\omega_2^2}}$$

and (15)

$$M = 1 + \frac{\omega_{pe}^2}{\omega_{ce}^2}\left[1+\frac{i\beta}{(\omega+i\eta)}\right] - \frac{\omega_{pi}^2}{\omega_0^2}\left[1+\frac{i\beta}{(\omega+i\eta)\delta}\right] - \frac{\omega_{pd}^2}{\omega_0^2}.$$

Right-hand side of Eqs. (12) and (13) when equated to zero represents linear response whose solutions are given by $\phi_{1n_1}$ and $\phi_{1n_2}$, respectively.

Expressing $\phi_1$ and $\phi_2$, i.e., solutions of Eqs. (12) and (13) in terms $\phi_{1n_1}$ and $\phi_{1n_2}$ (orthonormal functions), we get

$$\phi_1 = \sum_{n_1} A_{n_1}^{(1)}\phi_{1n_1} \tag{16}$$

and

$$\phi_2 = \sum_{n_2} A_{n2}^{(2)}\phi_{2n_2}. \tag{17}$$

In the absence of the pump wave, Eq. (12) becomes

$$\frac{\partial^2\phi_1}{\partial x^2} + K_{1dn_1}^2\phi_1 = 0 \tag{18}$$

Now subtracting Eq. (18) from Eq. (12),

$$[K_{1d}^2 - K_{1dn_1}^2]\phi_1 = \frac{e^2k_{0z}^4k^2\phi_0^*\chi_e}{4m_e^2\omega_0^2\varepsilon_d M}\left[1+\frac{i\beta}{(\omega+i\eta)}\right]$$
$$\left\{1+\chi_i\left[1+\frac{i\beta}{(\omega+i\eta)\delta}\right]+\chi_d\right\}[\phi_0\phi_1 + \phi_0^*\phi_2]$$

Substituting the values of $\phi_1$ and $\phi_2$ from Eqs. (16) and (17), we get

$$[K_{1d}^2 - K_{1dn_1}^2]\sum_{n_1} A_{n_1}^{(1)}\phi_{1n_1} = \frac{e^2k_{0z}^4k^2\phi_0^*\chi_e}{4m_e^2\omega_0^2\varepsilon_d M}\left[1+\frac{i\beta}{(\omega+i\eta)}\right]$$
$$\left\{1+\chi_i\left[1+\frac{i\beta}{(\omega+i\eta)\delta}\right]+\chi_d\right\}$$
$$\times\left[\phi_0\phi_0^*\sum_{n_1} A_{n_1}^{(1)}\phi_{1n_1} + \phi_0^*\phi_0^*\sum_{n_2} A_{n_2}^{(2)}\phi_{2n_2}\right]$$

Multiplying both sides by $\phi^*_{1m_1}$ and integrating over '$x$,' we get

$$\int [K_{1d}^2 - K_{1dn_1}^2] \sum_{n_1} A_{n_1}^{(1)} \phi_{1n_1} \phi_{1m_1}^* \mathrm{d}x = \int \eta_1 \phi_{1m_1}^* \left[ \phi_0 \phi_0^* \sum_{n_1} A_{n_1}^{(1)} \phi_{1n_1} + \phi_0^* \phi_0^* \sum_{n_2} A_{n_2}^{(2)} \phi_{2n_2} \right] \mathrm{d}x \tag{19}$$

where

$$\eta_1 = \frac{e^2 k_{0z}^4 k^2 \chi_e}{4 m_e^2 \omega_0^2 \varepsilon_d M} \left[ 1 + \frac{i\beta}{(\omega + i\eta)} \right] \left\{ 1 + \chi_i \left[ 1 + \frac{i\beta}{(\omega + i\eta)\delta} \right] + \chi_d \right\}$$

Taking only one value $n_1 = m_1$, we get

$$\left[ K_{1d}^2 - K_{1dn_1}^2 - \eta_1 \int \phi_0 \phi_0^* \phi_{1n_1}^* \phi_{1n_1} \mathrm{d}x \right] A_{n_1}^{(1)} = \eta_1 \sum_{n_2} A_{n_2}^{(2)} \int \phi_0^* \phi_0^* \phi_{2n_2} \phi_{1n_1}^* \mathrm{d}x \tag{20}$$

Similarly, we will get

$$\left[ K_{2d}^2 - K_{2dn_2}^2 - \eta_1 \int \phi_0^* \phi_0 \phi_{2n_2} \phi_{2n_2}^* dx \right] A_{n_2}^{(2)} = \eta_1 \sum_{n_1} A_{n_1}^{(1)} \int \phi_0 \phi_0 \phi_{1n_1} \phi_{2n_2}^* dx \tag{21}$$

Multiplying Eqs. (20) and (21) and taking $n_1 = n_2 = n$, we get

$$\left[ K_{1d}^2 - K_{1dn}^2 - \eta_1 \int |\phi_0|^2 |\phi_{1n}|^2 dx \right] A_{n_1}^{(1)} \left[ K_{2d}^2 - K_{2dn}^2 - \eta_1 \int |\phi_0|^2 |\phi_{2n}|^2 dx \right] A_{n_2}^{(2)} = \eta_1^2 A_{n_2}^{(2)} A_{n_1}^{(1)} \int \phi_0^* \phi_0^* \phi_{2n} \phi_{1n}^* dx \int \phi_0 \phi_0 \phi_{1n} \phi_{2n}^* dx$$

or

$$[K_{1d}^2 - K_{1dn}^2 - \delta_1][K_{2d}^2 - K_{2dn}^2 - \delta_2] = \mu \tag{22}$$

where $\delta_1 = \eta_1 \int |\phi_0|^2 |\phi_{1n}|^2 dx$, $\delta_2 = \eta_1 \int |\phi_0|^2 |\phi_{2n}|^2 \mathrm{d}x$ and $\mu = \eta_1^2 \int \phi_0^* \phi_0^* \phi_{2n} \phi_{1n}^* \mathrm{d}x \int \phi_0 \phi_0 \phi_{1n} \phi_{2n}^* \mathrm{d}x$.

As we know for modulational instability $k_z \ll k_{0z}$, $\omega \ll \omega_0$, we can expand $K_{1d}^2$, $K_{2d}^2$ using Taylor's series for a function of two variable as

$$
\begin{aligned}
K_{1d}^2 = K_{1d}^2(-\omega_0, -k_0) + \omega \frac{\partial K_{1d}^2}{\partial \omega_1}\bigg|_{-\omega_0} + k_z \frac{\partial K_{1d}^2}{\partial k_{1z}}\bigg|_{-k_{0z}} \\
+ \frac{\omega^2}{2} \frac{\partial^2 K_{1d}^2}{\partial \omega_1^2}\bigg|_{-\omega_0} + \frac{k_z^2}{2} \frac{\partial^2 K_{1d}^2}{\partial k_{1z}^2}\bigg|_{-k_{0z}}
\end{aligned} \tag{23}
$$

$$
\begin{aligned}
K_{2d}^2 = K_{2d}^2(\omega_0, k_0) + \omega \frac{\partial K_{2d}^2}{\partial \omega_2}\bigg|_{\omega_0} + k_z \frac{\partial K_{2d}^2}{\partial k_{2z}}\bigg|_{k_{0z}} \\
+ \frac{\omega^2}{2} \frac{\partial^2 K_{2d}^2}{\partial \omega_2^2}\bigg|_{\omega_0} + \frac{k_z^2}{2} \frac{\partial^2 K_{2d}^2}{\partial k_{2z}^2}\bigg|_{k_{0z}}
\end{aligned} \tag{24}
$$

Now assuming $\omega = \omega_r + i\gamma$ and using the condition for modulational instability, i.e., $\frac{\omega_r}{k_z} \approx \frac{\partial \omega_0}{\partial k_{0z}}$, we get

$$
\omega_r = \frac{\omega_0}{k_{0z}} \left\{ 1 - \frac{\left[1 + \frac{\omega_{pe}^2}{\omega_{ce}^2}\left(1 + \frac{i\beta}{\omega + i\eta}\right)\right]\omega_0^2}{\omega_{pi}^2\left[1 + \frac{i\beta}{(\omega+i\eta)}\right]\frac{m_i}{m_e} + \omega_{pi}^2\left[1 + \frac{i\beta}{(\omega+i\eta)\delta}\right] + \omega_{pd}^2} \right\} k_z \tag{25}
$$

and

$$
\gamma = \frac{\sqrt{\mu - \delta_1\delta_2 + B_1(\delta_1 + \delta_2 - B_1)}}{A_1} \tag{26}
$$

where $A_1 = \frac{2\omega_{pi}^2 m_i}{\omega_0^3 m_e M}\left[1 + \frac{i\beta}{(\omega+i\eta)}\right]k_{0z}^2$ and $B_1 = \frac{3\omega_r^2\omega_{pi}^2 m_i}{\omega_0^4 m_e M}\left[1 + \frac{i\beta}{(\omega+i\eta)}\right]k_{0z}^2 + \frac{k_z^2}{M}\left\{\frac{\omega_{pi}^2 m_i}{\omega_0^2 m_e}\left[1 + \frac{i\beta}{(\omega+i\eta)}\right] - 1\right\}$.

Now, we will discuss two cases of interest:

**Case I**: when DCF are present, i.e., $\eta$ is finite,
**Case II**: When DCF are absent, i.e.,$Q_{d_1} = 0$ when $\eta \to \infty$. The dispersion relation of Konar et al. [23] (cf. pages 3799 and 3800) is obtained for $\delta = 1$and $\beta = 0$.

## 3 Results and Discussions

We solve Eqs. (25) and (26) numerically to obtain real frequency ($\omega_r$) and growth rate ($\gamma$) of the unstable mode using following parameters: $n_{i0} = 5.0 \times 10^{10}$ cm$^{-3}$, $n_{d0} = 2.0 \times 10^4$ cm$^{-3}$, $T_e = T_i = 0.2$ eV, $m_i/m_e \approx 7.16 \times 10^4$ (Potassium), $a = 10^{-4}$ cm, $\omega_0 = 7.0 \times 10^9$ rad/s, $k_{0z} = 3.25$ cm$^{-1}$ and $k_{1z} = 0.035$ cm$^{-1}$, and $B_s = 2$ kg. We vary $\delta$ from 1.0 to 5.0.

Figure 1 shows the variation of $\omega_r$ (rad/s) of the unstable mode with $\delta (= n_{io}/n_{eo})$ in the presence and absence of dust charge fluctuations. It can be seen from Fig. 1

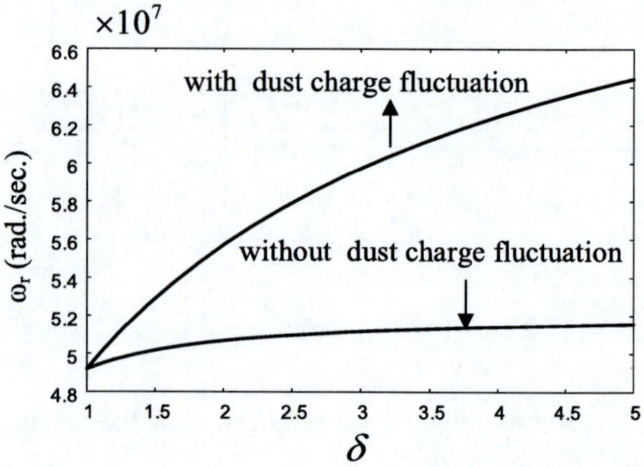


**Fig. 1** Unstable mode frequency as a function of $\delta(= n_{i0}/n_{e0})$ in presence and absence of DCF

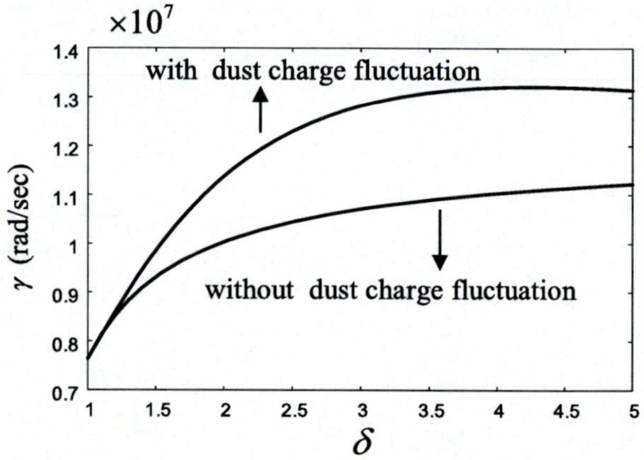


**Fig. 2** Growth rate $\gamma$ (rad/s) of the unstable mode as a function of $\delta(= n_{i0}/n_{e0})$ and in presence and absence of DCF

that $\omega_r$ increases with $\delta$ in both the cases and gets saturated for higher values of $\delta$ and increase is more significant in case of dust charge fluctuations. Figure 2, shows enhancement of growth rate $\gamma$ ($s^{-1}$) for different values of δ when pump amplitude is $\phi_0 = 0.023$ esu. It can be seen from Fig. 2 that $\gamma$ is enhanced by a factor ~1.73 (for $B_s = 2$ kg) [in the presence of DCF] as $\delta$ is increased from 1 to 4. The growth rate validates Barkan et al. [16] experimental finding when growth rate is almost doubled for appro. same parameters. In Eq. (26), $\mu \approx \delta_1\delta_2$ and since $B_1$ is positive, the growth is only possible when $\delta_1 + \delta_2 > B_1$ and this condition is satisfied when $\omega_r^2 > \omega_{pi}^2 I$, where $I = 1 + \frac{i\beta}{(\omega+i\eta)\delta}$. The growth rate is found proportional to pump amplitude as $\delta_1 \approx \delta_2$ and $B_1 < 2\delta$. Thus, the presence of dust charge fluctuations (DCF) makes the lower hybrid pump more modulationally unstable to low frequency quasimode for a particular value of $\delta$.

## 4 Conclusion

Non-local theory for investigation of Modulational instability in complex plasma slab is developed. The wave frequency and growth rate increase with $\delta$ and show significant increase in the presence of DCF. Modulational instability is proportional to pump wave amplitude while Landau damping has profound effect only for larger value of $\delta$.

## References

1. Wolf NS, Majeski R, Lashinky H, Tripathi VK, Liu CS (1980) Phys Rev Lett 45:799
2. Liu CS, Chan VS, Bhandari DK, Harvey RW (1982) Phys Fluids 48:1479
3. Liu CS, Tripathi VK, Chan VS, Stefan V (1984) Phys Fluids 27:1709
4. Akimoto K (1989) Phys Fluids B 1:1998
5. Stenflo L (1994) Phys Scr T50:15
6. Sharma RP, Ramamurthy K, Yu MY (1984) Phys Fluids 27:399
7. Saleem H, Murtaza G (1986) J Plasma Phys 36:295
8. Liu CS, Tripathi VK (1986) Phys. Reports 130:143
9. Tripathi VK, Liu CS, Grebogi C (1979) Phys Fluids 22:301
10. Kruer WL (1987) The physics of laser plasma interactions. Addision-Wesley, Reading, MA
11. Jain VK, Christiansen PJ (1981) Phys Lett A 82:127
12. Jain VK, Christiansen PJ (1984) Plasma Phys Controlled Fusion 26:613
13. Praburam G, Sharma AK (1992) J Plasma Phys 48:3
14. Whipple EC, Northdrop TG, Mendis DA (1985) J Geophys Res 90:7405
15. Jana MR, Sen A, Kaw PK (1993) Phys Rev E 48:3930
16. Barkan A, D'Angelo N, Merlino RL (1995) Planet Space Sci 43:905
17. Chow VW, Rosenberg M (1995) Planet Space Sci 43:613
18. Chow VW, Rosenberg M (1996) Planet Space Sci 44:465
19. Ma JX, Shukla PK (1995) Phys Plasmas 2:1506
20. Shukla PK, Vladimirov SV (1995) Phys Plasmas 2:3179
21. Vladimirov SV (1994) Phys Plasmas 1:2762
22. Annou R, Tripathi VK (1998) Phys Plasmas 5:60
23. Konar S, Jain VK, Tripathi VK (1989) J Appl Phys 65:3798

# A Comprehensive Survey on Different Routing Protocols and Challenges in Underwater Acoustic Sensor Networks

**B. Ragavi, V. Baranidharan, A. John Clement Sunder, L. Pavithra, and S. Gokulraju**

**Abstract** Underwater acoustic sensor networks are the most interesting and emerging technology in recent research. All are small-sized sensors with less memory space are deployed underwater to collect and transmit data from sender to sink node. Sensors consist of memory, CPU for storage process, and antennas to detect the signals. Underwater sensor networks are quite different from terrestrial networks because UASN isextremely working under low frequency but the terrestrial networks are working in high frequency. However, acoustic communication has many limitations such as poor data transmission, long end-to-end delay, limited bandwidth, and other parameters. These limitations lead to poor network lifetime and high power consumption. Nowadays, most researchers are mainly focused to design and implement an effective routing protocol to improve long network lifetime with energy consumption. This survey discussed various routing protocols and their behavior in underwater acoustic sensor networks. Finally, we compare the analytical performance metric tabulation of the different routing protocols. It is very useful for UASN researchers in the future to explore better research.

**Keywords** Underwater acoustic sensor network · Sensor nodes · Routing protocol · Characteristics · Limitations

## 1 Introduction

Ocean covers two-third of the earth's surface, and it is one of the most important ones for our human life. So, we gained a great interest to implement a technology called USWN [1]. It is one of the most interesting and promising technologies to enhance the ability in aquatic applications to monitor and control various parameters and gather data from underwater and reports periodically to sink nodes. Unmanned autonomous vehicles are also equipped with an underwater wireless sensor to find and detect various ocean environments such as ocean depth, void area, monitoring

B. Ragavi (✉) · V. Baranidharan · A. John Clement Sunder · L. Pavithra · S. Gokulraju
Department of ECE, Bannari Amman Institute of Technology, Sathy, Erode, India
e-mail: ragavi.co19@bitsathy.ac.in

 
S. Yadav et al. (eds.), *Recent Advances in Metrology*, Lecture Notes in Electrical Engineering 906, https://doi.org/10.1007/978-981-19-2468-2_34

an oil field exploitation, earthquake, and pollution. We deploy several underwater sensor nodes in the underwater environment to forward data from sender to sink node. After gathering these entire information, sender node forwards information to surface sonobuoys through a wireless radio channel. Wireless sensor networks are not suitable for USWN because all the terrestrial networks are worked under extremely high frequency [2]. Various signals are used to transmit data from source to a destination such as electromagnetic signal, optical signal, and acoustic signal. EM wave suffers severe attenuation because it has a very high frequency and optical signal which greatly suffer scattering and absorption. In complex environments, the acoustic signals are working under extremely low frequency, it is the most promising, and feasible technology in the underwater environments owe use many real-time applications and the speed of the acoustic wave is 1500 m/s. It monitors large areas than terrestrial WSN [3].

Acoustic signal supports transmitting data with an efficient packet delivery ratio and low-power noise. Routing and data forwarding are some of the most important to transmit data from source to sink node. So, we use a routing protocol to increase the performance and gain higher efficiency in the underwater environment. Generally, it faces many challenges such as the Doppler effect, temporary path loss, multipath effect in the acoustic channel, and other criteria. This cause leads to a high bit error rate and poor transmission. For instance, all the sensor nodes are always coped with limited bandwidth [4].

Hence, we need to improve the efficiency by using a routing protocol with a high packet delivery ratio with minimum bandwidth. Various routing protocols are used in an underwater scenario such as proactive, reactive, and geographic routing protocol. The proactive protocol establishes a routing table for each and every time based on network topology and mobility of dynamic nodes. If suppose the particular node failed during transmission (i.e., link loss, path loss, battery dies), then it changes the particular node routing table only. Reactive protocols are mostly used in a dynamic environment, but it incurs a high latency. If any node failed during transmission, it will change the entire topology. So, we move a geographic routing protocol (position-based protocol), a promising protocol for scalability and transmit data in greedy tragedy [5]. It estimates the current position of sensor accurate timing of signal and establishes a path efficiently to transmit a data without congestion and poor QoS. It cannot update any routing table and message in sensor networks and need not maintain the full dimension for discovery and maintenance. Therefore, we use large-scaled networks for long-term monitoring with many numbers of node and follows a greedy behavioral strategy to transmit a data packet from origin node (source node) to sonobuoys [6].

In this paper, we have to survey various routing protocols and their behavior reported in the literature, and the paper is organized as follows: In Sect. 2, we discuss a characteristic of underwater acoustic sensor networks. In Sect. 3, various existing routing protocols are taken under to reviewing the related paper. Finally, we compare the performance metric of various routing protocols. Finally, in Sect. 4 we conclude the conclusion and results.

## 2 Characteristic of Underwater Acoustic Sensor Networks

In this section, we discuss the characteristics of USWN. Already, we know UAWN is quite different from terrestrial networks. Terrestrial networks are work extremely under high frequency because it transmits data through the radio channel. So, it is impractical to implement in the underwater environment [7]. Thus, it is much more important to deploy and implement a routing protocol technique in acoustic communication considering all the characteristics.

### *2.1 Node Mobility and Three-Dimensional Architecture*

When compared to terrestrial network, acoustic communication gives a three-dimensional structure. In USWN, sensors are in a state of mobility because the movement of water in the surrounding environment frequently changes in the network topology [8]. Data transmission in a geographical three-dimension structure may create a hole in the routing path and lead to failure of data transmission and loss of more packets. When compared to the terrestrial network, we deploy a more number of mobile sensors in an acoustic environment to transmit data from sender to sonobuoys.

### *2.2 Speed and Propagation Delay*

They use various signal types such as electromagnetic waves, optical waves, and acoustic waves. In terrestrial networks, electromagnetic signals are used to transmit data through a wireless radio link. In acoustic sensor networks, we use all the above waves. Out of all these waves, we use the acoustic wave to transmit data over long-distance communication [9]. The speed of sound in underwater is 1500 m/s. It causes a higher propagation delay, which is a five-order magnitude of order higher than the radio frequency signal. The velocity of the sound wave varies based on different parameters such as temperature, depth, and salinity. So, we need to improve the design of the protocol to avoid those propagation delays and Doppler effect, etc.

### *2.3 Bandwidth Limitation*

Acoustic wave contains limited bandwidth depends upon a communication range and acoustic frequency [10]. Signal-to-noise ratio is one of the most important ones to transmit data over long-distance communication. So, we need more bandwidth to transmit data over a wireless channel. Therefore, we need to consider the bandwidth when designing a routing protocol.

### 2.4 Noise in Underwater Sensor Network

There are two types of noise in underwater sensor network which affects the underwater environment such as ambient noise and man-made noise [11]. Man-made noises are mainly created by human activities like shipping noise, motor noise, etc. Ambient noise also frequently occurs in ocean environments such as waves and tides.

### 2.5 Energy Consumption

Energy consumption is one of the major issues in both terrestrial and acoustic sensor networks because it is very difficult to replace all the sensor nodes. When compared to the terrestrial network, USWN consumes more energy during data transmission. Therefore, energy efficiency is the essential one to improve the performance of the underwater network [12].

## 3 Existing Routing Protocols in Literature

### 3.1 E2MR

Khalid et al. [13] proposed a protocol called efficient energy in multipath routing to reduce communication challenges. This problem frequently occurs in sensor nodes and transmission paths. Generally, routing consumes more power during the processing and forwarding of data packets. This novel scheme is designed to improve the network lifetime and also reduces the flooding of data packets. It creates a priority table based on the mobility of nodes and transmits data from source to sink node with the reference of a priority table. During this process, the formation of multiple copies in data packets can be able to avoid data flooding. We deploy many sensor nodes to collect and transmit data through a multi-hop fashion. In this type of transmission, we can use both acoustic modem and radiofrequency modem to collect all the information. In this energy-efficient method, a static node can continuously supply a power capable of receiving data from the source node to the sink node.

A small table is created by each and every node, and it calculates the priority value for all the active nodes. Under this calculation, the sensor knows depth information. It contains two phases such as setup phase and the initial phase. Firstly, the setup phase broadcasts a hello packet to all the sensor nodes, and Fig. 1 is depicted in E2MR protocol [13], which is shown below. Secondly, all the hello packets gain information such as residual energy (RE), shortest path, depth information, node ID, to all the sensor nodes. The node with the hello packet replies a hello message back to receiving the message. Having lower depth gives a higher priority to calculate the values and then transmit the data from source to sink node. This process continued

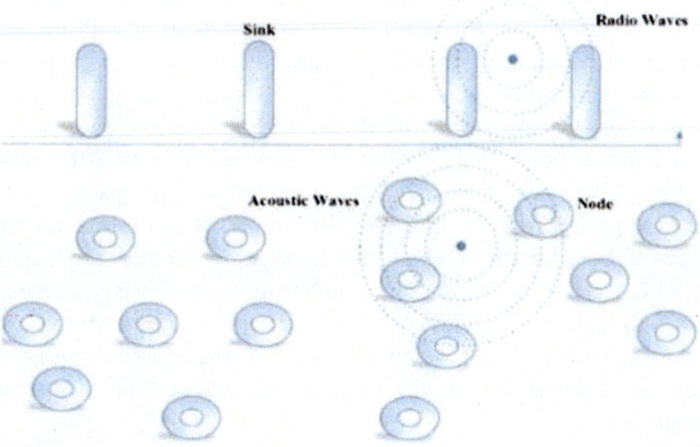


**Fig. 1** Architecture of E2MR

all the time and avoids data flooding with improved better efficiency within the range of communication. In the setup phase, we deploy a node in a random manner and initialize the broadcast message having some control message. After getting the hello packet from the sender node, it will send back a reply message to the source node and calculate the received signal strength indication (RSSI). After calculating RSSI, it transmits data based on lower depth with a higher priority table. The major limitation of this energy-efficient protocol takes a long time to calculate the distance based on sensor depth. It cannot find a void region. If suppose there may be any void in the sensor based on lower depth, there may be the chance to retransmit or drop a data from source to sink node.

### *3.2 Distance Vector-Based Opportunistic Routing Protocol*

Guan et al. [14] proposed on distance vector-based OR protocol to establish a distance vector with the least hop count toward the sink node. It coordinates all the packets based on distance vector and forwards all the packets from the sender node to receiver node. This opportunistic protocol improves QoS over a loss network. It requires a signaling exchange to all the sensor nodes by selecting a relay candidate set to forward data by coordinating the routing protocol. Distance vector-based OR protocol describes a distance based on the number of hops and hop count. It establishes the shortest path with the least hop count to deliver a packet. It works under a query mechanism for all the distance vector-based sensor nodes to calculate the shortest path with the least hop count. It exploits a distance vector through multi-hop forwarding and exchange information to all the sensor nodes. The sink node generates a query packet to all the sensor nodes having query ID or node ID and finds the distance to all the sensor nodes. The newest packets are also considered as the largest query ID. Each node first initializes the hop count from one to infinity. If the query packet arrives at node i, it compares the node ID along with the new query ID. If the query packet is greater than the recorded query ID, it is considered ad new packet. If the query packet is lesser than the recorded query id, it is considered as delay of the packet because of queuing. The packet with the same QID and QID arrived a node at several times due to creating multiple copies of the data packet. This process repeats until the packet reaches all the sensor nodes. In this process, all the sensor nodes can know the information about the distance vector with the least hop count toward the sink node. By selecting the number of nodes during transmission, this OR protocol will select the multiple relay candidate to transmit information with the least hop count to reduce a header overhead. DVOR source simply broadcasts all the data packets to relay candidates and creates a relay set to distribute information from source to sink node.

Each data packet contains a node ID, sequence number, and hop count, etc. By using this DVOR, the source node stops the unnecessary transmission because there is no available path to transmit data from source to sink node. At that time, this particular sensor node can change the hop count value as infinity.

So, all the sensor nodes cannot choose that particular sensor node and stop transmitting data from sender to sink transmission path, and Fig. 2 is depicted in DVOR protocol [14], which is given below. It can work under a waiting mechanism to forward data through OR protocol and adjust a waiting duration with the calculation of the shortest path with the least hop count. The node with $N = 1$ one-hop transmits data with 2 to short back off time. Node with $N = 2$ two-hop has zero waiting time period. DVOR can transmit data with the shortest path and least hop count based on limited time duration. The limitation of this DVOR is it introduces a long detour problem for extremely long detour problem, which leads to poor network lifetime.

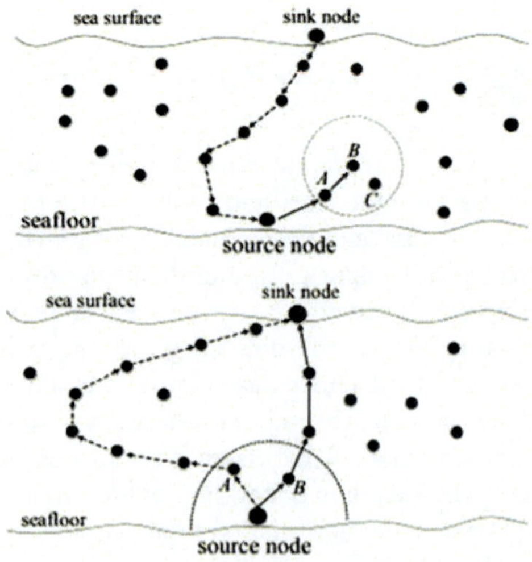


**Fig. 2** Establish a distance vector using a query mechanism

## 3.3 Radius-Based Courier Node Routing Protocol

Khalid et al. [15] presented a routing protocol called radius-based courier node in multipath communication. Generally, the acoustic channel has very limited bandwidth and higher end-to-end delay. Nowadays, most of researchers are mainly focused to design and implement various protocols to improve a long network lifetime and reduce latency. This proposed novel scheme are used a radius-based architecture with the combination of track ID, residual energy, and node ID to transmit data from sender to receiver. It is mainly designed to provide efficient energy with improved network lifetime. Both acoustic and radio frequency modem are equipped in the underwater environment. The courier node and static node are combined to perform a task (i.e., transmit data from the source node to sink node) in a multi-hop fashion. In this work, we consider the network area as a circle form. It is broadly divided into sink area and node area. The quaternary of the area is allocated for the sink area, and the outstanding areas are divided into node areas. In this form, nodes that are more static are deployed with a range of higher depth to improve the network lifetime. The node area is divided as equal segments into a triangle between static and courier nodes. Each triangle in the node area is further divided into the track and assigned a number called T-ID (track ID), and this is depicted from radius-based protocol [15] which is shown in Fig. 3. The physical address also assigned in the static node is called S-ID, and the courier nodes are named as C-ID. T-ID continuously changes under the movement of S-ID.

It focuses mainly on energy efficiency and power consumption and avoids multiple copies of data packets. Source node sends hello packets to all the nodes within the range of communication. After receiving hello packets to all the sensor nodes, it creates a small table by each and every node to calculate the depth information and residual energy. This hello packet contains track ID, node ID, and residual energy. If the static sensor node receives a data packet from the source, it decides to forward

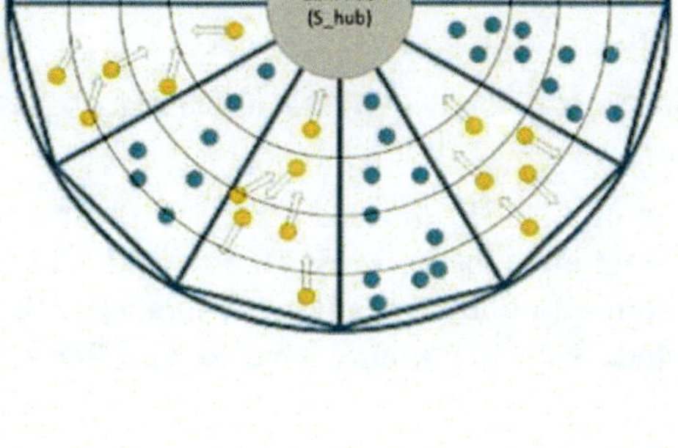


**Fig.3** Courier node and static node in radius-based protocol

data from the neighboring node to the sink node. If the static node is placed within the range of communication to transmit data based on residual energy from the source node to the sink node. If not in the range of communication, it searches another static node to forward a data packet within the range of communication. The major limitations of this RCMN are computational, and analytical complexity is too high. When compared to other protocols, it cannot able to identify the least depth region in the sensor node.

### 3.4 Evolutionary Routing Protocol

Faheem et al. [16] proposed an evolutionary routing protocol to improve the quality of service (QoS). Generally, QoS is the major challenging issue in USWN due to acoustic noise, low bandwidth, high bit error rate, and other interference. To overcome the above issues, we designed an evolutionary routing protocol based on clustering technique to improve packet delivery ratio and network lifetime and focussed to reduce an average end-to-end delay during data transmission. We deploy several sensor nodes with different depths followed by architectural properties. All the deployed nodes are having the same capability of transmission range and initial energy during initial transmissions. Based on geographical routing, each sensor node knows its own location. Every node moves only in the horizontal direction and slightly moves in the vertical position. All the sensors and their properties are worked under the CSMA mechanism, which is capable of receiving multiple copies at data packets at the same time. The routing path contains a sequence of non-negative integers. This integer denotes the CH nodes. The order of each cluster head represents the routing path of a first gene. After selecting the route, it initializes the initial population size and procedures due to the complexity of the initial population. We consider a random initialization after the deployment of sensor nodes; surface sonobuoys are responsible for evolutionary algorithm by using an individual random number generator.

After receiving a message, each receiving node can send an acknowledgment. Finally, it selects the parents by recombining the routing path to produce offspring. From the selected parent, it maintains a crossover operator to exchange parent chromosome information to get mutation with uniform mutation operator. The main limitation of the QoS-aware routing protocol is it consumes more power and it takes more time to general next cluster.

### 3.5 Latency-Aware Routing Protocol

Su et al. [17] proposed a latency-aware routing protocol based on deep Q-network to improve the network lifetime and avoid latency. Q-network-based energy adopts both off-policy and on-policy methods which depend on energy and depth rate. The node with maximum Q-value can able to transmit data in unicast and broadcast

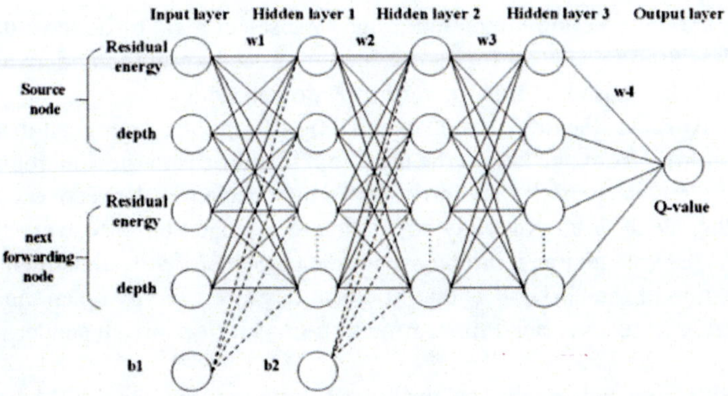


**Fig. 4** Hidden layers of QL and RL

communication. On policy, the method can be addressed by network topology based on a new routing mechanism. In such, lesser energy based optimization techniques, an autonomous agent are mainly concerned and utilized by reinforcement learning. RL and QL are the same but RL cannot need an environment model to handle all those problems. This QL can be worked under the Markov chain process, and Fig. 4 is depicted from latency-aware protocol [17].

Deep Q-learning techniques are always used to store the Q-value for too many states. These Q-values are widely used to predict the combination of the neural network to save energy and time. We deploy many nodes to collect information by applying a deep Q-learning technique. Each packet contains agents and actions. Agents can compromise a network, and it contains current information about sensor node, residual energy depth. A neighboring node compromises an action to perform various parameters to improve the network lifetime. The limitation of this proposed work is that we need more than three layers to transmit data from one hop to another hop and computational complexity is too high.

## 3.6 *Gedar*

Rodolfo et al. [18] proposed a geo-opportunistic routing protocol to recover a void region based on depth adjustment. Communication in underwater is one of the most challenging issues to transmit data with long end-to-end delay. This geo-opportunistic routing protocol transmits data through opportunistic forwarding from source to sink node based on depth and network topology. If the packet enters in a non-communicating region, the protocol should be attempting to route the packet by using some recovery method; in the work, the sensors are deployed in Euclidean space moving in the vertical direction. This sensor node can control the depth through sonobuoys. A system with an undirected graph contains a time and set of vertices corresponding to node and sonobuoys. The node having the same transmission power results in maximum communication range. We select a set of nodes with distance ($d$) which is used to select a forward selection procedure in the underwater model. We transmit data based on this forward selection procedure.

It greedily routes the information over a long-distance communication with short-term monitoring from one hop to the next hop. If a node is in the void region, it goes through the recovery mode. The arrival of a new packet in the recovery node goes through the queue through greedy, and data are rescheduled to forward data. If the node not in the void region, it directly sends data based on depth adjustment topology. The limitation of the GEDAR is that, it consumes more energy when the node is in a void region to recover need more time to retransmit data from sender to receiver (Table 1).

In this literature survey, the scalability and energy-efficiency of E2MR and DVOR are very high. When compared to RBCN, one of the main advantages is the residual energy in RBCN is very high when compared to E2MR and DVOR. End-to-end delay is comparatively low in DQELR when compared to all the above protocols. When compared to all the above protocols, opportunistic routing protocol achieves only a DVOR protocol. Out of all GEDAR and RBCN methods [19, 20] are gives an efficient packet delivery ratio during data transmission.

**Table 1** Comparison of various routing protocols

| Protocol | About protocol and its uses | Activeness | Packet delivery ratio | OR forwarding | Scalability | Residual energy | Energy efficiency |
|---|---|---|---|---|---|---|---|
| E2MR | Handlesenergy efficiency | Pro active | Medium | No | High | High | High |
| DVOR | To identify the shortest path with the least hop count | Pro active | Medium | Yes | High | Low | Medium |
| RBCN | Sends data based on radius courier node | Pro active | High | No | Medium | High | Medium |
| QERP | Improves the quality of service | Pro active | High | No | Low | Low | Low |
| DQELR | Improves the network lifetime and reduce latency | Pro active | Low | No | Medium | Low | Low |
| GEDAR | Improves QoS | Geographic | High | No | High | Low | Medium |

## 4 Conclusion

Different types of routing protocols in underwater acoustic sensor networks are discussed and investigated in this article. The characteristics and routing procedures of various routing protocols are also discussed. These protocols describe major issues to solve various network interrupts and provide an efficient platform to find a suitable routing protocol. The best part of this work is we surveyed various protocols and their challenges advantages and major limitations, which are useful for the USWN system for future reference. Finally, we compare a complete comparison of various routing protocols that are tabulated. This survey sheds some light on the effective design of routing protocols for underwater acoustic sensor networks.

## References

1. Scissile K (2003) Acoustic modems for underwater communications. In: Encyclopedia of wireless communications
2. Rice J, Green D (2008) Underwater acoustic communications and networks. SENSORCOMM
3. Yan H, Shi ZJ, Cui J-H (2008) DBR: depth-based routing for underwater sensor networks, pp 72–86
4. Hu T, Fei Y (2010) QELAR: a machine-learning-based adaptive routing protocol for energy-efficient and lifetime-extended underwater sensor networks, pp 796–809
5. Domingo MC (2012) An overview of the internet of underwater things. J Network Comput Appl, 1879–1890
6. Chen W, Yu H, Guan Q, Ji F, Chen F (2017) Reliable and opportunistic transmissions for underwater acoustic networks, pp 93–99
7. Cui Y, Qing J, Guan Q, Ji F, Wei G (2015) Stochastically optimized fountain-based transmissions over underwater acoustic channels, pp 2108–2112
8. Kao CC, Lin YS, Wu GD, Huang CJ (2005) A comprehensive study on the internet of underwater; Akyildiz IF, Pompili D, Melodia T (2005) Underwater acoustic sensor networks: research challenges. Ad Hoc Networks, pp 257–279
9. Khalid M, Ullah Z, Ahmad N et al (2017) A comparative simulation-based analysis of location-based routing protocols in underwater wireless sensor networks, pp 1–5
10. Khalid M, Ullah Z, Ahmad N et al (2017) A survey of routing issues and associated protocols in underwater wireless sensor networks, pp 134–141
11. Lin J, Yu W, Zhang N, Yang X, Zhang H, Zhao W (2017) A survey on internet of things: architecture, enabling technologies, security, and privacy, and applications, pp 1125–1142
12. Faheem M, Tuna G, Gungor VC (2018) QERP: quality-of-service (QoS) aware evolutionary routing protocol for underwater wireless sensor networks, pp 2066–2073
13. Lv T, Lin Z, Huang P, Zeng J (2018) Optimization of the energy-efficient relay-based massive IoT network, pp 3043–3058
14. Khalid M, Ahmad F, Arshad M, Khalid W, Ahmad N, Cao Y (2018) E2MR: an energy-efficient multipath routing protocol for underwater wireless sensor networks, pp 321–328
15. Guan Q, Ji F, Liu Y, Yu H, Chen W (2019) Distance-vector-based opportunistic routing for underwater acoustic sensor networks, pp 3831–3839
16. Khalid M, Cao Y, Ahmad N, Khalid W, Dhawankar P (2018) Radius-based multipath courier node routing protocol for acoustic communications, pp 341–347
17. Wang Z, Han G, Qin H, Zhang S, Sui Y (2018) An energy-aware and void-avoidable routing protocol for underwater sensor networks, pp 7792–7801

18. Coutinho RWL, Boukerche A, Vieira LFM, Loureiro AAF (2014) GEDAR: geographic and opportunistic routing protocol with Depth Adjustment for mobile underwater sensor networks, pp 251–256
19. Su Y, Fan P, Fu X, Jin Z (2019) DQELR: an adaptive Deep Q-network-based energy-and latency-aware routing protocol design for underwater acoustic sensor networks, pp 9091–9104
20. Aslam S, Ejaz W, Ibnkahla M (2018) Energy and spectral efficient cognitive radio sensor networks for internet of things, pp 3220–3233

# Substrate Integrated Waveguides Analysis for Different Dielectric Materials—A Comprehensive Survey

**S. Saranya, V. Baranidharan, S. Harikirubha, K. J. Kishore Kumar, and R. Sharan**

**Abstract** In this paper, we analyze and design an antenna model for a substrate integrated waveguide (SIW) to examine the impacts of dielectrics on performance parameters. The parameters calculated in this analysis are the transmission gain, the electrical field and the return loss (RL). Polymethyl methacrylate (PMMA), mica, glass fiber and silicon (Si) were used as various dielectrics to test the results in the 6–20 GHz frequency range. In accordance with the finite element method (FEM), formulated on the casting of SIW structures, an ordered design stage was carried out to optimize the geometric dimensions. The outcome has been obtained; it shows the highest change in the electrical field which was noticed in PMMA, the gain of transmission is increased by increasing up to 9.751 GHz frequencies, and RL was minimal at this frequency accordingly. The transmission was possible in two situations, but the improved results were noticed with silicon as dielectric substrates.

**Keywords** Substrate integrated waveguide (SIW) · Silicon · Polymethyl methacrylate (PMMA) · Mica · Glass fiber · Finite element method (FEM)

## 1 Introduction

In recent years, wireless components and systems have received increased interest as new applications are being introduced and developed for millimeter waves (mm-waves). In fact, in the frequency range over 60–94 GHz, a variety of applications have recently been proposed, including automotive radars, wireless networks, biomedical devices and imaging sensors. The deployment of mm-wave technologies is critical for the evolution of wireless systems, as the use of mm-waves naturally supports broadband and high-resolution techniques.

Success depends primarily on the availability of a cost-effective technology suitable for the mass production of components and systems in most of these systems. High-density integration techniques should be expected to be able to offer widespread

S. Saranya (✉) · V. Baranidharan · S. Harikirubha · K. J. Kishore Kumar · R. Sharan
Bannari Amman Institute of Technology, Sathy, India
e-mail: saranya.co19@bitsathy.ac.in



S. Yadav et al. (eds.), *Recent Advances in Metrology*, Lecture Notes in Electrical Engineering 906, https://doi.org/10.1007/978-981-19-2468-2_35

solutions for mm-wave commercial applications, combined with a low-cost manufacturing process. The core of these systems is linked to the active component, which includes, among others, components such as local oscillators, mixers and possibly low noise amplifiers. Such components can now be integrated at a reasonably low cost in the form of chipsets. Currently, several semiconductor companies are working to develop chipsets operating at 60 GHz or even higher frequencies.

The small integrated radio frequency (RF) system field's recent advancement has measured wavelengths to micron level. These devices have been developed and optimized for production and integration in RF circuits. The dielectric waveguide is the basic building blocks of integrated circuits. It is not only a communication medium for limiting and transmitting signals, but also used in couplers and filters as a circuit basis. Traditional waveguide structures have the advantage of more power efficiency and large Q factor, but they also have the drawbacks of large, huge and affordable mass production techniques. In trend, SIW has attracted increasing importance in microwave community due to its versatility in implementation of μwave circuits. In fact, many of the tools developed for the waveguide technology can be created using substrates integrated waveguide. The SIW equipment which is implemented using dielectric substrates facilitates integration in another planar circuit, for example, the micro-strip technology. The existence of a dielectric also makes it possible to reduce measurements with a scaling factor that is equal to the square root of relative permittivity (Fig. 1).

The existence of dielectrics within the waveguide enhances the loss, based on tangent loss of materials used. The loss is also caused by the comparatively low level of the substrate waveguide, which will increment the loss of the conductor. The dielectric loss is minimized by using low loss tangential substrates, but these substrates are costlier. The conductor losses are minimized by the use of thicker substrates, which affect cost, weight and scale of the unit. There is no desideratum for the transfer between components made up of different techniques, which reduces damage and parasitic. This work involves modifying the basic structure of SIW and evaluating the execution of the simulated antenna for two dielectric materials.

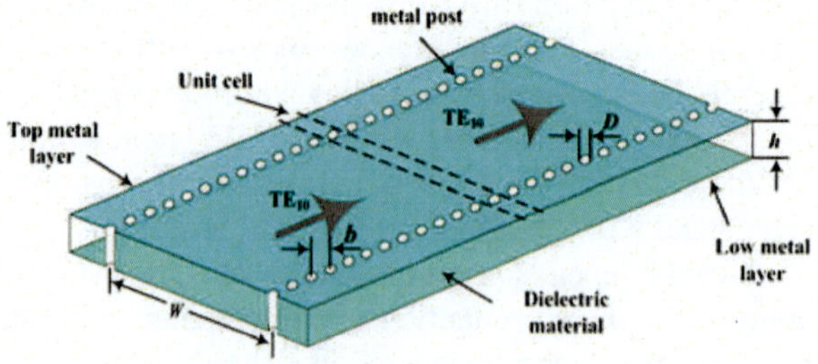


**Fig. 1** Substrate integrated waveguide using dielectric material [1]

## 2 Literature Review

Some of the related works have been reviewed, and the major purpose of the literature is explained briefly in this literature review section.

The authors Dmitry Zelenchuk and Vincent Fusco are giving the analysis of the use of SIW resonators for the description of printed circuit board materials (PCB) [2]. Simple technical considerations for dielectric characterization provided with the rectangular and the circular SIW resonators are given. Rectangular and spherical cavities with different types of coupling have been developed and measured to characterize substrates of low and high permittivity.

In the proposed system, the authors, Jiangling Tong et al., achieved the same result because they used a glass material as the insulator and showed a similar pattern of rectangular filled waveguides [3]. In addition, the simulated S-parameters show the insertion loss of 0.67 dB/cm at 20 GHz, and they got 100% relative bandwidth.

The authors, Yasser Arafat et al., have discussed various dielectrics on the SIW. Finally, he saw the highest change in the electrical fields in polymethyl methacrylate [4]. In this paper, four various dielectric materials are compared in the frequency range 1–10 GHz. They finalized that PMMA is more effective compared to other dielectric materials. The authors, Antonio Morini et al., have reviewed a substrate integrated waveguide used in beam forming networks [5]. The device consists of four ports and six ports along with the magic tree. They explained the design process of all devices and found a possible solution for the prototype.

The authors, Jagmeet Kour et al., discussed the effects of dielectric materials on their performance [6]. They specify that the improved results were noticed with silicon as dielectric substrates. In this paper, three various dielectric materials are compared in the frequency range 6–11 GHz. The results of their study have shown that transmission is possible in all three situations, but improved results have been seen for silicon as a dielectric substrate.

## 3 Substrate Integrated Waveguide

### *3.1 Overview*

The structure shown in the figure is known as SIW or substrate integrated waveguide. It has three layers, i.e., the upper and lower main planes and the dielectric layer placed in between these two planes. The metal is located between the two planes through an array of holes. Microwave devices, such as filters, antenna arrays, power dividers, directional couplers, phase shifters, couplers and junctions, are designed and manufactured using SIW structures due to many advantages (Fig. 2).

SIW is a new transmission line type. Adding metal to the ground plane creates a rectangular guide inside the substrate and relays the structure with rows of vias inside. The dielectric filled waveguide is transformed to a SIW by means for the sidewalls of

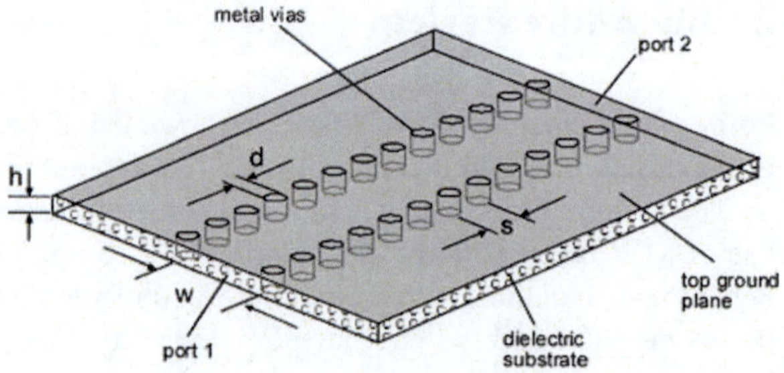


**Fig. 2** SIW [7]

the waveguides. The SIW is one of dielectric filled waveforms (DFW). The substrate thickness does not influence the frequency cutoff. The substrate therefore has no thickness range. The dielectric losses are inversely proportional to the thickness of the substrate. The characteristic propagation of SIW waveguides is same as traditional rectangular waveguides. The field structure and dispersion feature are also same. In addition to these similitudes, structures of SIW have a large Q factor and an excellent ability to handle power with a self-coherent electrical shield. This is an advantage of these structures over traditional metal waveguides.

## 3.2 *Geometry*

For SIW structures, dielectric substrates, generally adopted for micro-strip and coplanar waveguide circuits, can be used. Compared to the width of the waveguide, the thickness of the dielectric layer is typically small and can range from 0.2 to 1 mm. The relative dielectric constant $\varepsilon_r$ is typically from 2 to 10 range.

The SIW geometry is fully defined by three parameters: the holes diameter d, the waveguide width w and the longitudinal spacing s between the holes as shown in Fig. 3. The width w is mainly related to the cutoff frequency of the basic SIW mode, as the SIW structure is similar to the rectangular waveguide. The holes' diameter

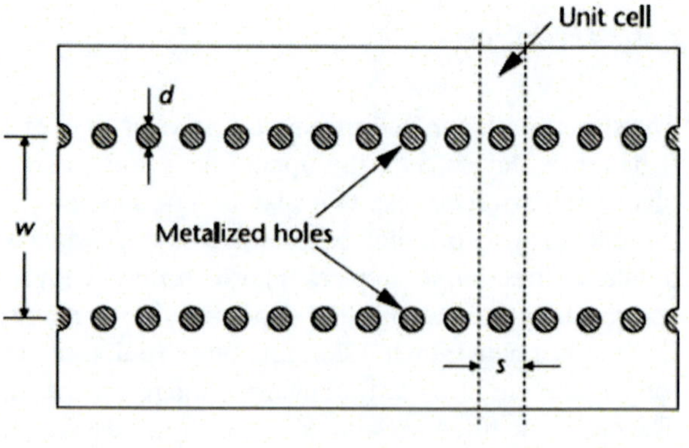


**Fig. 3** SIW geometry

$d$ is typically a small fraction of $w$: A value $d < w/8$ is recommended due to the periodic modulation of the waveguide width to avoid possible band-gap effects. The longitudinal spacing s affects electromagnetic field containment, as it determines the gap width ($s - d$) between pairs of metal vias. The minimum value is $s = d$, meaning there is no gap and the shielding is perfect. If the value of s increases, the gaps are greater and the containment of the field deteriorates. The golden rule is $s < 2.5d$, and the typical choice is $s = 2d$.

### 3.3 *Principle of Operation*

The propagation of waves in SIW structures is similar to the propagation in classical rectangular waveguides: The basic SIW mode is similar to the $TE_{10}$ mode of a rectangular waveguide as shown in Fig. 4. With this mode configuration, the surface current flows exactly like in a rectangular waveguide along the top and bottom metal planes of the SIW, and on the sides, it can flow vertically along the metalized surface of the cylinders, being minimally disturbed by the gaps (provided the gaps are small). For this reason, the electromagnetic field is confined within the SIW, and radiation leakage is not present. This operating mechanism applies only to rectangular waveguide $TE_{n0}$ modes, where the surface current flows in vertical direction on the sidewalls.

The surface current on the sidewalls has a longitudinal component for the other modes of the rectangular waveguide, namely TM modes and $TE_{n0}$ modes with p hierarchy, and thus, it would be severely disturbed by the gaps. For this reason, the SIW structure does not support these modes. In conclusion, those similar to the rectangular waveguide $TE_{n0}$ modes are the only modes supported by SIW structures.

SIW components can be classified as H-plane waveguide structures, due to their geometry and mode pattern. The electric field is normal to the wide wall, and the vertical direction does not vary in its amplitude. In addition, the thickness h of the substratum plays no role in the wave propagation characteristics, except for loss of the conductor.

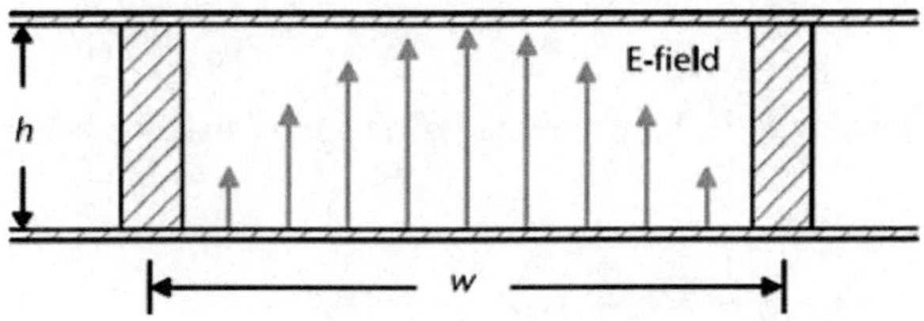


**Fig. 4** SIW mode's electric field pattern

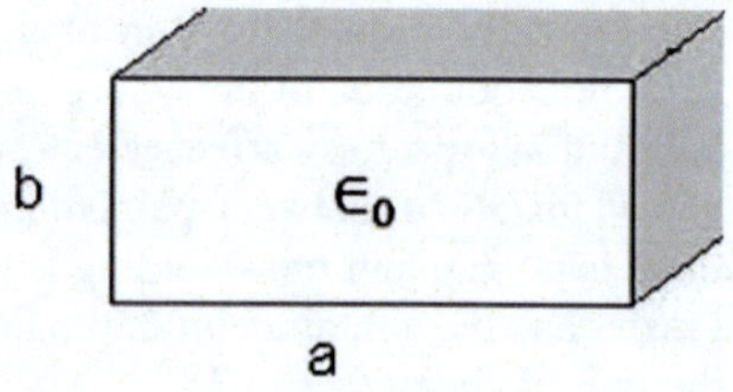


**Fig. 5** Rectangular waveguide dimension [7]

# 4 Design Equations

SIW devices can be considered dielectric filled waveguides (DFW), so DFW can be the starting point. The parameter "*b*" is not significant as it does not influence the waveguide cutoff frequency in mode $TE_{10}$. Therefore, the thickness of substrate can vary; dielectric loss is affected (Fig. 5).

For a rectangular waveguide, the arbitrary mode of frequency cutoff is found using the following formula:

$$fc = \left(\frac{c}{2\pi}\right)\sqrt{\left(\frac{m\pi}{a}\right)^2 + \left(\frac{n\pi}{b}\right)^2} \quad (1)$$

where *c* is light speed, (*m*, *n*) is mode numbers, and (*a*, *b*) is waveguide dimensions.

The simplified formula for $TE_{10}$ mode is

$$fc = \frac{c}{2a} \quad (2)$$

Same cutoff frequency for DFW, dimension "$a_d$", can be found by

$$a_d = \frac{a}{\sqrt{\varepsilon R}} \quad (3)$$

The design equations for SIW

$$a_s = a_d + \left(\frac{d^2}{0.95p}\right) \quad (4)$$

where "*d*" is diameter and "*p*" is pitch (distance between the vias).

# 5 Design Structure of SIW

The design of the SIW comprises the upper and lower surfaces of the substrate and the two parallel ones by means of the substrate fences. Via is designed in such a

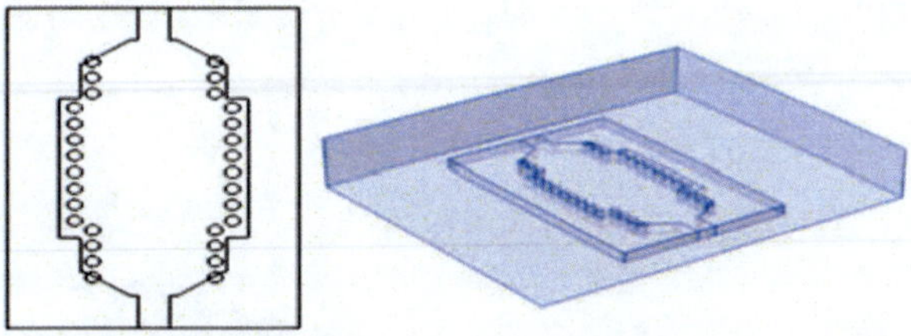

**Fig. 6** Design structure of SIW [6]

way that only vertical current patterns presented on the sidewalls will survive in the SIWS (Fig. 6).

The $TE_{10}$ mode is supported by the SIW because the current path cannot be broken through the fence. This is true for all $TE_{m0}$ modes because the sidewall current distribution is same. At the other hand, the perpendicular to the vertical component of surface zero exists in the sidewalls of all nonzero $TE_{mn}$ and TM modes. These current paths are cut from the SIW structure, and radiation is generated. Therefore, it can be concluded that the $TE_{m0}$ mode only exists in the SIW structure. In this experiment, different substrates were used to study the effect of SIW on the diffusion model. The model was designed using the electromagnetic frequency domain solvers. The 6.0–20 GHz frequency range is applied through a centralized port to analyze the results of various SIW designs.

## 6 Results and Its Comparison

The various dielectric substrates were taken into account in order to examine their effect on the propagation patterns of the SIWs as given in the table. The model was developed using HFSS. The frequency of 12.0 GHz is used via wave ports to evaluate the performance. The pitch distance $p = 1.90$ mm, the diameter of the lanes shall be held as $d = 0.80$ mm, and $W_{siw} = 4.0$ mm.

### *6.1 Silicon Dielectric Material on SIW*

The radiations generated by the electric field for substrates taken as silicon are shown in Fig. 7.

The maximum value (700 V/m) is for silicon which is clear from the line bar corresponding to these graphs (Fig. 8).

A reduction in RL is seen at 9.31 GHz, and the gain of transmission increases to 9.31 GHz and then it decreases, likewise, the silicon transmission loss and RL. For silicon, the return loss showed a major dip at frequency of 6.6 GHz, but the transmission gain is negligible and remains constant.

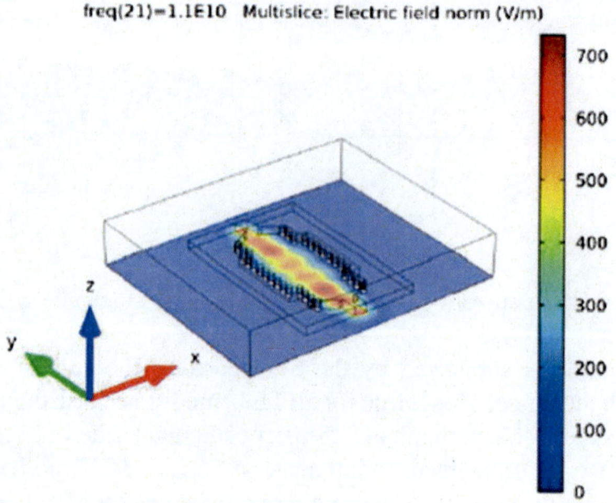


Fig. 7 Silicon substrate electric field [6]

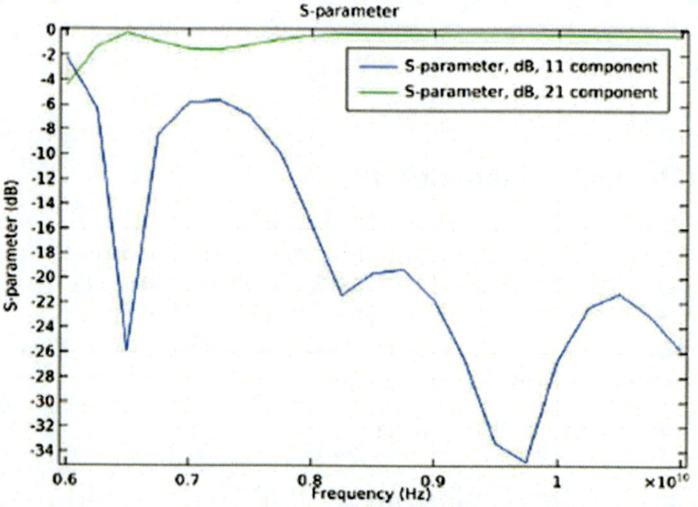


Fig. 8 Silicon substrate *S* parameter [6]

## 6.2 PMMA Dielectric Material on SIW

The bar line next to these diagrams shows the highest value (989.21 V/m) of the electrical fields applied to PMMA, when the lowest value (723.80 V/m) applied to silicon is clear from comparing Figs. 7 and 9.

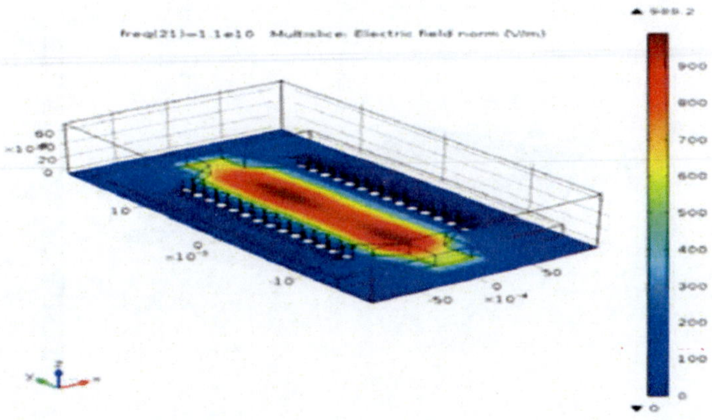


Fig. 9 PMMA substrate electric field [4]

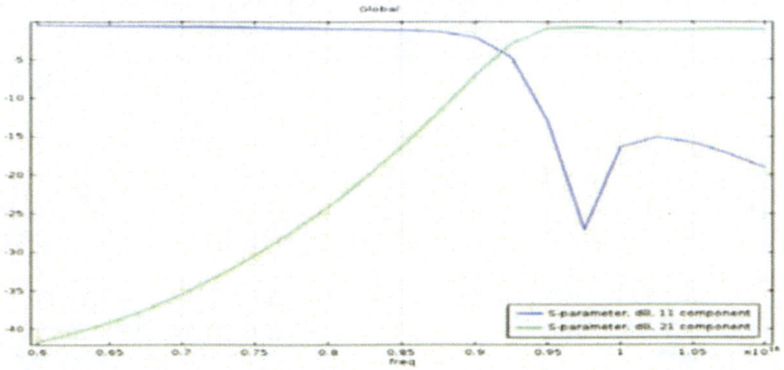


Fig. 10 PMMA substrate S parameter [4]

The PMMA substrates used in the simulations show an increment in gain of transmission up to a frequency of 6.0–9.50 GHz. The transmission gain is unity from 9.50 to 11.0 GHz. A decrease in RL value of −27 dB at a frequency of 9.75 GHz (Fig. 10).

## *6.3 Glass Fiber Dielectric Material on SIW*

Figure 11 demonstrates the difference between the S parameter and the glass fiber substrate frequency. The first drop in RL of −34 dB is seen at 10.4 GHz. The gain of transmission is up to 10 GHz, after which it is saturated.

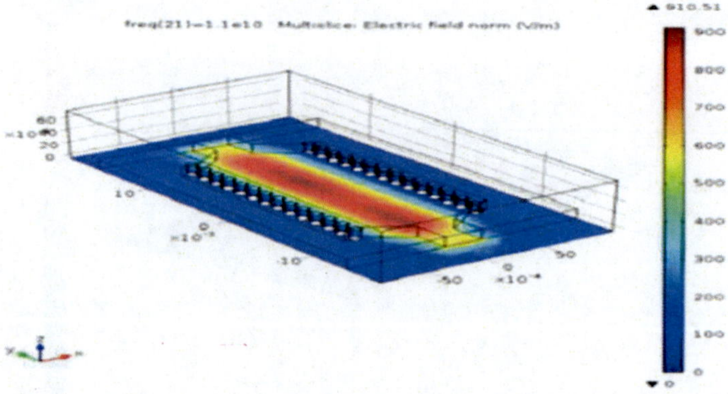


**Fig. 11** Glass fiber substrate electric field [4]

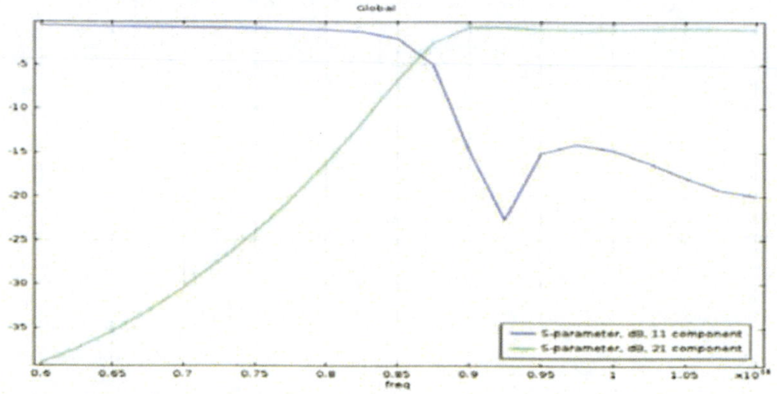


**Fig. 12** Glass fiber substrate S parameter [4]

Figure 12 shows the characteristic curve between the parameters of S and the frequency. RL and gain of transmission were plotted for each dielectric substrate used in this analysis.

## *6.4 Mica Dielectric Material on SIW*

The electrical field distribution of the mica substrate remains at 726.68 V/m (Figs. 13 and 14).

The parametric curve for RL and gain of transmission is observed in the case of mica as a substrate. The RL is nearly unit with a frequency of 7.0 GHz, and the gain of transmission shows dips of −21 dB, −26 dB and −28 dB at 7.0 GHz, 8.5 GHz and 10.0 GHz.

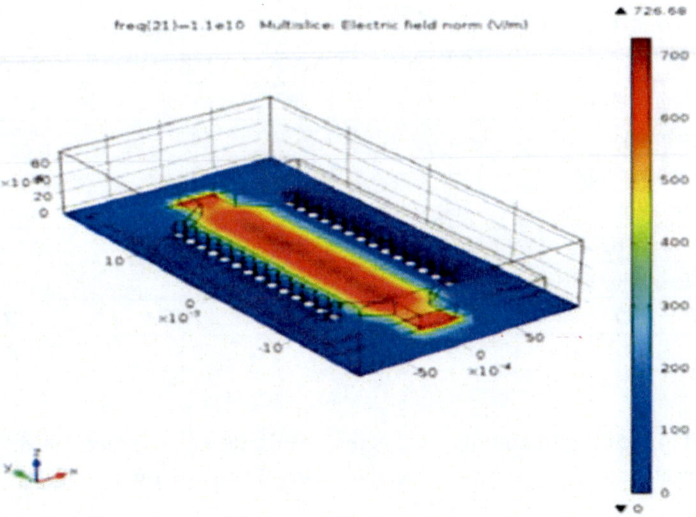

Fig. 13 Mica substrate electric field [4]

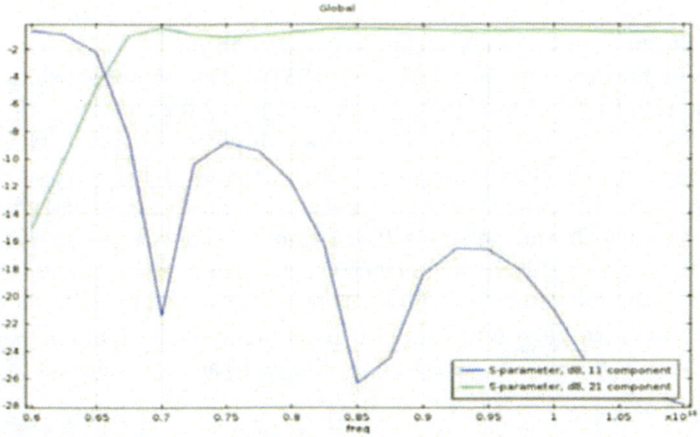


Fig. 14 Mica substrate S Parameter [4]

## 6.5 Comparison of Different Dielectric Materials on SIW

The various materials used as substrates in SIW and their properties are given in Table 1.

The relative permeability of all the four dielectric materials is 1, and the permittivity of silicon is higher than the other three dielectric materials. The electrical conductivity of glass fiber material is zero, and the electric field of PMMA is high

**Table 1** Properties of dielectric materials

| Materials | Silicon | PMMA | Glass fiber | Mica |
|---|---|---|---|---|
| Relative permeability | 1 | 1 | 1 | 1 |
| Relative permittivity | 11.9 | 3.7 | 4.4 | 5.7 |
| Electrical conductivity | $10^{-12}$ | $10^{-19}$ | 0 | $2 \times 10^{-15}$ |
| Electric field (V/m) | 700 | 950 | 1.7k | 1k |
| Frequency range (GHz) | 6.3–6.8 | 6–9.5 | 9.2–9.4 | 8.5–9.7 |
| Resonant frequency (GHz) | 6.6 | 11 | 9.3 | 9 |
| Bandwidth (MHz) | 500 | 250 | 200 | 1200 |
| Gain (dB) | 0.75 | 1.2 | 2.3 | 0.95 |

compared to the other materials. In this test result, we can see that PMMA and glass fiber materials on SIW are more effective than silicon and mica.

## 7 Conclusion

An experimental work is conducted to analyze the impact of various dielectric substrates on the propagation of EM wave of SIW. The various substrates, such as PMMA, mica, glass fiber and silicon, were used to test the dielectric effect through the experiment. S-parameters like RL and gain of propagation were calculated for frequencies in the 6.0–20 GHz. It can be concluding that the SIW works effectively at about 9.0 GHz. The outcome has been obtained; it shows the highest change in the electrical field which was noticed in PMMA, the gain of transmission is increased by increasing up to 9.751 GHz frequencies, and RL was minimal at this frequency accordingly. The test results of various dielectric materials on SIW show that the result evaluation for glass fiber and PMMA is higher than Si and mica as dielectric substrates. In future, we will analyze different parametric effects of dielectric material on SIW.

## References

1. Tan Q, Guo Y, Zhang L, Lu F, Dong H, Xiong J (2018) Substrate integrated waveguide (SIW)-based wireless temperature sensor for harsh environments. Sensors 18(5):1406
2. Zelenchuk D, Fusco V (2010) Dielectric characterisation of PCB materials using substrate integrated waveguide resonators
3. Tong J, Tummala R, Shorey A, Sundaram V (2015) Substrate-integrated waveguides in glass interposers with through-package-vias. ©2015 IEEE Electronic Components & Technology Conference, 978-1-4799-8609-5/15

4. Khan S, Arafat Y, Arya S, Singh SR (2010) Modelling, design and parametric considerations for different dielectric materials on substrate integrated waveguide. WSEAS Trans Commun 13. E-ISSN 2224-2864
5. Morini A, Venanzoni G, Farina M, Mencarelli D, Prudenzano F (2019) Review of substrate integrated waveguide circuits for beam-forming networks working in X-band. Appl Sci 9:1003. https://doi.org/10.3390/app9051003
6. Singh G, Kour J, Arya S (2017) Design and modeling of substrate integrated waveguide based antenna to study the effect of different dielectric materials. IRJET 04(04). net p-ISSN 2395-0072, e-ISSN 2395-0056
7. Bozzi M, Xu F, Deslandes D, Wu K (2007) Modeling and design considerations for substrate integrated waveguide circuits and components. In: 2007 8th international conference on telecommunications in modern satellite, cable and broadcasting services, pp VII–XVI

# Significance and Implications of Noise Mapping for Noise Pollution Control

**S. Kumar, B. S. Chauhan, and N. Garg**

**Abstract** The paper presents the pros and cons associated with noise mapping studies for understanding the noise scenario, analyzing the hot spots, and devising appropriate control measures for noise control. Noise mapping has been widely used exercise in Europe, and in India, gradually, some studies have been reported recently. The paper analyzes the various aspects pertaining to the development of noise maps, identification of hot spots, consideration of noise control measures, and futuristic policies for noise pollution control. The study analyzes the various noise control measures that can be helpful for noise pollution control in metropolitan cities of India. A case study of one such site in Delhi city with simulative implementation of noise control measures is presented.

**Keywords** Noise mapping · Interpolation technique · Sampling strategy · Noise control action plan

## 1 Introduction

Noise mapping is a vital tool for the assessment, management, and control of ambient noise levels. In past decades, there had been many studies carried out that analyze and correlate the increased ambient noise levels with health hazards [1–3].

The environmental noise management and control is one of the important aspects of preserving the environment for sustainable development goals and promoting health and quality of life. The European Noise Directive 2002/49/EC (END) had introduced two aspects for urban planning and noise management: strategic noise maps and action plans [4–6]. As such in Indian perspectives, these strategies have to be also implemented on a wider scale, especially for the metropolitan cities so as

S. Kumar (✉) · B. S. Chauhan · N. Garg
CSIR-National Physical Laboratory, New Delhi 110012, India
e-mail: 23saurabhkumar@gmail.com

N. Garg
e-mail: ngarg@nplindia.org

Academy of Scientific and Innovative Research (AcSIR), Ghaziabad 201002, India



S. Yadav et al. (eds.), *Recent Advances in Metrology*, Lecture Notes in Electrical Engineering 906, https://doi.org/10.1007/978-981-19-2468-2_36

to control the noise pollution levels. The Central Pollution Control Board (CPCB) of India has earlier initiated the establishment of National Ambient Noise Monitoring Network (*NANMN*) project with an objective of monitoring, recoding, and analyzing the continuous ambient noise levels at seventy sites in seven major cities of India. Consequently, the noise-monitored data are indispensable to formulate control measures for the ambient noise levels at noisy spots. The National Green Tribunal (NGT), Delhi, in his orders dated 15th March, 2019, has advocated the development of noise maps of the various cities in India so as to initiate remedial action plans. However, noise mapping is at very primitive stages in India as compared to that in European Union scenario. This is attributed to the fact that various technical issues require further subtle investigations so as to develop and devise a harmonized procedure that can be followed by all the State Pollution Control Boards, Pollution control committees, and other stake holders. The present paper highlights some of these technical aspects and suggests the optimum solution in order to develop a harmonized and cost-effective solution with lowest levels of uncertainty [7–10]. The study shall be indispensable for reducing the ambient noise levels in metropolitan cities of India.

## 2 Methodology

Noise map is a graphical representation of the noise levels of a particular site or area under consideration. These maps may be plotted based on the actual observations acquired via noise monitoring of the site or area as per the legal protocols or by using a validated noise model. The use of the geographical information systems (GISs) is an important aspect in order to signify the ambient noise levels and analyzing the effect of noise exposure and consideration of noise control measures. Figure 1 shows a typical illustration of a noise map developed for CSIR-National Physical Laboratory, New Delhi site based on the actual measurements conducted during the working hours in daytime. Eventually, various considerations are involved when it comes to developing the noise maps of various cities and on a large scale. The accurate information pertaining to the geometrical coordinates of the site, number of buildings and residents, road surfaces, vehicular type and flow, sound absorption characteristics of facades, etc., that are required prior developing an accurate noise map with minimal uncertainty levels. The exclusive details of all such input parameters and methodologies available had been discussed in details earlier by Garg et al. [9].

In Indian perspectives, the establishment of continuous noise monitoring terminals at various sites of the cities is a very cumbersome and expensive process. The use of validated model shall be indispensable in this regard for developing reliable noise maps. Apart from the technical aspects such as sampling strategies, use of suitable spatial interpolation algorithms, validation of the simulation results, color code scheme, and initiation of suitable cost-effective noise control action plans stands a big challenge.

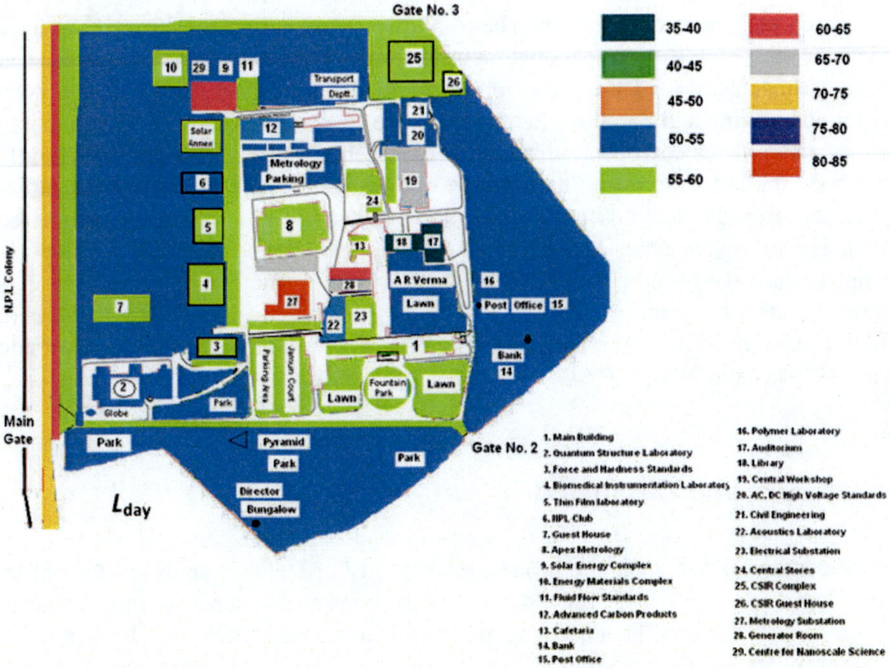


**Fig. 1** Noise map of CSIR-National Physical Laboratory, New Delhi

Also, in case of actual noise monitoring at various sites in the cities, development of an appropriate sampling strategy either categorization based or grid based [11], noise monitoring in various seasons, etc., are the issues for which further studies and deliberations among expert community are required to arrive at a final consensus.

## 3 Noise Control Action Plans

The noise management and control is an important aspect that needs to be considered, especially for the noisy hot spots. While considering the initiation and implementation of suitable noise control measures, the economic considerations, cost–benefit analysis plays an important role. An action plan should define actions in a 'smart' way (specific, measurable, achievable, realistic, time related). So, it should not merely be a policy statement but should define the targets to be achieved and must be accountable [12]. There had been various studies that reported the benefits of noise control measures. Dintrans and Préndez reported the best average noise emission reduction by 2.8 dB(A) for combined measures of traffic flow reduction (by 30%), speed reduction (maximum speed limit of 40 km/h), and road surface improvement or replacement [13]. Ausejo et al., [14] reported that in order to achieve the best long-term solution, global noise abatement measures, i.e., traffic management and local measures, i.e., noise screens, should be combined. This study emphasized the tunnels to be the most effective noise reduction measure by 30 dB(A). The noise screens had been estimated to yield the reduction up to 15 dB(A). A similar study had been reported by Bunn and Zannin, whereby the noise control measures: speed reduction, pavement replacement, decreased flow of heavy vehicles and all the measures jointly had been studied. The restriction of heavy vehicle traffic was considered as an effective measure and reductions in A-weighted sound levels of 6–7 dB had been estimated [15]. The European experience revealed diverging approaches to noise action planning between the Member States. Noise barriers and traffic management measures were the most commonly reported. The cost–benefit analysis revealed that where action plans have been implemented; the directive was efficient with a favorable cost–benefit ratio of 1:29 [16].

## 4 Results and Discussion

A case study of one such site in metropolitan city of Delhi is analyzed for understanding the benefits and application of noise control measures in ambient noise levels. Figure 2 shows the noise map of a roundabout near Siddhartha Hotel in Delhi city during the daytime when traffic density was maximum. The noise map has been generated using the CadnaA model version 4.1.137 (32 bits). FHWA TNM model had been employed for noise predictions during the peak hours in daytime. The bitmap was imported from the Google Earth platform and then calibrated for reliable predictions. The noise predictions were validated with the actual measurements at various locations, and prediction results were measured to be within ± 3 dB(A).

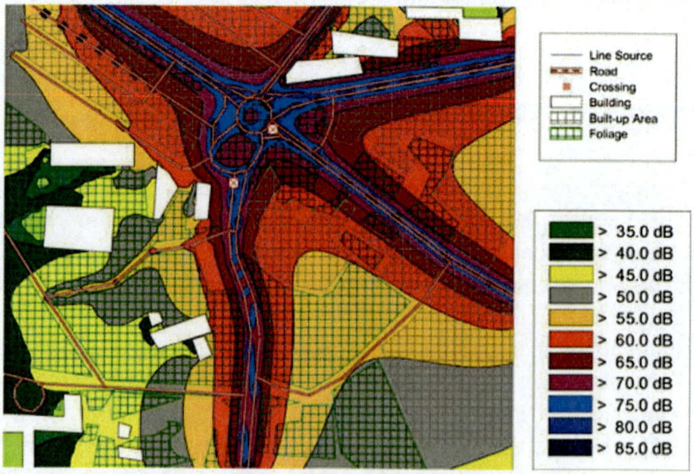


**Fig. 2** Noise map of roundabout near Siddhartha Hotel in Delhi

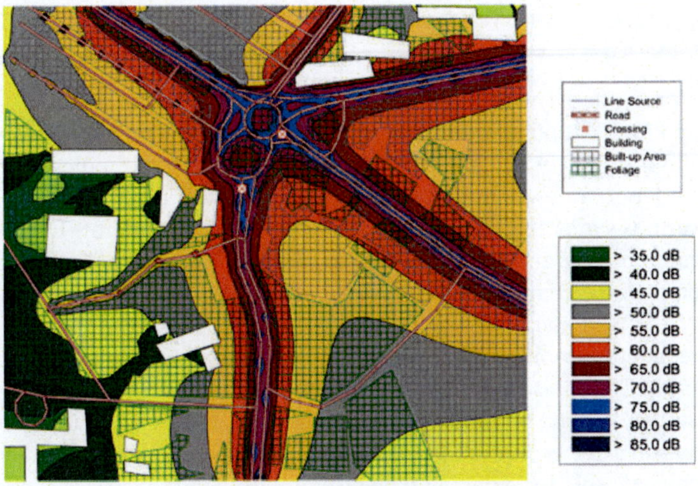


**Fig. 3** Simulated noise map of roundabout near Siddhartha Hotel with noise control measures

Eventually, the noise control measures were considered to develop a simulative platform for understanding the overall effect of all the noise control measures at the noise exposure levels at various receptor locations.

The maximum average speed measured was up to 40–45 km/h, and maximum percentage of heavy vehicles (30%) was observed. Figure 3 shows the simulated noise map considering all the possible noise control measures. It was observed from the simulation studies that around 2–4 dB(A) reductions can be possible due to the various control measures such as: restricting the heavy vehicles (up to 10%), speed reduction (up to 25 km/hr), and changing the road surface to dense graded asphalt concrete. However, in order to achieve higher noise-level reductions, erection of noise barrier near the shops and hotel premises (up to 10 dB(A)) can be a good solution. However, sometimes, erection of noise barriers has certain issues being the cost's incurred and the disturbance of the urban landscape attributed to blockage of the view and wind's stream [17]. The inclusion of green vegetation, reduction in honking activity, and enhancing the urban soundscape by the addition of a musical water fountain at the roundabout can be a good solution [18]. Further, the noise control at the receiver ends particularly the hotel premises and shops, the use of enhanced acoustic balcony, improved insulation of facade, and window shall be helpful in realizing the acoustic comfort [19]. Also, considerable noise-level reductions can be realized by adopting land-use management measures such a development of a highway for diverting the traffic and heavy vehicles and prohibiting the honking noise. Eventually, a simulative noise mapping exercise for each case shall be helpful in analyzing the benefits of all such measures. However, the cost–benefit analysis of all the measures like erection of 4 m noise barrier (Euro 400 per $m^2$) and porous asphalt double-layer road surface (Euro 10.5 $m^2$) as highlighted in a European report [20] should be prior undertaken to find the optimum solution [21].

## 5 Conclusions

The paper presents the various aspects and issues associated with noise mapping and noise pollution control in Indian perspectives. It is evident that noise mapping is at primitive stages in Indian scenario. However, the legal bodies and some of the research laboratories in India are now focused on working on developing noise maps of various cities for noise pollution prevention and control. Eventually, the concept shall be more implicit with the reporting of various studies carried out in different parts of the country in future scenario. The role of CSIR-National Physical Laboratory, the National Metrology Institute (NMI) of India shall be pivotal in this regard for conducting noise mapping and control studies in collaboration with various State Pollution Control Boards and other stake holders. It is thus imperative that noise mapping studies carried out in different parts of the country and initiation of noise control measures for bringing the ambient noise standards within the legal limits shall be indispensable for reducing the noise pollution in India.

## References

1. van Kempen E, Casas M, Pershagen G, Foraster M (2018) WHO environmental noise guidelines for the European Region: a systematic review on environmental noise and cardiovascular and metabolic effects: a summary. Int J Environ Res Public Health 15(2):379
2. WHO-JRC (2011) Burden of disease from environmental noise-quantification of healthy life years last in Europe. European center for environment and health JRC EU
3. Münzel T, Sørensen M, Schmidt F, Schmidt E, Steven S, Kröller-Schön S, Daiber A (2018) The adverse effects of environmental noise exposure on oxidative stress and cardiovascular risk. Antioxid Redox Signal 28(9):873–908
4. Directive 2002/49/EC of the European Parliament and of the Council of 25th June 2002, relating to the assessment and management of environmental noise. Official J Euro Commun
5. Kok A (2019) Refining the CNOSSOS-EU calculation method for environmental noise. In: Proceedings Internoise, Madrid
6. The Noise Pollution (Regulation and Control) rules (2000) Ministry of Environment and Forests, India
7. Vogiatzis K, Remy N (2019) Environmental noise mapping as a smart Urban tool development. Chapter in Smart Urban Development. https://doi.org/10.5772/intechopen.88449
8. Garg N, Maji S (2016) A retrospective view of noise pollution control policy in India: status, proposed revisions and control measures. Curr Sci 111(1):29–38
9. Garg N, Chauhan BS, Singh M (2020) Normative framework of Noise mapping in India: strategies, implications and challenges ahead. Acoustic Australia
10. Garg N, Surendran P, Dhanya MP, Chandran AT, Asif M, Singh M (2019) Measurement uncertainty in microphone free-field comparison calibrations. MAPAN-J Metrol Soc India 34:357–369
11. Gozalo GR, Morillas JMB (2016) Analysis of sampling methodologies for noise pollution assessment and the impact on the population. Int J Environ Res Public Health 13(5):490
12. Luzzi S (2019) Noise reduction and control in Urban planning: the strategic action plan of florence. Lecture, St. Petersburg 22nd Mar 2011
13. Dintrans A, Préndez M (2013) A method of assessing measures to reduce road traffic noise: a case study in Santiago, Chile. Appl Acoust 74:1486–1491

14. Ausejo M, Tabacchi M, Recuero M, Asensio C, Pagán R, Pavón I (2011) Design of a noise action plan based on a road traffic noise map. Acta Acust Acust 97:492–502
15. Bunn F, Zannin PHT (2015) Urban planning-simulation of noise control measures. Noise Control Eng 63(1):1–10
16. ETC/ATNI (2019) Noise Action Plans. Managing exposure to noise in Europe, No ETC/ATNI Report No 8/2019. https://www.eionet.europa.eu/etcs/etc-atni/products/etc-atni-reports/etc-atni-report-8-2019-noise-actionplans-managing-exposure-to-noise-in-europe. Accessed 10 Feb 2021
17. Jo Jaechang, Effective traffic noise policy of urban development projects in South Korea. Capstone fall, 2017, MPA/MPP Capstone Projects. 292, https://uknowledge.uky.edu/mpampp_etds/292. Accessed 25 Feb 2021
18. Jeon JY, Lee PJ, You J, Kang J (2010) Perceptual assessment of quality of urban soundscapes with combined noise sources and water sounds. J Acoust Soc Am 127(3):1357–1366
19. Cheung FSM, Li LYC, Lai AKY, Chan HCK (2019) Case studies of innovative window and balcony design for traffic noise mitigation. In: Proceedings of Acoustics 2019, 10th–13th Nov 2019 Cape Schanck, Victoria, Australia
20. Conference of European Directors of Roads (CEDR) (2013) Value for money in road traffic noise abatement. CEDR project group road noise. http://www.carreteros.org/explotacion/cedr/3_CEDR.pdf. Accessed 24 Feb 2021
21. Vaibhav K, Saini P, Garg N (2020) Development of smartphone app for environmental noise measurements. In: Proceedings of international conference on industrial and manufacturing systems (CIMS-2020), Dr. B. R. Ambedkar National Institute of Technology, Jalandhar, Punjab, India, CIMS-358

# Electrical Properties of PU/CdS Nanocomposites

**Ajay Pal Indolia, Malvika Chaudhary, M. S. Gaur, and Sobinder Singh**

**Abstract** Polyurethane/CdS nanocomposite films were prepared by a solution mixing method. The surface morphological and electrical properties of PU/CdS nanocomposite films were investigated by characterization techniques such as scanning electron microscopy (SEM) and thermally stimulated discharge current (TSDC). The SEM images show the dispersion of CdS nanoparticles in PU matrix. In order to observe the effect of nanofillers in electret state, charge relaxation, and charge storage properties of nanocomposites, TSDC is employed to understand the mechanism of charge storage. TSDC data are an evidence of different types of relaxation process. The activation energy, released charge, and charge carrier mobility decrease while relaxation time and peak current increase with CdS nanoparticles concentration.

**Keywords** PU · CdS · TSDC · SEM · Electret · Nanocomposites

## 1 Introduction

Organic–inorganic nanocomposites are attracting considerable interests in recent years due to not only the combined properties of organic polymers and inorganic materials [1–10] but also the creation of nano-interface between organic polymers and inorganic nanofillers and therefore widely used in many fields such as plastics, rubbers, and coatings. Generally, organic–inorganic nanocomposites are of two typical kinds depending upon the strength or level of interaction between organic and inorganic phases. Some polymer nanocomposites, especially containing nano-$SiO_2$

A. P. Indolia (✉) · M. Chaudhary
Department of Applied Science, Delhi Technical Campus (GGSIPU Delhi), Greater Noida, UP, India
e-mail: pajay81@gmail.com

M. S. Gaur
Department of Physics, Hindustan College of Science and Technology Farah, Mathura, UP, India

S. Singh
Department of Applied Science, Maharaja Surajmal Institute of Technology (GGSIPU Delhi), Delhi, India

 
S. Yadav et al. (eds.), *Recent Advances in Metrology*, Lecture Notes in Electrical Engineering 906, https://doi.org/10.1007/978-981-19-2468-2_37

or nano-$TiO_2$, were prepared by the sol–gel technique and investigated the influence of nanoparticles on the mechanical, thermal, and optical properties of the polymer and the relationship between structure and properties [11–17].

The micro-scale and larger inorganic fillers incorporated into organic polymers are scientifically well explored; the decrease in size of the inorganic component into the nano-dimension and the simultaneous increase of the interfacial area results in new extraordinary materials properties which require further investigations. The material parameters, such as charge storage properties, surface voltage stability, determination of mean depth of internal charge, activation energies of traps, and trap structure of the material, have been investigated by TSDC. The research in this area is not only essential for the successful application of existing materials but also for the developments of new and improved materials. In view of above background, we have chosen to study the electrets in CdS nanoparticles-embedded polyurethane samples. The CdS nanoparticles are semiconducting material forming an interface with polymer molecules, thus provide the new trapping site for charge localization. Therefore, it is well expected that the embedding of CdS nanoparticles plays an important role in electret state of polyurethane.

The main goal of this study is to prepare PU/CdS nanocomposite samples by the sol–gel technique or solution mixing method and prepare thermo-electrets and study the thermally stimulated discharge current (TSDC) to calculate the activation energy, charge released, and relaxation time to understand the relaxation and polarization mechanism in PU/CdS nanocomposites.

## 2 Experimental

Cadmium acetate, cadmium chloride, carbon disulphide, and *N*, *N*-dimethylformamide were purchased from Merck India ltd. Polyurethane (MW = 466.61078/mol) was supplied by Redox India. The thin film of polyurethane and PU/CdS nanocomposites was prepared by sol–gel technique or solution mixing method given below:

(1) 5 g of polyurethane dissolved in 100 ml of *N*, *N*-dimethylformamide (DMF) and stirred for the period of 3 h at 333 K. The homogeneous milky solution was obtained. This was used to prepare thin film of pristine PU. The glass plates were cleaned in hydrofluoric acid, rinse with acetone and distillated water, and coated by dipping in the PU solution. The PU-coated glass plates were vacuum dried at 313 K. After 24 h, the samples of pristine PU have now been easily peeled off from glass plates.

(2) In order to prepare PU/CdS nanocomposite thin films, the cadmium acetate (Cd $(CH_3COO)_2$), ammonium thiocynate ($NH_4CNS$) have been taken according to weight percent ratio and dissolve in DMF under stirring at 313 K using magnetic stirrer for 1 h. This solution was added drop by drop in a solution of pristine PU as prepared in step 1. Now, the final solution of nanocomposite samples

was kept in sonicator at 20 kHz frequency for 10 min for good dispersion of nanoparticles in polymer matrix. The remaining method for nanocomposite sample preparation is the same as discussed above. The PU/CdS polymer nanocomposite films were prepared by Zelner et al. [18] applying the same method successfully.

## 3 Results and Discussion

SEM (Fig. 1) shows the morphology of CdS nanoparticles dispersed with different weight ratio into polyurethane. For the PU/CdS nanocomposite samples, at low CdS concentration, CdS particles can be dispersed homogeneously into polyurethane regardless of the particle size of CdS, while some small aggregation of CdS nanoparticles occurs at relatively high concentrations. This aggregation of nanoparticles is forming the cluster. The interaction of CdS nanoparticles with macromolecular chains causes the modification in chain dynamics. During preparation, some of the polyester segments were chemically bonded with CdS nanoparticles, which can prevent aggregation of nanoparticles.

In the present experiment, PU and PU/CdS nanocomposites of 1%, 5%, 9%, 13% weight ratios have been prepared. The TSDC of all samples was observed with poling field of 25, 75, and 150 kV/cm at 373 K. The TSDC for PU samples polarized at 373 K with polarizing field of 25, 75, and 150 kV/cm is characterized by peak appeared at $450 \pm 2$ K (Figs. 2, 3 and 4). However, nanocomposite samples are characterized by one unstable peak (i.e., say $\beta$ peak) in the temperature range from 366 to 388 K and other stable peak (i.e., say $\alpha$ peak) in the temperature range from 443 to 458 K. $\alpha$ peak is strong as compare to $\beta$ peak.

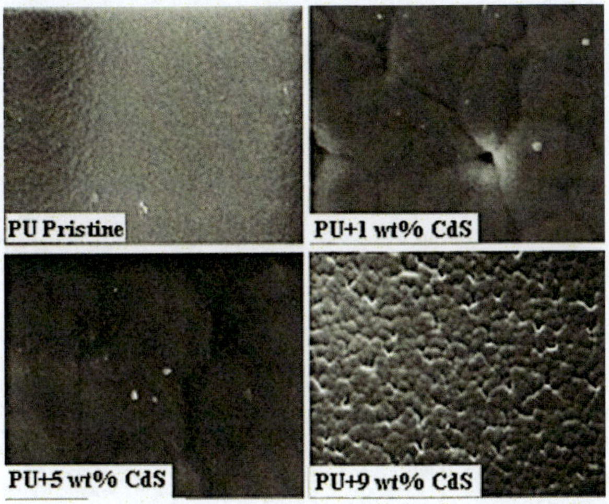


**Fig. 1** SEM images of PU/CdS nanocomposites

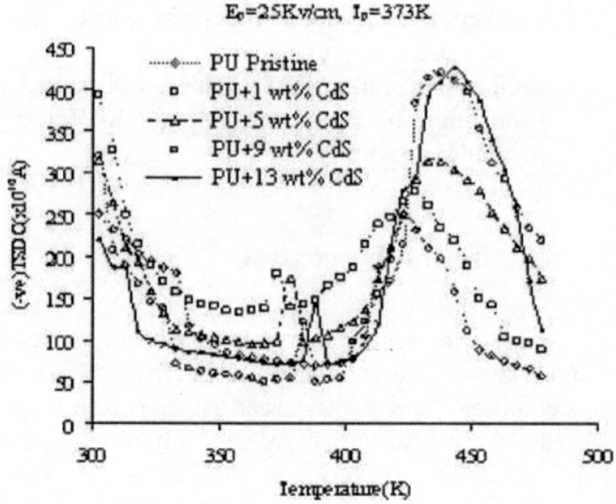


**Fig. 2** TSDC characteristics of PU pristine and PU/CdS nanocomposites polarized with polarizing field ($E_p$) of 25 kV/cm at polarizing temperature ($T_p$) of 373 K

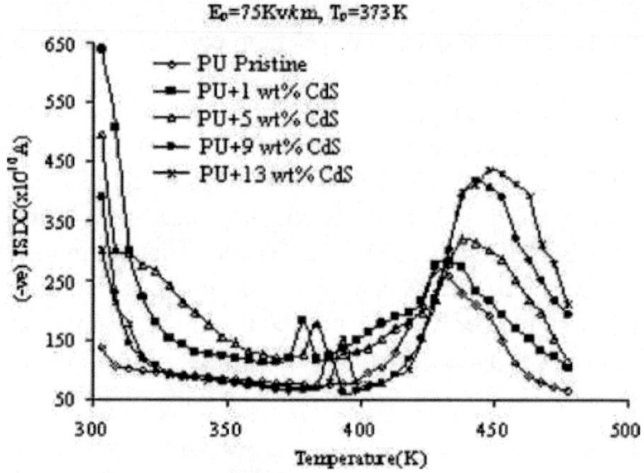


**Fig. 3** TSDC characteristics of PU pristine and PU/CdS nanocomposites polarized with polarizing field ($E_p$) of 75 kV/cm at polarizing temperature ($T_p$) of 373 K

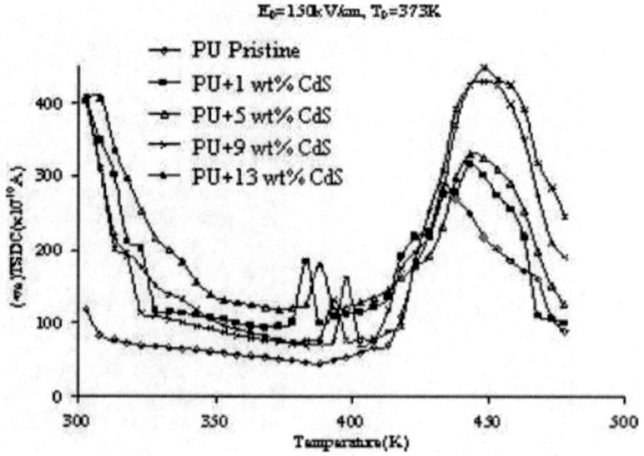


**Fig. 4** TSDC characteristics of PU pristine and PU/CdS nanocomposites polarized with polarizing field ($E_p$) of 150 kV/cm at polarizing temperature ($T_p$) of 373 K

For pristine PU, single peak is observed at high temperatures, designated as α peak; actually, low temperature peak of pristine PU is not observed, which is attributed to crankshaft motion of the $(CH_2)$ n sequences and with motion of the polar carbonyl groups with attached water molecules, respectively [19]. The high temperature peak is well observed in present study. The shoulder and the peak at higher temperatures are assigned to the α-relaxation associated with the glass transition of the soft phase of PU and the interfacial Maxwell–Wagner–Sillars relaxation (MWS peak). The accumulation of charge carriers at the interfaces between hard micro-domains and soft microphase is also the possible reason of this peak [19]. This peak in nanocomposite samples is stronger. It is assigned to rotation of the CdS nanoparticle-attached phenyl ring around the carbon atom and rotation of larger units between crosslink's around the ether segment [20].

The appearance of TSDC MWS peak, characteristic of the microphase-separated morphology of PU is appeared in the nanocomposite samples at 450 ± 2 K, suggesting that the microphase separation into hard-segment microdomains and interaction of CdS with PU affects the soft-segment microphase. Interestingly, similar effects are observed in epoxy resin/layered silicate nanocomposites, where the results to be reported elsewhere so that the long-range heterogeneity of epoxy resin is destroyed in the nanocomposites [21].

The TSDC peak at high temperature (i.e., α-relaxation peak) is corresponding to glass transition temperature. The theoretically reported $T_g$ of PU is well agreed with the difference of ± 5 K. In addition, a broad shoulder appeared on the high-temperature side of the peak in the nanocomposites, which was attributed to the α-relaxation in an interfacial region where interaction with the particle surface increases the glass transition.

When the PU/CdS nanocomposite samples were polarized at room temperature, a current reversal is obtained. The magnitude of current varied from experiment to experiment. This fact is due to change in poling condition of samples as well as density of CdS in PU samples. During the cooling from high temperature to room temperature, it is expected that large temperature gradient will exists within the sample, which results in the disturbance of trapped charges. Therefore, sign of the TSDC was opposite to that of the charging current (−ve). The TSDC characteristics for samples polarized above room temperature are qualitatively similar. In PU/CdS, nanocomposites samples show an unstable new peak in the temperature range from 366 to 388 K; however, this new peak is not appeared in pristine PU for any condition of polarization. In other words, the TSDC characteristics are similar for all samples at room temperature polarization and also similar above room temperature polarization for all samples. It has been noticed that the current is smaller for low-temperature polarization than the high-temperature polarization. It is noticed that the peak is shifted with respect to the polarizing field.

The relationship between the peak current and polarizing field is shown in (Fig. 5). The peak current in some cases falls on a straight, and extrapolations of this curve pass through a point at no vanishing applied field $E_o$. From the present experiment, it may be possible that the charge injected into PU/CdS nanocomposites in polarization at temperature above the room temperature for all polarizing field is ionic in nature. The

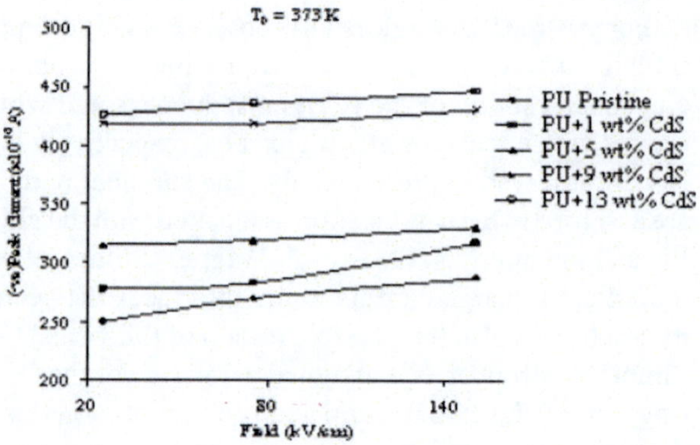


**Fig. 5** Peak current ($\alpha$ peak) versus polarizing field characteristics of PU pristine and PU/CdS nanocomposites with polarizing temperature ($T_p$) of 373 K

polarization of insulating polymers can be produced by dipole alignment, migration of ion within the material, or charge injection from the electrode or accumulation of charge careers in interface. In the case of uniform volume polarization, for example, dipole alignment, a linear dependence of peak current on applied field should be obtained, and the extrapolation of this linear plot passes through the origin [22].

In contrast to the case of dipole alignment, which is corresponding to first peak (β peak) observed in all PU/CdS nanocomposites at all condition of polarization, however, the dipole polarization is no fully observed in pristine PU because activation is varying between 0.23–1.01 eV, which corresponding to both dipolar (β-relaxation) relaxation and space charge relaxation (α-relaxation). The activation energy calculated for $\beta$ peak (i.e., dipolar relaxation) of PU/CdS nanocomposites samples is well agreed with value reported for other polymers [23]. The activation energy was calculated by using initial rise method [24]. The α-relaxation peak (high temperature peak) of nanocomposite samples is not only alone due to α-relaxation process. The dipolar—relaxation and MWS polarization cannot be ignored fully because (i) peak current of α peak is the function of field shown in (Fig. 5), and (ii) the calculated activation energy is not fully agreed with α-relaxation process these results suggested that the existence of multiple relaxation process in PU/CdS nanocomposite. Since no any complete theory for relaxation process in polymer nanocomposites is developed, therefore, we inference that multiple relaxation processes are operative in nanocomposite samples for which the present study is an evidence.

There are several concepts of traps in polymers have been suggested (i) Van-Roggen has proposed that the amorphous regions in polymers have different energy levels as compare to crystalline regions and therefore can act as traps [25]. This model should give the single trap depth. (ii) There are various models of molecular motion in polymer such as α, β, and γ dispersion have been observed in many polymers. However, there is no more report on that how these motions affected by introducing the nanofillers in polymer matrix. The nano-size particle provides energy to polymeric chain for motion; some of motion (i.e., relaxations) may disappear, and some of the motion or relaxations are strongly appeared. Gaur et al. [26] recently pointed out that the trap structure of polymers is modified by incorporation of nanofillers due to which large variation is obtained in activation energy.

In present case, the semiconductor behavior of CdS nanoparticle may have quite less energy band gap as compare to traditional CdS. This low-energy band gap of CdS is very less as compare to polyurethane subsequently when molecular species of CdS attached with polymer molecules and forming interface of nanomatric order.

These interfaces itself act as new trapping levels formed at different depth. The high-temperature TSDC peaks may be controlled by the trap concentration. This trap concentration is the function of wt % of CdS nanoparticles in PU matrix.

The trapped charge carrier concentration was calculated using following relation [27]

$$n_t = 2.7\,\mathrm{Jm\,kTm^2/ed}\gamma A$$

where $J_m$ is the maximum value of the current density, e the electronic charge, k the Boltzmann constant, $T_m$ the peak temperature, $\gamma$ the heating rate, A the activation energy, and d the sample thickness. It has been found that carrier concentration $n_t$ (i.e., varying from $3 \times 10^{18}\ m^{-3}$ to $5 \times 10^{18}\ m^{-3}$) increases with concentration of CdS nanoparticles, which is an evidence for new trapping sites introduced by incorporation of nanoparticles. It is evident from characteristic behavior of TSDC that the new traps introduced in PU matrix can act as electrons and hole traps with various energy levels. The current start increasing at a certain temperature due to frequent trapping of charge carriers and exponential decrease of current after attaining the maximum value is caused by trapping of charge carrier at deeper and deeper level. It is evident from higher value of activation energy of α peak, (i.e., 1.5 to 3.5 eV) for nanocomposite samples, which is normally not obtained in case of polymer like polyurethane. These results suggested that the structure of pristine PU is modified by filling of CdS nanoparticles which causes the behavior of TSDC and modified the various parameters like charge released, relaxation time, and activation energy particularly shown in Table 1. In other hand, the TSDC of nanocomposite samples seems to be due to space charge captured in shallow traps is released and neutralized by recombining with the carriers responsible for the ohmic conduction during the depolarization of the sample. Under such situation, the observed TSDC is mainly due to the carriers trapped in deeper traps [28].

In our present work, we have demonstrated that nanoparticles additives modified the TSDC behavior as it is also reported in literature [29–33].

**Table 1** TSDC data of PU/CdS nanocomposites

| S.no | Poling field kV/cm | Samples | Peak position (K) | Poling temperature 373 K | | |
|---|---|---|---|---|---|---|
| | | | | activation energy (eV) | Relaxation time(s) $X10^{-16}$ | Charge released (c) |
| 1 | 25 | PU | I 423 | 0.27 | 1.00 | $9.6 \times 10^{-8}$ |
| | | | II | | | |
| | | 1wt% CdS | I 373 | 0.13 | 1.00 | $3.4 \times 10^{-8}$ |
| | | | II 428 | 0.51 | 1.01 | $5.3 \times 10^{-8}$ |
| | | 5wt% CdS | I 378 | 0.13 | 1.00 | $5.0 \times 10^{-8}$ |
| | | | II 433 | 0.93 | 1.02 | $1.2 \times 10^{-7}$ |
| | | 9wt% CdS | I 383 | 0.24 | 1.00 | $4.7 \times 10^{-8}$ |
| | | | II438 | 1.64 | 1.04 | $8.0 \times 10^{-8}$ |
| | | 13wt%CdS | I 388 | 0.27 | 1.00 | $5.4 \times 10^{-8}$ |
| | | | II 443 | 3.08 | 1.08 | $8.1 \times 10^{-8}$ |
| 2 | 75 | PU | I 428 | 0.56 | 1.01 | $1.0 \times 10^{-7}$ |
| | | | II | | | |
| | | 1wt% CdS | I 378 | 0.16 | 1.00 | $3.5 \times 10^{-8}$ |
| | | | II 433 | 0.60 | 1.01 | $5.4 \times 10^{-8}$ |
| | | 5wt% CdS | I 383 | 0.10 | 1.00 | $5.1 \times 10^{-8}$ |
| | | | II 438 | 0.66 | 1.01 | $1.2 \times 10^{-7}$ |
| | | 9wt% CdS | I 388 | 0.20 | 1.00 | $4.8 \times 10^{-8}$ |
| | | | II 443 | 1.50 | 1.03 | $8.0 \times 10^{-8}$ |
| | | 13wt%CdS | I 393 | 0.09 | 1.00 | $5.9 \times 10^{-8}$ |
| | | | II 448 | 3.02 | 1.08 | $1.2 \times 10^{-7}$ |
| 3 | 150 | PU | I 433 | 0.61 | 1.01 | $1.1 \times 10^{-7}$ |
| | | | II | | | |
| | | 1wt% CdS | I 383 | 0.13 | 1.00 | $3.5 \times 10^{-8}$ |
| | | | II 443 | 1.18 | 1.03 | $6.1 \times 10^{-8}$ |
| | | 5wt% CdS | I 388 | 0.10 | 1.00 | $5.2 \times 10^{-8}$ |
| | | | II 443 | 0.55 | 1.01 | $1.2 \times 10^{-7}$ |
| | | 9wt% CdS | I 393 | 0.26 | 1.00 | $5.1 \times 10^{-8}$ |
| | | | II 448 | 0.80 | 1.02 | $8.2 \times 10^{-8}$ |
| | | 13wt% CdS | I 398 | 0.18 | 1.00 | $6.2 \times 10^{-8}$ |
| | | | II 453 | 1.87 | 1.04 | $1.2 \times 10^{-7}$ |

## 4 Conclusion

It was found that a strong correlation between CdS wt % and TSDC parameters exists. The charge release is directly related with charge storage capability of polymer nanocomposites. The new trapping sites introduced by nanoparticles demonstrate the potentiality of nanoparticles to improve the charge storage and electret behavior of polymer.

## References


1. Yan C, Jude OI (1999) Chem Mater 11:1218
2. Wang YT, Chang TC, Hong YS, Chen HB (2003) Thermochim Acta 397:219
3. Ogoshi T, Itoh H, Kim KM, Chujo Y (2002) Macromolecules 35:334
4. Shang XY, Zhu ZK, Yin J, Ma XD (2002) Chem Mater 14:71
5. Matˇejka L, Duek K, Pleštil J, Kríz J, Lednický F (1999) Polymer 40:171
6. Huang ZH, Qiu KY (1997) Polymer 38:521
7. Yu YY, Chen CY, Chen WC (2003) Polymer 44:593
8. Zhou SX, Wu LM, Sun J, Shen WD (2002) Prog Org Coat 45:33
9. Yu YY, Chen WC (2003) Mater Chem Phys 82:388
10. Pu ZC, Mark JE (1997) Chem Mater 9:2442
11. Chen Y, Iroh JO (1999) Chem Mater 11:1218
12. Chan CK, Peng SL, Chu IM, Ni SC (2001) Polymer 42:4189
13. Hsiue GH, Kuo WJ, Huang YP, Jeng RJ (2000) Polymer 41:2813
14. Gao Y, Choudhury NR, Dutta N, Matisons J, Reading M, Delmotte L (2001) Chem Mater 13:3644
15. Xiong MN, You B, Zhou SX, Wu LM (2004) Polymer 45:2967
16. Mascia L, Kioul A (1995) Polymer 36:3649
17. Chen XC, You B, Zhou SX, Wu LM (2003) Surf Interface Anal 3:369
18. Zelner NF, Minti H, Reisfeld R (1997) Chem Mater 9:2541
19. Pissis P, Apekis L, Christodoulides C, Niaounakis M, Kyritsis A, Nedbal J (1996) J Polym Sci B 34:1529
20. Georgoussis G, Kyritsis A, Bershtein VA, Fainleib AM, Pissis P (2000) J Polym Sci B 38:3070
21. Vanderschueren J, Gasiot J, Braunlich P (1980) Applied physics, Berlin, Springer vol 37, p 205
22. Bucci C, Fieschi R (1996) Phys. Rev 2:148, 816 (1966), 148
23. Garg M, Quamara JK (2006) Nucl Instrum Meth Phys Res B 246:355
24. Van Turnhout J (1975) Thermally stimulated discharge current of polymer electrets. Elsevier, Amsterdam
25. Van Roggen A (1965) Ann Rep Comf Elect Insulation Not Res Council Public 3:1356
26. Shukla P, Gaur MS (2008) Iran Polym J 17:745
27. Neagu E (1994) Mater Lett 21:119
28. Jones JF (2003) Appl Environ Microbiol 69:6515
29. Okumoto S (2005) FRET nanosensors. PNAS 102 :8740
30. Benkekaa N (2004) Thermochim Acta 413:39
31. Garg M, Quamara JK (2005) Nucl Instrum Meth Phys Res B 355:246
32. Correia NT, Ramos JM (1999) J Polym Sci 37:227
33. Khare PK, Sahu DK, Verma A, Srivastava RK (2004) Indian J Pure Appl Phys 42:693

# IoT-Enabled Sensor Node for Environmental Monitoring

**Parag Chourey, Nirbhow Jap Singh, Kirti Soni, and Ravinder Agarwal**

**Abstract** In the environmental monitoring system, the atmospheric boundary layer (ABL) plays an important role. It is a major source of concern for the scientific community, and it necessitates the implementation of an effective and quick monitoring system to make better decisions. SODAR is a critical tool for continuous determining the environmental parameters in real time. The ABL height is a critical parameter for determining the air quality in a given area. The use of the well-connected SODAR network to determine ABL height from remote locations in real time is beneficial for analyzing and planning the environmental monitoring system. The Internet of Things (IoT) is a ground-breaking technology, which allows various environmental monitoring stations to communicate with one another. The interconnection of several IoT-enabled sensor nodes is known as an IoT network. Multiple sensors, a data acquisition system, and a data communication system make up the sensor node. It accomplishes the goal of monitoring local environmental parameters and sharing information with an IoT network. A sensor node is composed of a SODAR, a wind sensor, a temperature, and a relative humidity sensor. The indigenous SODAR design, integrated into common data acquisition software, with other meteorological sensors was used. For the reliable performance of the proposed system, the calibration of meteorological sensors was important. Before installation, all meteorological sensors used were calibrated. In addition, the data from the sensor node-enabled by IoT were also validated with comparisons to the local data acquisition system.

**Keywords** Internet of Things · Meteorological sensors · Mixing height · Planetary boundary layer features · Meteorological sensors · Sensor node

P. Chourey (✉) · N. J. Singh · R. Agarwal
TIET, Patiala, Punjab, India
e-mail: choureyparag@gmail.com

K. Soni
CSIR-NPL, Delhi, India



S. Yadav et al. (eds.), *Recent Advances in Metrology*, Lecture Notes in Electrical Engineering 906, https://doi.org/10.1007/978-981-19-2468-2_38

# 1 Introduction

Environmental monitoring is important in various aspects of life including the monitoring and control of air pollution, meteorological forecasts, water quality control, and monitoring. Environmental monitoring aims to facilitate the appropriate application under favorable environmental conditions. Advanced technologies such as IoT and sensor networking play a critical role in sustainable growth and a healthy society. The monitoring and meteorological parameters on the atmospheric borders (ABL) provide conceptual knowledge about the critical parameters affecting the air quality in an area [11, 19]. The ABL lies within the 1–2 km of the lower atmosphere where all living entities exist. ABL monitoring has a key role in estimating pollutant emissions from the atmosphere. Collective information on weather and ABL height helps to determine the atmospheric dispersion and the air particles' emission process. This process shows the carrying capacity to the atmosphere of air pollutants [13]. So, it becomes essential to monitor the ABL structure collectively and some important weather parameters effectively in environmental monitoring.

The ABL monitoring techniques are broadly classified into two categories, i.e., direct or in situ measurements and remote sensing techniques. The remote sensing techniques have significant advantages over in situ techniques due to their wide monitoring range and cost-effectiveness. The SODAR is one of the best tools for continuous monitoring of the ABL structure in real time [1] and recommended by the regulation authorities like Environmental Impact Association (EIA) USA and Central Pollution Control Board (New Delhi) India for ABL structure monitoring [6].

In the decade, the Internet of Things (IoT) has been evolved as a technology. With real-time data accessibility, cloud storage, and the ability to make autonomous decisions, it revolutionized the way objects interact with one another [7, 12, 17]. In this chapter, the parameters such as temperature, relative humidity, wind speed, and wind directions are addressed with the ABL structure for the collective measurement of important air pollution meteorological parameters. The networking between the sensors is required for the collective measurement and monitoring of these parameters on a single platform. The proposed method integrates SODAR in the design of a sensor network with meteorological sensors like wind sensors, temperature sensors, and relative humidity sensors. A modern signal processing unit, data acquisition unit, data networks, and data storage facilities are also provided in the sensors network for the creation of an IoT-enabled sensor node. Figure 1 shows the block diagram of the proposed IoT capable node.

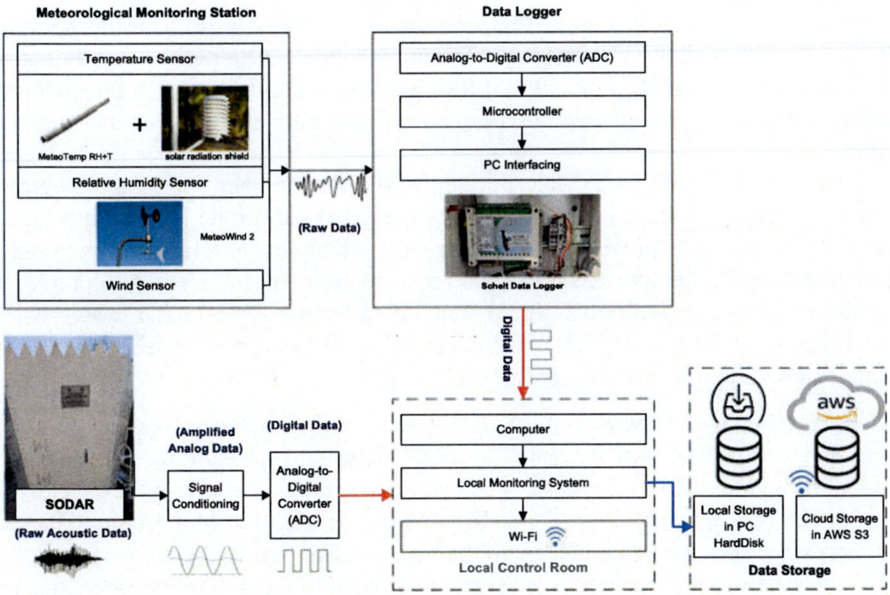


**Fig. 1** Block diagram of IoT-enabled sensor node consisting of SODAR and meteorological sensors (temperature, relative humidity, and wind)

## 2 System Design

The IoT-enabled sensor node consists of sensors, data gathering systems, data tracking and communication, and storage facilities. The following are the main components of the proposed IoT-enabled sensor-node system.

(a) Meteorological Sensors

  i. MeteoTemp (RH + T) for temperature and relative humidity measurements
  ii. MeteoWind2 for wind speed and wind direction measurement
  iii. Data logger for meteorological sensors

(b) Signal processing unit (SPU) for monostatic SODAR for ABL height measurement
(c) Common data acquisition software
(d) Data storage and data communication software.

**Meteorological Sensors**: The MeteoTemp (RH+T) and MeteoWind2 sensors manufactured by Barani design technologies, Slovakia was used. An anemometer with a wind vane had the excellent accuracy of World Meteorology Organization (WMO) and MEASNET standards (MEASNET/IEC 61400-12-1:2005 calibrations and ISO/IEC 17025:2005 calibrations) for meteorology and wind resource assessment [18]. It had a very low power consumption and integrated lightning and surge protection for reliable operation. Temperature and humidity sensor consumes ultra-low power with accurate measurement of air temperature and humidity as per the standard of WMO (ISO/IEC 17025:2005) for long-term stability [10]. These meteorological sensors can provide reliable continuous monitoring in adverse environmental conditions [3].

Schelt Technology, India, has developed the data logger for meteorological sensors. It supplies the fundamental analog-to-digital converters, microcontrollers, and interfaces for PC processing transmits and communicates with the local monitoring station for the raw analog signal received from sensors. The program was designed to process additional digital signal data for use with the hardware board for monitoring purposes. The block diagram of the meteorological sensor data acquisition system is shown in Figure 2.

**Monostatic SODAR**: Monostatic SODAR was used as a sensor for creating the sensor node. The purpose of installing SODAR was to measure the ABL height in real time. This SODAR was designed and developed indigenously in the CSIR-National Physical Laboratory (CSIR-NPL), New Delhi India. Monostatic SODAR works on the acoustic principle where the high acoustic signal was transmitted into the atmosphere, and a backscattered signal was received from the atmosphere due to temperature turbulence [6]. The received signal was further processed through the signal conditioning stage. The signal conditioning stage consists of a preamplifier and a narrow band-pass filter. Figure 3 depicts the gain and noise characteristics of an indigenous preamplifier design. The single low-noise amplifier with state-variable (SV) narrow band-pass filter configuration was used in the pre-amplification stage.

Fig. 2 Block diagram of meteorological sensor data acquisition system

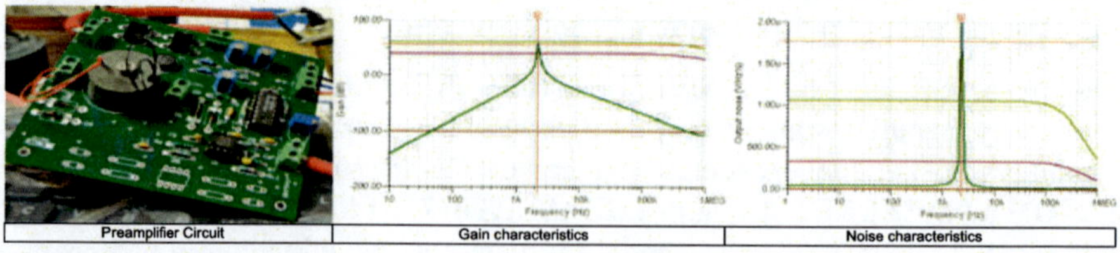


Fig. 3 SODAR preamplifier gain and noise characteristics

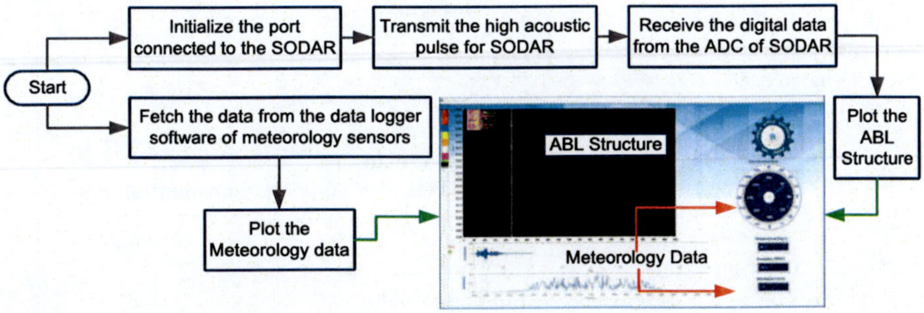


**Fig. 4** Flowchart of the common data acquisition software

The total gain and output noise of the design were 54 dB and 2μV/√Hz. Further, the processed analog signal was converted into the digital signal using the built-in sound card of the computer [4].

**Common Data Acquisition Software**: The digital signal from the sensors was sent to the local monitoring station. The common data acquisition software provides further data processing and manipulation for local data visualization in a desirable format. This software also controls the transmission and receiving of the acoustic signal for SODAR. Figure 4 shows the software's flowchart. This software helps meteorology and SODAR information to be combined in real time with visualization.

**Data Storage and data communication Software**: The local monitoring station has a data storage facility for offline monitoring of historical data. This system also includes Wi-Fi for Internet access. Each local monitoring station worked as a gateway for the IoT networking and shares the real-time data with the AWS-S3 cloud storage system in near real time. The data communication from the monitoring station to the cloud was established using the NI-LabVIEW-based software. Figure 5 depicts the flowchart of the AWS-S3 data communication software. The new bucket of each day was created automatically from the software. The data were updated using the put-object function of AWS-S3 after every 20 s. The identity and access management function provides secure data access as well as privacy, consent, and identity verification while data communication [2, 5]. Further, the data were stored in the AWS-S3 cloud storage. The sensors, signal conditioning unit, data acquisition system, and data communication system work together to create an IoT-enabled sensor node to visualize and share data on its own.

## 3 Results and Discussion

The results of the planned IoT-enabled sensor-node system are discussed. In the first part, the data analysis of SODAR and meteorological sensors, i.e., ABL height, temperature, relative humidity, wind speed, and direction, is described. In the second

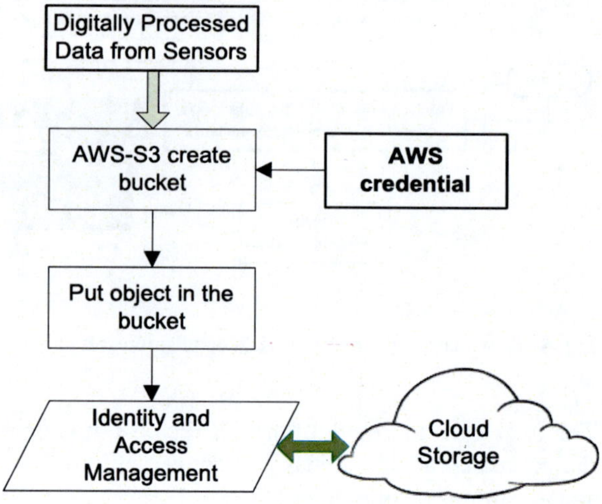


**Fig. 5** Flowchart of the LabVIEW and AWS-S3-based data communication software

part of the analysis, the statistical analysis has been done on the IoT-enabled sensor-node data and uncertainty and measurement accuracy.

**Data Analysis**: The detailed study of ABL structures and corresponding meteorological data is presented in Fig. 6. The data have been plotted using the LabVIEW-based offline software. This software provides a single platform for simultaneously visualizing all sensor data. The data from 19th October 2020 have been selected for the demonstration purpose. Figure 6a presents the 24-h data analysis of ABL structure, wind direction, hourly mean wind speed, hourly mean relative humidity, and hourly mean temperature data. This software also cumulates the ABL height information after manual calculation. The graph shows that wind speed and temperature have a proportional relationship, whereas relative humidity has an inverse relationship with temperature and wind speed. The variation in ABL structure is also visible in the graph.

Figure 6b–d presents the special structures found during the different time frames on 19th October 2020. Figure 6b depicts the thermal plumes with periodical spikes during daytime. These structures show a more common phenomenon of atmosphere and formed due to rise in the temperature at daytime. The higher-thermal plumes were considered favorable for higher pollutant dispersion [8, 9, 15]. In this period, the temperature and wind speed shows an increasing trend, whereas the relative humidity decreases with time. Figure 6c demonstrates the rising layer structure of ABL. The rising layer structure was formed during the transition period of day (from early morning to afternoon period). The transition period may vary from season to season. In this graph, it was observed that the ABL height starts increasing and split from the bottom structure of ABL in the morning time [14]. With the start of the rising layer, the temperature, wind speed graph was increasing, and the opposite trend was visible with relative humidity. Figure 6d presents the inversion period of the day. It is visible from the graph that during the inversion time, a stable ABL structure

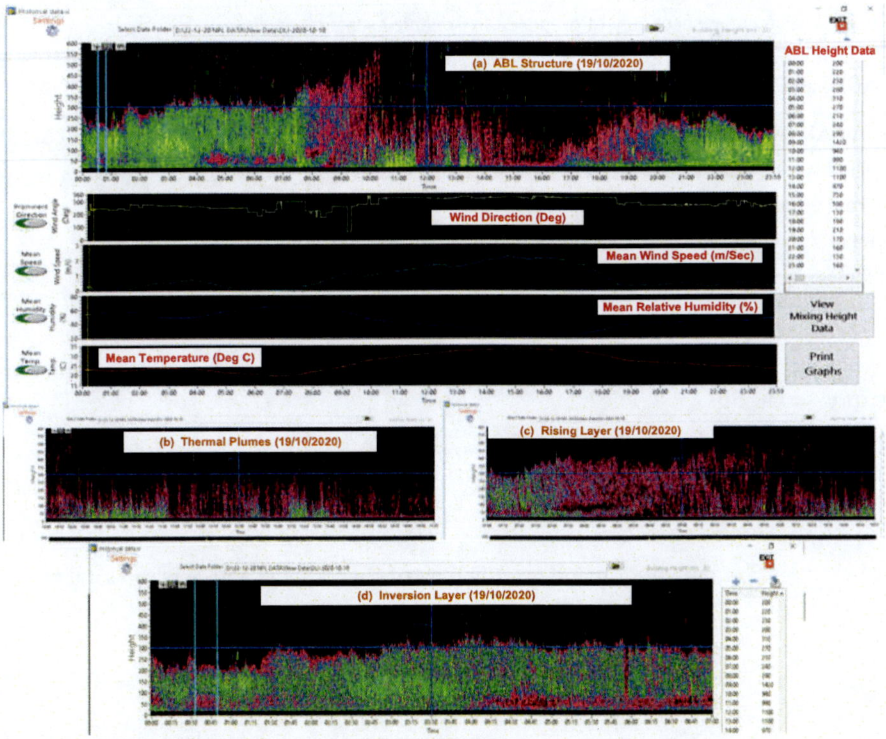


**Fig. 6** Various ABL structures of 19th October 2020 and corresponding meteorological data, **a** 24-h ABL structure **b** thermal plumes structure, **c** rising layer structure, **d** inversion layer structure

is formed [16]. In this period, most of the meteorological parameters were settled down, and less variation was visible in comparison to the thermal plumes period.

**Statistical Analysis of IoT-Enabled Data Communication System**: The accuracy of the IoT-enabled sensor node was verified. In this, the data communication accuracy was checked by comparing the received data points with transmitted data points. The one-month data of the proposed sensor node were collected from the local monitoring station and cloud. The communication error was classified into two categories, i.e., system error and communication error. The sluggish network, malfunction in software, routine maintenance, and power interrupts were the major cause of the system error. Whereas, the error due to the data transmission system was considered as a communication error. The data communication percentage accuracy and percentage error are calculated using following equations.

$$\%\text{Accuracy} = \frac{\text{Total Received Data Points(Rx)}}{\text{Total Transmitted Data Points(Tx)}} \times 100 \tag{1}$$

$$\%\text{Error} = \frac{\text{Tx} - \text{Rx}}{\text{Tx}} \times 100 \tag{2}$$

**Table 1** Percentage accuracy of proposed IoT-enabled sensor node

| Uncertainty estimation of IoT-enabled sensor-node system | Data from (Oct 01, 2020–Oct 30, 2020) | | | | |
|---|---|---|---|---|---|
| | Accuracy (%) | Error (%) | $\sigma$ (%) | $u$ (%) | $u_c$ (%) |
| SODAR (ABL structure) | 99.999 | 0.00005 | 0.00104 | 0.00019 | 1.06766 |
| Wind speed (WS) | 99.994 | 0.00542 | 2.14246 | 0.39784 | |
| Wind direction (WD) | 99.964 | 0.03569 | 1.79420 | 0.33317 | |
| Relative humidity(RH) | 99.976 | 0.02344 | 4.74598 | 0.88130 | |
| Temperature (T) | 99.964 | 0.03590 | 1.65038 | 0.30646 | |

The standard deviation ($\sigma$), standard uncertainty, and combined uncertainty are expressed in Eqs. 3, 4, and 5.

$$(\sigma) = \sqrt{\frac{(x_i - \mu)^2}{N}} \tag{3}$$

where $x_i$ is %accuracy of one day, $\mu$ is mea of %accuracy, and N is the total numbr of days.

$$\text{Standard Uncertainty}(u) = \frac{\sigma}{\sqrt{N}} \tag{4}$$

$$\text{Combined Uncertainty}(u_c) = \sqrt{u_{\text{ABL}}^2 + u_{\text{WS}}^2 + u_{\text{WD}}^2 + u_{\text{RH}}^2 + u_T^2} \tag{5}$$

Table 1 shows the proposed system's estimated uncertainty results. The statistical analysis of the IoT-enabled sensor node reveals that the accuracy was greater than 99.9 % for all sensor data, with an error of less than 0.1 %. The sensor-node system's standard deviation and standard uncertainty were also evaluated. All sensors showed less variation and uncertainty, according to the data. The ABL data have the lowest standard uncertainty. The suggested sensor-node system was a combined uncertainty of 1.067 %. Overall, the statistical analysis indicates that the IoT-enabled sensor-node system was considerably high accuracy and extremely little uncertainty.

## 4 Conclusion

This chapter describes a sensor node enabled for the monitoring of the environment by SODAR and meteorological sensors. The sensor node with IoT measures and shares with the cloud in almost reality the ABL height, wind speed, wind direction, relative temperature, and humidity data. The suggested sensor-node data collection program concurrently monitors all parameters on a single platform. Data transmission was a difficult challenge because of the enormous amount of data. The described data transmission system achieves 99.99% accuracy with a total error rate of less than 0.1%. Research on the uncertain estimation shows that the total uncertainty of the system is around 1%. The suggested IoT-enabled sensor node is extremely useful for the scientific community working in atmospheric research since it allows them to remotely monitor critical environmental parameters in real time. This sensor node may be placed at numerous places in the future to create an IoT-based environment monitoring system. In the future, this sensor node can be installed at multiple locations for creating an IoT-based environment monitoring network.

**Acknowledgment** This research was carried out at the CSIR-NPL. The authors are grateful to the Director of the CSIR National Physical Laboratory (NPL).

## References

1. Asimakopoulos DN, Helmis CG, Michopoulos J (2004) Evaluation of SODAR methods for the determination of the atmospheric boundary layer mixing height. Meteorol Atmos Phys 85:85–92
2. Brantner M, Florescu D, Graf D, Kossmann D, Kraska T (2008) Building a database on S3. In: Proceedings of the 2008 ACM SIGMOD international conference on Management of data. Pp 251–264
3. Chourey P, Singh NJ, Soni K, Agarwal R (2022) SODAR based meteorological sensor network for air pollution monitoring in Northern India. MAPAN 1–15
4. Chourey P, Soni K, Singh NJ, Agarwal R (2022) IoT-Sodar network for airshed management planning. IETE J Res 1–15
5. Garfinkel S (2007) An evaluation of Amazon's grid computing services: EC2, S3, and SQS. Harvard Comput Sci Gr Tech Rep TR-08–07
6. Gera BS, Raghavendra T, Singh G, Ojha VK, Malik J, Gera N, Gupta NC (2011) Instrumentation and computer capabilities for improving sodar data acquisition. Int J Remote Sens 32:4807–4817
7. Khan MA, Salah K (2018) IoT security: review, blockchain solutions, and open challenges. Futur Gener Comput Syst 82:395–411
8. Kumar N, Soni K, Agarwal R (2021) Prediction of temporal atmospheric boundary layer height using long short-term memory network. Tellus A Dyn Meteorol Oceanogr 73:1–14
9. Kumar N, Soni K, Garg N, Agarwal R, Saha D, Singh G (2017) SODAR pattern classification and its dependence on meteorological parameters over a semiarid region of India. Int J Remote Sens 1161:3466–3482
10. Leštinská L, Danc IJ (2016) Temperature and humidity calibration procedure ISO/IEC 17025:2005
11. Li Z, Guo J, Ding A, Liao H, Liu J, Sun Y, Wang T, Xue H, Zhang H, Zhu B (2017) Aerosol and boundary-layer interactions and impact on air quality. Natl Sci Rev 4:810–833
12. Madakam S, Lake V, Lake V, Lake V et al (2015) Internet of things (IoT): a literature review. J Comput Commun 3:164
13. Pleim JE (2007) A combined local and nonlocal closure model for the atmospheric boundary layer. Part II: application and evaluation in a mesoscale meteorological model. J Appl Meteorol Climatol 46:1396–1409
14. Singal SP, Lewthwaite EWD, DSW (1989) Estimating atmospheric stability from monostatic acoustic sounder records. Atmos Environ 23:2079–2084
15. Singal SP, Aggarwal SK, Pahwa DR, SG B (1984) Stability studies with the help of acoustic sounding. Atmos Environ 19:221–228

16. Singal SP, Gera BS, Kallistratova MA, Petenko IV (1997) Sodar aspect sensitivity studies in the convective boundary layer. Int J Remote Sens 18:1809–1819
17. Talavera JM, Tobón LE, Gómez JA, Culman MA, Aranda JM, Parra DT, Quiroz LA, Hoyos A, Garreta LE (2017) Review of IoT applications in agro-industrial and environmental fields. Comput Electron Agric 142:283–297
18. Westermann D, Westermann H (2018) Measnet anemometer calibration procedure ISO/IEC 17025:2005
19. Zhou W, Jia Y, Peng A, Zhang Y, Liu P (2018) The Effect of IoT new features on security and privacy: new threats, existing solutions, and challenges yet to be solved. IEEE Internet Things J 4662:1–11

# Review and Metrological Evolution of Primary Pressure Standard

**Chanchal, Afaqul Zafer, Renu Singh, Ajay Kumar, and Sanjay Yadav**

**Abstract** In metrology, pressure measurement and calibration have been a focus for many scientific researches. In the field of pressure, the advancement in technology is mainly based on two factors: (i) geometrical parameters (fabrication of large diameter cylinders and piston) (ii) on dimension and high accuracy. There are many devices for pressure measurement like barometer, manometer, deadweight tester, etc. However, the accuracy in measurement for deadweight testers is highest. Therefore, the main focus of this paper is on deadweight tester, precise and accurate measurements of the pressure and calibration and measurement capabilities (CMC) of parameter 'pressure' for CSIR-NPL (National Physical Laboratory). CSIR-NPL provides various services to the industry of India by generating the CMC value for different standards, including for the pressure metrology.

**Keywords** Deadweight tester · Calibration · National metrology institute · Pressure · Deadweight tester

Chanchal (✉) · R. Singh
GD Goenka University, Gurugram 122103, India
e-mail: schanchal038.cs@gmail.com

Chanchal
Government College for Women, Faridabad. Haryana 121001, India

A. Zafer · S. Yadav
CSIR-National Physical Laboratory, New Delhi, India

Academy of Scientific and Innovative Research (AcSIR), CSIR- NPL Campus, Dr. K. S. Krishnan Road, Delhi 110012, India

A. Kumar
Maharaja Surajmal Institute of Technology, New Delhi 110058, India



S. Yadav et al. (eds.), *Recent Advances in Metrology*, Lecture Notes in Electrical Engineering 906, https://doi.org/10.1007/978-981-19-2468-2_39

## 1 Introduction

Pressure and its measurement play an important role in industries. Due to the advancement of the environmental and semiconductor-based industries, there is an increase in demand of precise measurement of pressure at widespread pressure. The instruments used in industries for the measurement of pressure require high accuracy. To fulfil the above requirements, deadweight testers, piezoelectric transducers, silicon resonance gauges (SRGs), capacitance diaphragm gauges (CDGs), etc. are used. For this, the instrument needs to be calibrated.

Calibrations of such devices are done in national metrological institute and various calibration laboratories. For this purpose, deadweight tester (DWT) is considered as an evolution in primary pressure standard. DWT is a popular hydromechanical device which makes use of Bernoulli's principle for achieving the maximum accuracy of calibration. This uniqueness makes DWT a primary standard and is used internationally in calibration laboratories. The calibration of pressure is done in two modes depending on the type of fluids. The two modes are hydraulic (use oil as fluid) and pneumatic mode (use air as fluid). The cross-floatation method is used for the calibration of device in hydraulic mode. It is a mechanical-driven device and requires large amount of human efforts/man power to operate on high range of pressure. Taking observation during the experiment is a difficult task during equilibrium (when the mass is rotated on the piston cylinder assembly). The erroneous in the readings may occur because of leakage. For the remedy, root cause analysis of leakage is required. At high pressure, sometimes, leakage occurs due to compression of fluid. This problem is resolved by putting the system back to the service.

## 2 A Historical Take on DWT

Pressure measurement and its control is the most demanding variable in almost all industries. For many years, pressure measurement attracted the interest of science. It is considered that evolution in pressure measurement started with invention of mercury barometer in 1643 by Evangelista Torricelli which is the simplest devices for the measurement of atmospheric pressure of a particular location [1–3]. The concept of atmospheric pressure was introduced by Otto Von Gucricke through an experiment known as Magdeburg hemispheres [4]. This experiment was demonstrated on 8th May 1654 in Rosenberg in front of Emperor Ferdinand III. In this demonstration, 30 horses were deployed for the separation of hemisphere later on in 1656 this demonstration with 16 horses (8 on each side bracket) in Magdeburg, and this experiment became so popular that it is used as an ideal experiment to explain the concept of air pressure/atmospheric pressure. In this experiment, two Copper hemispheres were joined with the help of grease, and vacuum was created inside them due to the high-power atmospheric pressure even the team of horses could not separate them. In the middle of eighteenth century, various researchers introduced their inventions which help in pressure measurement in various fields. In 1844, Lucien Vidi invented the aneroid barograph which records barometric pressure versus time in the graphical form and also used to make continuous readings of atmospheric pressure. In this apparatus, there is a pressure-sensitive element which is a partially evaluated metallic cylinder that is linked to a pen in a way that atmospheric pressure is noted as the vertical displacement of the pen. There was another development in barograph in 1845 and is based on photography by Francis Ronald's. In1849, Bourdon gauge (Manometer) was developed by Euge'ne Bourdon which became the most prominent invention. For measuring low pressure down to $10^{-6}$ torr, McLeod invented the McLeod gauge in 1874. The McLeod gauge consists of a vacuum system and the tube which is filled with mercury; the sample of the gas is filled in the vacuum system. By tilting the tube which is filled with mercury, the mercury rushes towards the sample when the pressure is created, and thus, created pressure is measured with the help of mercury manometer. By knowing the ratio of initial and final volume (compression ratio), the pressure of original volume can be determined by using Boyle's law. This measurement is most suitable for non-condensable gases like nitrogen and oxygen and is not an appropriate method for condensable gases like ammonia and carbon dioxide water vapour because these gases condense when compressed by McLeod gauge. The condensed gases show pressure much lower than the actual pressure which reserve with error reading. For the remedial purpose, cold trap can be used with McLeod gauge for condensation of vapours before entering into the gauge. The McLeod gauge provides the same calibration for all non-condensable gases and is easy to use. In 1881, Samuel Seigfried Karl Ritter Von Basch designed a sphygmomanometer to measure blood pressure [5]. With the beginning of nineteen's, German physicist Marcello Pirani developed Pirani Vacuum gauge based on heat loss measurement [6]. An important invention in the field of metrology for the calibration of pressure gauges, i.e. deadweight tester which will be discuss in detail in this paper [7].

## 3 Deadweight Tester and Working Principles

The DWT setup consists of oil reservoir, a collector port (which is used to provide the pressure for the device to be calibrated), piston cylinder assembly, mass sets and hydraulic compressor with rotary piston [8]. To operate this device, human effort is required as it is mechanically driven device.

The pressure generated is calculated by the equation [9]:

$$P = \frac{\Sigma_i m_i \cdot g(1 - \rho_a/\rho_i) + \gamma C}{A_0\left(1 + b_1 p_n + b_2 p_n^2\right)\left[\left(\alpha_c + \alpha_p\right)(T - T_c)\right]} \pm \Delta p \tag{1}$$

$m_i$ = Mass of the weight
$\rho_a$ = Density of air at laboratory condition

$\rho_i$ = i[th] weight density
$\gamma$ = Surface tension of fluid
$C$ = Circumference of the piston emerging out from the fluid
$A_0$ = Piston—cylinder's effective area at Zero pressure
$\alpha_c$ and $\alpha_p$ = Thermal expansion coefficients of cylinder's and piston's material
$T$ = Assembly temperature
$T_r$ = Temperature at which $A_0$ is referred
$b$ = Effective area's pressure coefficient
$d$ = Coefficient of change of area with respect to jacket pressure
$\Delta p$ = It is head correction in term of pressure (where $\Delta p = [(\rho_f - \rho_a). g. H]$, in this, equation H depicts height difference between two deadweight tester's reference level, and ($\rho_f$) is transmitted fluid density.

## 4 Invention Summary

In dead weight tester, the pressure measurement is based on two factors: (i) fabrication of large diameter cylinders and piston (ii) on dimensions and high accuracy measurement. The uncertainty in measurement of pressure in dead weight tester is very less $\cong 2.2 \times 10^{-6}$(1), when piston and cylinder assembly is round in shape of diameter ± 30 nm and straight within 100 nm over a substantial fractions of their height and diameter at 20 °C provided by PTB (Physikalish Technische Bundesanstalt) ± 15 nm and good geometry of piston and cylinder assembly. In every industry for measurement purpose, it is essential for every equipment to be accurate, precise, efficient and easy to use. All the above qualities are fulfilled by the deadweight tester for the calibration of pressure measurement. By avoiding the pressures undershoot and overshoot, the equilibrium of piston can be easily achieved in the measurement system of deadweight piston. It is a device to provide a better version of deadweight calibration/measurement. In this, regardless of the vertical position of the piston, speed of the piston rotation is accurately measured and controlled. Also, according to the requirement, the gap between the piston and cylinder can be adjusted to get accuracy in measurement/celebration of pressure. Whilst using an automated drive system in deadweight tester, the perturbation and interruptions that cannot be predicted easily during the measurement of pressure can be reduced. Deadweight testers are of three kinds on the basis of medium which is measure and lubricant used for measuring elements.

1. Oil—oil (oil-operated oil-lubricated pressure control unit)
2. Gas—oil (gas-operated oil-lubricated pressure control unit)
3. Gas—gas (gas-operated gas-lubricated pressure control unit).

In case of oil-operated oil-lubricated pressure control units, the device under test is sent with oil and connected with oil deadweight tester oil will not contaminate the deadweight tester. By rotating, the piston friction can be lowered and give high accuracy in the measured value. Whilst operating gas oil, the lubricant used for PCU (pressure control unit) leagues towards the gas circuit of deadweight tester; due to this reason, the reservoir is included in the system; the reservoir becomes full; then, it causes an uncontrollable oil head. In third case, i.e. gas–gas direct pressure control (DPC) use should be taking very important mobile operating that the PCU is very sensitive to contamination; also, device under testing does not introduce any foreign material in the system and does not contaminate the system.

## 5 Contemporary Techniques and Different Type of Assembly

See Figs. 1, 2, 3, 4, 5 and 6.

This paper [10] explains that the pressure transducers are the alternative to calibrate the pressure standard. They are also called reference pressure transducers. When these reference pressure transducers are used, then the pressure source should be

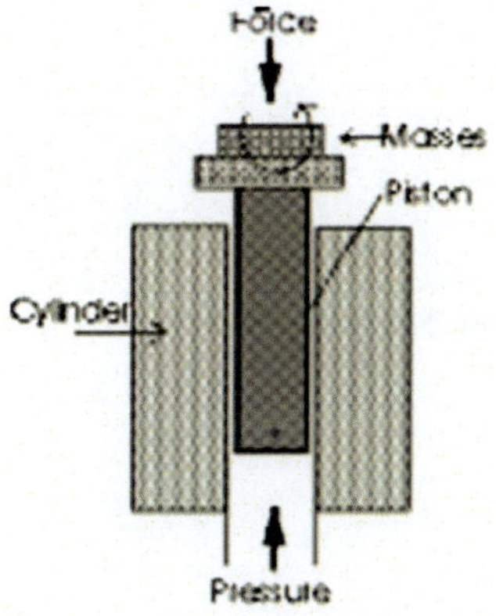


**Fig. 1** Simple assembly with top loading

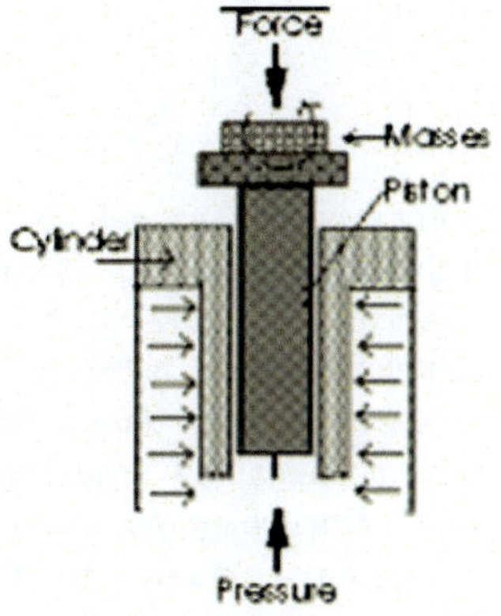


**Fig. 2** Re-entrant type with top loading

Fig. 3 Control clearance type with top loading

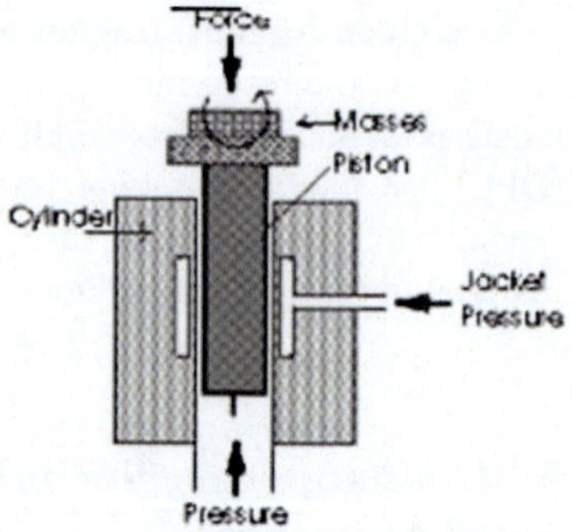


Fig. 4 Simple type with overhang loading

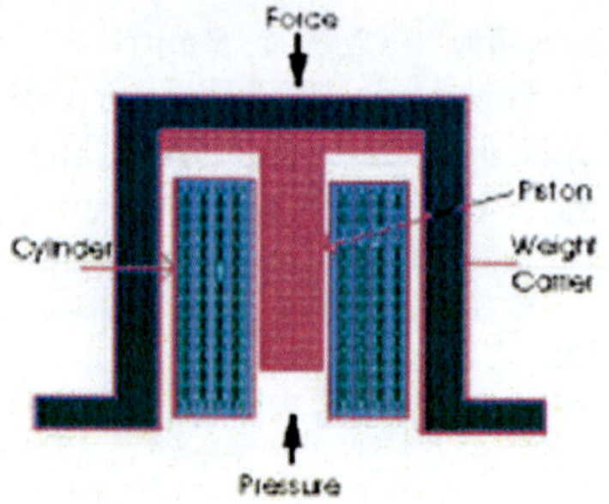


Fig. 5 Simple type dual-range assembly with

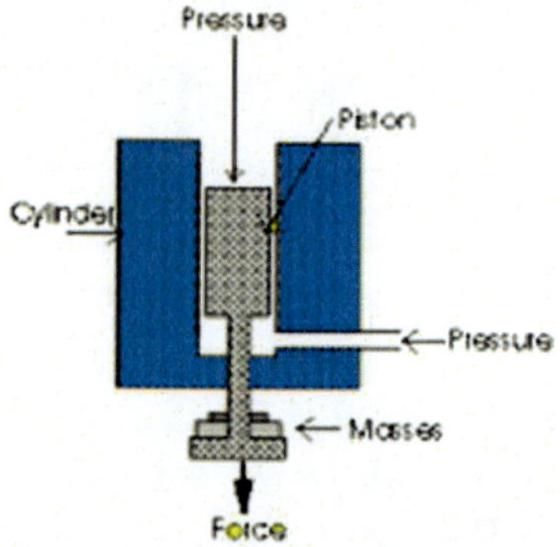


motor driven or manual. Using pressure transducer, actual pressure can be measured even if the test pressure is not produced precisely.

The author in the paper explains the development in the pressure standards [11]. This development is based on (a) largest diameter piston cylinder fabrication with good geometry (b) their capabilities to produce minimum relative uncertainty. This paper explains how uncertainty can be reduced from 10 to 5 PPM when dimension measurements are included in the gauges.

The paper explains the latest development in the National Institute of Standards and Technology [12]. These advances were on two fronts. (1) Fabrication of piston

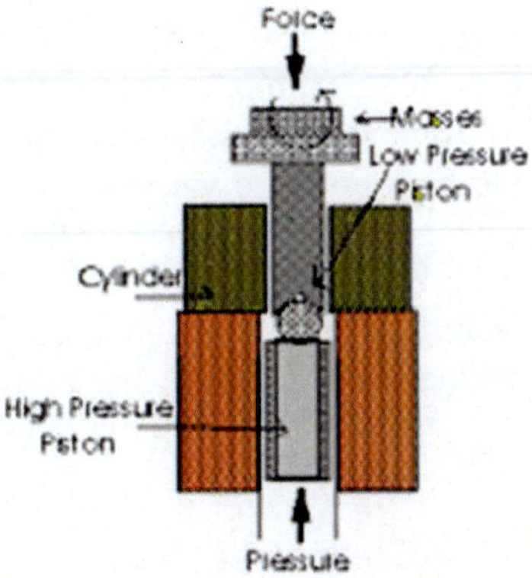


**Fig. 6** Differential type of assembly with suspended loading top loading

with good geometry and large diameter. (2) The combination of advanced geometry capability with some deadweight testers at NIST provides the capability of achieving total relative uncertainty in generated pressure near to 10 PPM. This paper explains that in generated pressure, use of latest dimension measurement in the newest gauge helps to reduce the total relative uncertainty to 5 PPM.

NIST [*Julia scherchlight* et al. *2017 Metrologia 54S125*] developed a standard which produces pressure in the range of 10^-6 Pa × 10^-10 Pa and possibly lower. This is possible by use of the cold atom vacuum standard (CAVS) which acts as an absolute sensor as well as primary standard. CAVS is based on the fact of cold loss that sensing atoms like alkali metals (Lithium) trap because of collision by gas like $H_2$ in vacuum. The value pressure can be obtained by averaging collisions (along cross-section). That is why, it is called primary standard as it uses the cross-section calculated by ab initio method for $Li + H_2$ system. And the first primary standard is fountain clock.

Wladimir Sabuga in his paper explains that five National Meterology jointly produce devices to measure high pressure in accurate and reliable manner [13]. As a result of this research, a new pressure measuring device was developed to measure a pressure of 1.6 Gpa measuring material elastic constant using resonant ultrasound spectroscopy, viscosity of pressure transmitting liquid at 1.4 GPa, density, hardness of material of high pressure; component and measurement of dimension of piston cylinder are the base of the characterisation. This project developed a device which has measuring capability of 1.6 Gpa and can be accurately used for calibrating a pressure measuring device up to 1.5 GPa.

In 2017, Karl Jousten with other co-workers realised the unit of pressure. He concludes that the optical methodology with thermodynamics definitions can be used for better accuracy then the traditional method force/area [14]. Necessary corrections are to be needed for the real-gas effect. This is the most promising to explain the pressure unit at the maximum level of metrology.

Rosendahl, in 2018, explained how differential pressure can be used in fluid flow processes like determination of flow rate, fluid velocity (e.g. Venturi, Pitot tube, orifice plate). Generate the pressure up to range from 500 kPa to 20 MPa, with the help of twin pressure balance which is gas operated. Various parameters of differential pressure manometer were evaluated [15].

Tomoya Muramoto in his paper explained the capillary method used to measure the high-pressure viscosity. For this, he makes the use of two pressure balances. Change in the fall rate of the balance is used to calculate the volume flow. For the pressure, the range up to 20 MPa viscosity of bis (2-ethylhexyl) Sebacate was recorded. Thus, concluded that the system developed was accurately helped in the evaluation of pressure dependence of viscosity.

## 6 Differential Pressure

Differential pressure is the difference between the two points difference. Differential pressure measurement instrument consists of two input ports [16–18]. An unknown pressure is applied to both inputs. The output is difference between the two, and differential pressure is obtained by measuring each port pressure with separate instrument then subtract the two readings. For the measurement of differential pressure, two deadweight testers can be used. Differential pressure is used in the following applications:

i. Flow of measurement of gaseous and liquid medium
ii. Level measurement in close vessels
iii. Monitoring of filter contamination
iv. Overpressure measurement in clean rooms
v. Control of pumping plants

It is widely used to calibrate all kind of pressure gauge. Examples: pressure transmitter, industrial pressure gauges, etc. Some of the advantages are as follows: (i) Simple in construction (ii) Easy to use (iii) Widely used to calibrate wide range of pressure gauge (iv) Shape and size of piston cylinder can be varied according to the requirement. There are many other accessories other than deadweight tester are used to calibrate pressure [19, 20]. Some of the important ones include: reference pressure monitor which is used to measure the pressure in pressure line connecting to different components of the calibrator, needle valve (for opening and closing of pressure line), barometric pressure monitor, thermometer to measure room temperature, humidity measurement instrument, platinum resistance thermometer (to measure the temperature of piston and cylinder assembly in the deadweight tester), split level, oil reservoir, pressure transmitting fluid (Sebacate oil), hydraulic screw pump (used to generate pressure) (Table 1).

After observing the different uses of differential pressure applications and due to its national importance, it is of at most significance that the work should be preceded in the direction of establishing the CMC for differential pressure measurement and to establish the calibration and measurement capabilities by using the deadweight tester to measure the differential pressure more than the maximum range [21].

**Table 1** CMC (Calibration and measurement capability) values for differential pressure in different countries

| Service provider | Institute service code | Comments | Quantity | Value CMC | Expanded uncertainty CMC |
|---|---|---|---|---|---|
| Russian Federation<br>VNIIM | VNIIM/11 | Working fluid: ethanol | Differential pressure, liquid medium | [0.1 to 100] Pa | 0.1 Pa |
| Russian Federation<br>VNIIM | VNIIM/12 | Working fluid: water | Differential pressure, liquid medium | [100 to 5000] Pa | 0.34 Pa |
| Korea, Republic of<br>KRISS | KRISS/C204112 | | Differential pressure, oil medium | [0 to 20] MPa | |
| Croatia<br>FSB-LPM | | | Differential pressure | [0 to 1.00E7] Pa | |

## 7 Conclusion

Deadweight tester is one of the best methods for precise evaluation of the parameter 'pressure' and is very suitable as a primary pressure standard. With its long-term stability and simple working principle, deadweight tester can be considered as the standard for 'all time and all people'.

## References


1. Benedict R (1984) Fundamentals of temperature, pressure, and flow measurements
2. West JB (2013) Torricelli and the ocean of air: The first measurement of barometric pressure. Physiology 28:66–73. https://doi.org/10.1152/physiol.00053.2012
3. Woo SY, Choi IM, Lee YJ et al (2004) Calibration apparatus for precise barometers and vacuum gauges. J Vac Sci Technol A Vacuum Surfaces Film 22:1816–1819. https://doi.org/10.1116/1.1735935
4. Lambert LB (1979) A history of pressure measurement. Meas Control 12:116–124. https://doi.org/10.1177/002029407901200305
5. Noyes B Jr (1936) The history of the thermometer and the sphygmomanometer. Bull Med Libr Assoc 24:155
6. Shie J, Chou BCS, Chen Y (1995) High performance Pirani vacuum gauge. J Vac Sci Technol A Vacuum, Surfaces, Film 13:2972–2979. https://doi.org/10.1116/1.579623
7. Grossman J (2009) The electronic deadweight tester—a modern replacement for the conventional deadweight tester. Cal Lab 1–4

8. Yadav, Sanjay. "Characterization of dead weight... - Google Scholar. https://scholar.google.co.in/scholar?hl=en&as_sdt=0%2C5&q=+Yadav%2C+Sanjay.+%22Characterization+of+dead+weight+testers+and+computation+of+associated+uncertainties%3A+a+case+study+of+contemporary+techniques.%22+Metrology+and+Measurement+Systems+14.3+%282007%29%3A+453-469.+&btnG=. Accessed 6 Jul 2021
9. Yadav S, Standards V Prssure Measurements , Measuring Instruments and Their Calibrations
10. Zafer, Afaqul, and Sanjay Yadav. "Design and development... - Google Scholar. https://scholar.google.co.in/scholar?hl=en&as_sdt=0%2C5&q=Zafer%2C+Afaqul%2C+and+Sanjay+Yadav.+%22Design+and+development+of+strain+gauge+pressure+transducer+working+in+high+pressure+range+of+500+MPa+using+autofrettage+and+finite+element+method.%22+International+Journal+of+Precision+Engineering+and+Manufacturing+19.6+%282018%29%3A+793-800&btnG=. Accessed 6 Jul 2021
11. Schmidt J, Cen Y, Driver R, et al A primary pressure standard at 100 kPa. iopscience.iop.org
12. Vern E (1994) NIST Pressure Calibration Service
13. Sabuga W, Rabault T, Wüthrich C et al (2017) High pressure metrology for industrial applications. Metrologia 54:S108. https://doi.org/10.1088/1681-7575/AA895A
14. Jousten K, Hendricks J, Barker D, et al (2017) Perspectives for a new realization of the pascal by optical methods. iopscience.iop.org 54:146. https://doi.org/10.1088/1681-7575/aa8a4d
15. Rosendahl M, Nazareth RS, Magalhaes MR et al (2018) New calibration procedure for differential pressure using twin pressure balances for flowrate measurement. J Phys Conf Ser 1044. https://doi.org/10.1088/1742-6596/1044/1/012053
16. 1. Dilawar, Nita, et al. "Characterization of a pneumatic... - Google Scholar. https://scholar.google.co.in/scholar?hl=en&as_sdt=0%2C5&q=1.%09Dilawar%2C+Nita%2C+et+al.+%22Characterization+of+a+pneumatic+differential+pressure+transfer+standard.%22+Metrologia+40.2+%282003%29%3A+74.&btnG=. Accessed 6 Jul 2021
17. Dogra S, Yadav S, Bandyopadhyay AK (2010) Computer simulation of a 1.0 GPa piston-cylinder assembly using finite element analysis (FEA). Meas J Int Meas Confed 43:1345–1354. https://doi.org/10.1016/j.measurement.2010.07.010
18. Woo S-Y, Choi I-M, Song H-W (2009) A low differential pressure standard in the range of 1 Pa to 31 kPa at KRISS. Metrologia 46:125–128. https://doi.org/10.1088/0026-1394/46/1/016
19. Yadav, Sanjay, and A. K. Bandyopadhyay. "Proficiency... - Google Scholar. https://scholar.google.co.in/scholar?hl=en&as_sdt=0%2C5&q=14.%09Yadav%2C+Sanjay%2C+and+A.+K.+Bandyopadhyay.+%22Proficiency+Testing+Program+Under+NABL+in+the+Pressure+Range+7-70+MPa+Using+a+Dead+Weight+Tester.%22+Med.+J.+Meas.+Contrl+1.3+%282005%29%3A+138-151.&btnG=. Accessed 6 Jul 2021
20. Chauhan J, Vijayalakshmi V, Muralidharan V, Sreedhar S (2020) Automation of Hydraulic Dead Weight Tester. Int Conf Electr Electron Eng ICE3 2020 236–239. https://doi.org/10.1109/ICE348803.2020.9122991
21. Yadav S, Kumar A, Dilawar N et al (2018) Role of national pressure and vacuum metrology in Indian industrial growth and their global metrological equivalence article in MAPAN. J Metrol Soc India Springer 33:347–359. https://doi.org/10.1007/s12647-018-0270-8

# Prospects of Digitalizing Dimensional Metrology

**Mahammad Arif Sanjid, K. P. Chaudhary, Sanjay Yadav, Mrinal Sen, and Sanjoy K. Ghoshal**

**Abstract** Industry 4.0 is developed combining a variety of new technologies like IoT, AI, and additive manufacturing, etc. Production is accelerated employing Internet that decentralizes monitoring of the physical processes using smart sensors by a virtual copy of the physical world. Digitization offers significant opportunities to formulate conformity assessment of quality of products, processes, and services. Metrology is one of the pillars of quality infrastructure that ensure all the measurements traceable to SI units. The International Committee for Weights and Measures (CIPM), started work to transform the SI unit for digitizing. The CIPM task group and PTB, Germany is pursuing the task of Digital SI project. Digital transformation of metrological services refers to digitalizing the quality infrastructure aiming to conformity assessments. Generation of digital calibration certificate is ultimate product of digital transformation of metrology. Virtual modelling of measurement process and accumulated data forms the digital twins. Digitalization of dimensional metrology needs development of universally acceptable models of dimensional calibration services. Recently, simulation methods are used to improve the diameter measurement of piston—cylinder of pressure balances. Multiple sensors are used to generate redundant coordinate data referring to a common datum. These works pave

M. A. Sanjid (✉)
Length, Dimension and Nanometrology, CSIR—National Physical Laboratory, New Delhi 110012, India
e-mail: sanjid@nplindia.org

K. P. Chaudhary
Maharaja Surajmal Institute of Science and Technology, New Delhi 110058, India

S. Yadav
Pressure, Vacuum and Ultrasonic Metrology, CSIR—National Physical Laboratory, New Delhi 110012, India

M. Sen
Department of Electronics Engineering, Indian Institute of Technology (Indian School of Mines), Dhanbad, Jharkhand 826004, India

S. K. Ghoshal
Department of Mechanical Engineering, Indian Institute of Technology (Indian School of Mines), Dhanbad, Jharkhand 826004, India

 
S. Yadav et al. (eds.), *Recent Advances in Metrology*, Lecture Notes in Electrical Engineering 906, https://doi.org/10.1007/978-981-19-2468-2_40

way to design a framework for the digital dimensional metrology. A generic model is proposed considering the recommendations of ISO 17450.

**Keywords** SI units · Traceability · QUDT · FAIR · Simulation

# 1 Introduction

Advent of secured Internet technologies enabled data driven Industrial 4.0 that involves (i) automation using IoT (ii) encrypted data access (iii) decentralized decision-making (iv) process control using AI. Industry 5.0 is imminent through the augmentation of a growing consciousness of machines in line with human [1].

An organization's qualifies to Industry 4.0 depends on its capability of continuous innovation to respond to change, and quickly mastering the products, processes in response to customer's demand. The challenges of Industry 4.0 are to upgrade their infrastructure, work force, and culture to digital formats. The objective of setting a comprehensive framework for the conceptual and structural design of Industry 4.0 systems to organize the standard resource and environmental data is took up by Reference Architectural Model Industrie 4.0 [2].

FAIR abbreviated as Findability, Accessibility, Interoperability, and Reuse of digital assets explained the principles for scientific data management by emphasizing machine-action ability [3]. Similarly, a trilateral between Alliance Industrie du Futur, the National Plan "Transition 4.0" (former Impresa 4.0), and Platform Industrie 4.0 is working on three core topics: standardization, SME integration and test beds, and governmental support for the digitalization processes in the European manufacturing sector [4]. A digital quality infrastructure includes generation—exchanges of digital calibration certificates, legal metrology and authentic traceability [5]. It is obvious to implement this digital transformation lie mainly in the following fields:

- Capabilities in the field of IT and software;
- Modelling and virtual measurement process or "digital twins";
- Real-time data storage and cloud services;
- Autonomous systems;
- Development and integration of apps, and
- Establishing a link between the virtual and the physical world (CPS).

The attribute of digitalizing dimensional metrology for the Industry 4.0 includes in process measurement of dimension product specification with feedback, measurement models to evaluate uncertainties, and accessing the data to enable decision-making for conformity assessment. Digital transformation of metrological services, dealing the high dimensional data with secured transmission, and developing analytical methods and standardized procedures for interconnected and virtualized measuring systems are focus of digital transformation of metrology.

Traditionally, experts (e.g. metrologist, quality inspector, and manufacturer) interpret the results of the analogue calibration certificates for the production, the quality control and the decision-making. On the contrary, digital calibration certificate (DCC) serves for the electronic storage, the authenticated, encrypted, and signed transmission, and the uniform interpretation of results [6]. Specifically, there is a scope of research for the digitalization of dimension metrology and its standardization. The metrology community is working rigorously to choose appropriate data format to expresses the results of the given dimensional calibrations [7].

Authors proposed a process graph to generate digital calibration certificate of dimensional calibration. This format as a data structure, contain all the measurement related data. The proposed format also includes the standard operator of ISO 17450 of dimensional measurement process to enable simulation of measurement process [8]. Accordingly, the simulation of dimensional calibration services along with data could form the desirable digital twins.

## 2 Digitalizing the Calibration Results

Measurement values, data, algorithms, mathematical, and statistical procedures as well as communication and security architectures are the basis of digital expansion and transformation. The plan is to create an internationally recognized DCC format for the entire field of metrology. This format should be machine readable as well as independent of the software platform. Based on the exchange formats of DCC, one should able to assess the risk and opportunity for legal metrology applications and intelligent means of digital type examination of the certificates for the "digital twin" (DT).

Some metrological norms includes the SI units, the International vocabulary of metrology (VIM), the GUM [9], the CODATA, and ISO/IEC 17,025 [10]. IEC TS 6270 "Identification of units of measurement for computer-based processing" shows the handling of the SI units and derived units in XML. XML-based data structures support robust Cryptographic methods. A major advantage of XML is its machine readability. The digital calibration certificate contain four areas:

- Administrative data (regulated area),
- Measurement results (partially regulated area),
- Comments (not regulated area), and
- Document (additional area).

In the present article, the segment of expression of measurement results is worked out. The representation of the measurement results is one of the most challenging tasks, because of the variety of measurements results that are in use by different scientific and strategic sector. At the same time, a simple structure can be presented for those who do not already have an existing data exchange format. Therefore, this area cannot be generally regulated. However, representation of the basis of the SI unit of measurements must be strongly regulated. They can be represented as a scalar, vector, matrix or tensor. A complete measurement result includes the following data:

- Identifier,
- Measurement value,
- Expanded measurement uncertainty,
- Coverage factor,
- Unit, and
- Time.

The use of an individual data structure is possible provided the results are complete and take the SI into account. In the case of the individual data structures, it must be considered that often descriptive components and information on measurement uncertainties are not available. In these cases, adjustments must be made. In addition, tools must be provided to visualize the content. The universal data structure is a linear structure. It consists of blocks of text and blocks with measurement results. The text blocks serve as an explanation of the results of the measurements and can be carried out, in addition to English, in other languages.

**Digital twin**

Measurement process simulation in combination along with the data forms the digital twin of the calibrated measuring instrument. These results significantly increase the quality of the production, since the behaviour of the measuring device is precisely known.

## 3 Dimensional Metrology

Manufacturer develops products according to customers demand. This increases the complexity in the manufacturing systems and is one of the main challenges for future production. The geometric product specification are described in ISO 1101 [11]. The geometric characteristics of products are measured in dimensional metrology. Gauge practise can be found in IS 3455 [12].

The calibration services are classified and given at BIPM website [13]. The six categories of measurement are radiation frequency, linear, angular, form, complex, other. The radiation measurement are traceable to time and frequency standards and He–Ne-based primary laser source at National physical laboratory (NPL), India. Gauge blocks are artefact standards have nanometre range uncertainty. Mostly, model equations of dimension metrology represent linear time invariant systems. As the length of material depends on the temperature at which it has been measured, an uncertainty of thermal measurement 10 mK is desirable to calibrate the thermometers. Refraction index of air changes 0.97 ppm per unit change of atmospheric pressure. Whenever, laser interferometer is used for displacement measurement the pressure sensors that are calibrated in the pressures standards division of NPLI provide the correction value to the wavelength. Table 1 shows some calibration services and their respective standards. A comprehensive discussion of uncertainty estimation of dimension measurement is given in literature [14].

**Table 1** Classification of some dimension calibration services

| Dimensional calibration services | | Measurand |
|---|---|---|
| **Recommended radiations** | • Laser sources of stabilized frequency<br>• Isotopic optical sources | • Vacuum wavelength;<br>• Radiation frequency |
| **Linear measures** | Displacement<br>• 1D length comparator<br>• Gauge block comparator<br>• Liner voltage differential transformer | • The error of indicated in the displacement |
| | **Artefact standards**<br>• Graduated scales<br>• Length of gauge blocks<br>• Size of ring gauge<br>• Size of plug gauge<br>• Size of sphere | • Deviation from size from reference length |
| **Form standards** | • Flatness | • Flatness; parallelism; wedge angle, |
| | • Roundness | • Roundness; amplitude and phase harmonic content |
| | • Straightness | • Straightness; |
| | • Cylindricity | • Cylindricity |

# 4 Digitalizing Dimensional Measurements

The objective of digitalizing dimension measurement is to generate DCC. The result expressed by the DCC is categorically divided into six different technical domains. Figure 1 shows structure of these technical aspects.

Foremost, the geometric parameter intended to measure needs unambiguous definition. Such definitions are electronically retrievable from the database of digital document standard (e.g. ISO 1101, ISO 80000, etc.). All the dimension specification are traceable to digital representation of SI standard i.e. metre. The operators of the skin model discussed in ISO 17450 are acceptable to plan the measurement

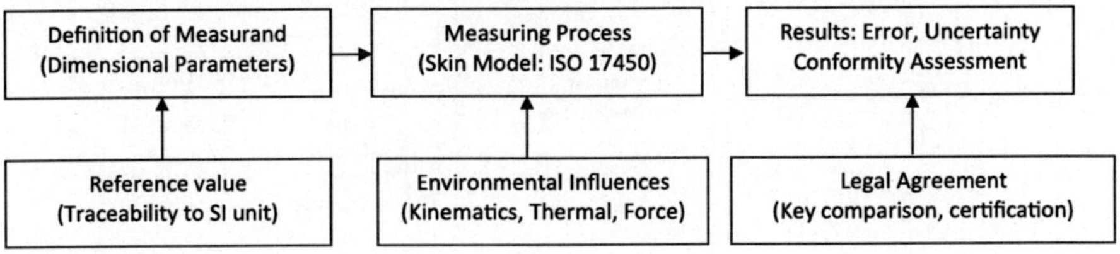


**Fig. 1** Six technical aspect of results of calibration certificate

procedure. When and wherever it is required, we can introduce the influences of environmental condition, material properties, and process parameter into the measuring process.

Ultimately, the numerical expression of results includes the measured deviation from the nominal value, uncertainty of measurement, and appropriate conformity statements. We noted that the digitized key comparison data is useful for the conformity analysis despite of the legal requirement of tolerances. This scrutiny leads to propose a linear data structure useful for the digitalization of dimension metrology. The following section discusses the details.

## 4.1 Data Structure for Dimensional Measurement

A simple data format is devised for the generation of DCC considering the interoperability issue and fulfilling the requirements recommended in ISO 17025. According to Robert Brown, Mitutoyo, interoperability is defined as " *the ability of machine, device, sensor, software, and people to connect and communicate with each other*". Figure 2 depicts proposed format useful to develop the data structure. The data struc-

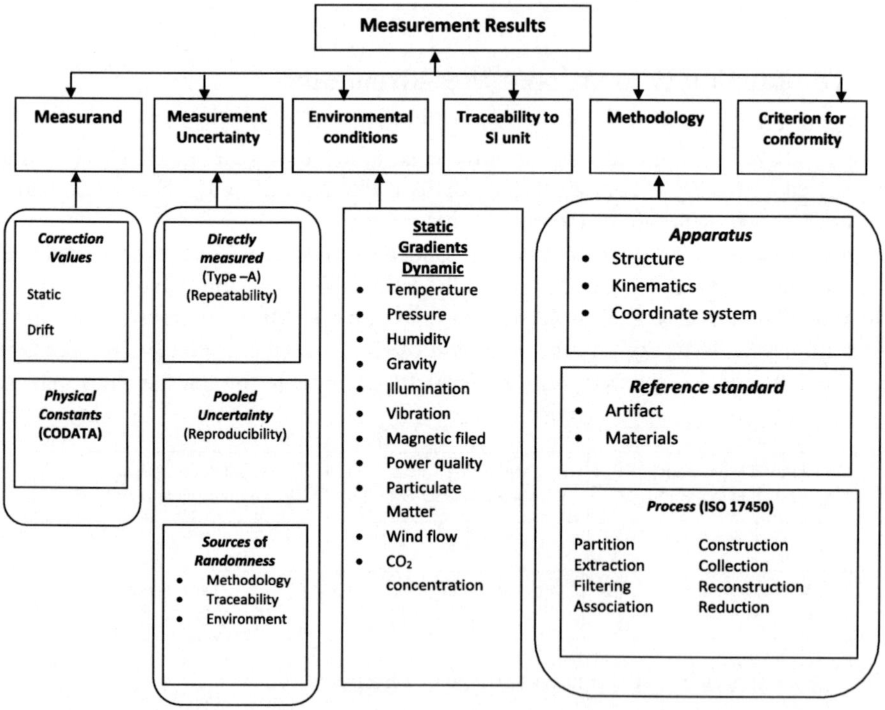


**Fig. 2** Data format devised for the digitalization of dimensional calibrations

ture can be store different numerical values that are acquired during the calibration process.

Usually, a calibration and test certificate contain at least six technical data filed viz. (i) estimate of measurand, (ii) measurement uncertainty (iii) environmental conditions (iv) traceability to SI unit (v) measurement methodology (vi) criterion for the conformity of geometric specifications of product to the given quality traits.

Usually, a calibration and test certificate contain at least six technical data filed viz. (i) estimate of measurand, (ii) measurement uncertainty (iii) environmental conditions (iv) traceability to SI unit (v) measurement methodology (vi) criterion for the conformity of geometric specifications of product to the given quality traits.

The measurand filed contain corrected reading that are obtained by direct measurement of the specimen. The correction values includes the constant and monotonically increasing time depended. A mathematical relationship give the numerical value of the measurand accounting correction values, physical constants, environmental parameters, and material properties. Some physical quantities can be called from CODATA website, and some constant from standards (e.g. the reference temperature recommended in ISO 1).

One can use the mathematic model to estimate the measurement uncertainty using the standard deviation of the directly measured qualities and pool uncertainty of measuring apparatus. GUM advocate statistical analysis (Type-A) of directly measured qualities for calculating the repeatability; and the pooled standard deviation of previously measured value of similar / same artefact to estimate reproducibility.

Nevertheless, all the environment condition like temperature of air/specimen, air pressure, humidity can be estimated as static, their gradients in space, and temporally dynamic—linear. Let us say, temperature of air is measured, the proposed data structure can encodes the measured result. Correction values and uncertainty contribution due to the influence of different environmental conditions can be calculated according to the measurement model equation.

The SI units has been for the human, but they needs to be represented for machines following the recommendation of ISO 11240 [15]. Their format must be usable in the software (e.g. LaTex, MathML) by scripting languages (e.g. python, java, PHP), and natural language processor to print documentation (pdf).

The SI representation by the QUDT (abbreviated as Quantity, Unit, Dimension, and Types ontology) worked out schema (OWL, SHACL) to comprise the requirements of encoding, ontology to enable transmission through URI on web [16]. It grammar adopted the vocabulary that include units, kind of quantity, dimensional vector. Thus, the traceability of SI units links to QUDT through query language SPARQL of resource design framework.

Digitalization methodology is still pending even though the dimensional calibration services are classified for universal use. The measuring apparatus needs to self-calibrate instruments to develop virtual infrastructure. However, the method of calibration can be sub-divided as process parameters, contribution of error due to reference standard, and performance of machine. Specific and critical analysis is required to bifurcate different uncertainty components into these categories. The concept of inherent uncertainty suggested in the literature is useful in the cause effect analysis [17].

Ironically, analysis of risk, opportunity is inevitable for the conformity assessment. Apex level calibration may not need conformity criterion. However, statement of conformity is essential an essential legal requirement. Technically, the proposed data format is devised following the prospects of digitalization of dimensional metrology. In first instance, the format covers all the aspects of the DCC including interoperability.

## 5 Conclusion

The importance of digitalization of metrology in line with Industry 4.0 is essential to operate the feedback during the machining. Artificial intelligence-based feedback contrary to manual statistical process control enable additive manufacture as well as increase the production. Globally, national metrology institutes mainly PTB, Germany, NIST–USA, and NPL, UK are working to develop the digital infrastructure. Several communities together with BIPM, ILAC, and ISO are digitalizing their database KCDB, conformity assessment, and ISO documents, respectively. We have already discussed the progress of representation of SI unit for the generation of the digital calibration certificates. Representatives from FAIR, QUDT, and Mitutoyo are organizing workshop to invite stakeholders from cyber security community, legal metrology inspectors, and metrologist.

Digitalization of dimension metrology is emerging through the development of virtual calibration measuring machines. Computer simulation of standard methodologies of calibration, self-calibration of machine, and inherent traceability are key technologies to be accelerated [18, 19]. Focused effects at National Physical Laboratory, New Delhi in developing algorithm of simulation inspired to devise a data structure of digital calibration certificate. The data format is compatible with ongoing digitalization projects according to ISO standard guidelines. NPL, India is active to participate in prospective digitalizing process.

## References

1. Value-oriented and ethical technology engineering in industry 5.0: a human-centric perspective for the design of the factory of the future
2. Wang Y, Towara T, Anderl R (2017) Topological approach for mapping technologies in reference architectural model Industrie 4.0 (RAMI 4.0). In: Proceedings of the world congress on engineering and computer science 2017, WCECS 2017, vol 2. San Francisco, USA, 25–27 Oct 2017

3. Wilkinson M, Dumontier M, Aalbersberg I et al (2016) The FAIR guiding principles for scientific data management and stewardship. Sci Data 3:160018. https://doi.org/10.1038/sdata.2016.18
4. Digital platforms in manufacturing industries, federal ministry for economic affairs and energy (BMWi) (2021)
5. The digital twin in Industrie 4.0—a short introduction to properties, submodels & asset administration shells (AAS), Presentation
6. Hackel S, Härtig F, Hornig J, Wiedenhöfer T The digital calibration certificate. https://doi.org/10.7795/310.20170403
7. Wilkinson MD (2016) The FAIR guiding principles for scientific data management and stewardship, Sci Data 3:160018. https://doi.org/10.1038/sdata.2016.18
8. ISO 17450:2011 Geometrical product specifications (GPS)—general concepts
9. GUM: guide to the expression of uncertainty in measurement
10. ISO/IEC 17025:2017, General requirements for the competence of testing and calibration laboratories
11. ISO 1101:2017(en) Geometrical product specifications (GPS)—geometrical tolerancing—tolerances of form, orientation, location and run-out
12. IS 3455: Gauging practice for plain work pieces, Bureau of Indian Standards
13. www.bipm.com/L services
14. Res J (1997) Natl Inst Stand Technol 102:647
15. ISO 11240:2012, Health informatics—identification of medicinal products—data elements and structures for the unique identification and exchange of units of measurement
16. www.qudt.org
17. Arif Sanjid M, Ghoshal SK, Sen M (2020) Reviving the inter-laboratory comparison measurement results. Trans Inst Meas Control 42(4):823–831. https://doi.org/10.1177/0142331219879817
18. Metrology for the digitalization of the economy and society, Report of the German council of science and humanities (2017)
19. Liu M, Cheung C, Cheng C, Lee W (2016) A gaussian process data modelling and maximum likelihood data fusion method for multi-sensor cmm measurement of freeform surfaces. Appl Sci 6:409